Leap 4.0: African Perspectives
on the Fourth Industrial Revolution

EDITED BY

Zamanzima Mazibuko-Makena
and Erika Kraemer-Mbula

THIS PROJECT WAS SUPPORTED BY:

science & innovation

Department:
Science and Innovation
REPUBLIC OF SOUTH AFRICA

MAPUNGUBWE
INSTITUTE FOR STRATEGIC REFLECTION (MISTRA)

First published by the Mapungubwe Institute for Strategic Reflection (MISTRA) in 2021

142 Western Service Road
Woodmead
Johannesburg

ISBN 978-1-928509-16-5

© MISTRA, 2021

Production and design by Jacana Media, 2021
Text editor: Terry Shakinovsky
Copy editor: Megan Mance
Proofreader: Lara Jacob
Designer: Sam van Straaten

Set in Stempel Garamond 10.5/15pt
Printed and bound by Print on Demand

Please cite this publication as follows:
MISTRA. 2021. *Leap 4.0: African Perspectives on the Fourth Industrial Revolution.*
Zamanzima Mazibuko-Makena and **Erika Kraemer Mbula** (eds).
Johannesburg: Mapungubwe Institute for Strategic Reflection

Contents

Contents

Preface

For all the critiques of the notion of a Fourth Industrial Revolution (4IR), the idea does capture the enormity of changes currently taking place in many areas of human endeavour. The impact that this revolution will have on humanity is still a matter of educated conjecture. However, with many social activities in various parts of the world already well into the depths of its application, the 4IR's contours are already taking shape.

This revolution straddles production and exchange, services, communication, entertainment, governance and many other areas of social intercourse. In this way, it affects human agency and has major implications for social wellbeing. As machine learning and Artificial Intelligence permeate and merge with human activity, they do enhance capabilities; but they can also undermine livelihoods in profound ways.

In developing countries, particularly in Africa, questions remain about whether this revolution will happen soon, assume different paths compared to advanced economies or percolate the continent in a disjointed fashion.

Given that the first three industrial revolutions were not a universal occurrence, there is much that should concern Africans about the application of 4IR technologies. For South Africa and other African countries, industrialisation was a slow process that started after more advanced economies had embarked on that journey. It was largely imported, along with the cannon and forced acculturation. Out of this emerged economies with a subordinate relationship with the colonisers and with socioeconomic path dependencies that endure to this day. Most of Africa is still catching up with basic industrialisation and the utilisation of information and communication technologies of previous revolutions.

However, in Southeast Asia, developmental states seem to have embarked on a modernisation drive to raise not only rates of growth, but also to improve social wellbeing.

This volume, *Leap 4.0: African Perspectives on the Fourth Industrial Revolution*, examines the challenges of the 4IR in the context of social realities in South Africa and the rest of the continent. The authors proceed from the premise that new technologies seep into social endeavours because they contain the possibility of improving the human condition. As their positive attributes become evident and as they generate large returns for the inventors and practitioners, they become globally pervasive. However, if their introduction is poorly managed, they can destroy human livelihoods and disrupt ecological balances.

The challenge for Africa is not whether, but how and with what effect, these technologies will find systemic application. If the process is haphazard, the continent will find itself dictated to by experiences that are not attuned to its social contexts. Africans will thus be at the mercy of conglomerates that break down national barriers in search of profits, and will end up as objects of autonomous machines and processes programmed elsewhere. It is therefore the focus of this volume to identify the challenges and opportunities the 4IR presents, especially to workers and other marginalised sectors of society.

The authors examine the prerequisites for the successful introduction of 4IR, including infrastructure, skilled personnel and appropriate regulation. They underline the importance of inclusive innovation, with a deliberate objective to create new jobs and reduce inequality.

The idea of *Leap 4.0* is used to underscore the possibility of leapfrogging various stages of development – that it is not necessary to go through the full gamut of previous revolutions to become part of the 4IR. The utilisation of mobile technologies and cryptocurrencies in financial transactions is a case in point. These changes should, however, not happen by mere coincidence. Deliberate intent, clear policy and effective regulation are required.

Besides case studies on mining and banking, passenger transport and electrification, the authors also examine broader issues such as indigenous knowledge systems, language processing, preservation of the environment and intellectual property rights.

So ubiquitous are 4IR technologies that a single book cannot examine all the issues that it poses. The central message from this volume is about courageous leadership as well as inclusivity in conceptualisation and application – all rooted in a transdisciplinary approach.

MISTRA hopes that this contribution to the 4IR discourse will help inform strategies required for South Africa and Africa at large to benefit from a step-change in humanity's technological development. We express our gratitude to the authors and the editorial team for their efforts, and to the Department of Science and Innovation for its generous support.

Joel Netshitenzhe
Executive Director

Acknowledgements

The Mapungubwe Institute for Strategic Reflection (MISTRA) would like to express its earnest gratitude to the project leaders and co-editors of this volume, Zamanzima Mazibuko-Makena and Professor Erika Kraemer-Mbula, who provided oversight of, and editorial contributions to, this book. Thank you to the project coordinator, Anelile Gibixego, for providing valuable support, and to Lerato Mahlangu for her assistance. Thanks also to the contributing authors for the time and energy they dedicated to producing these chapters. Appreciation is also extended to the subject specialists who reviewed these chapters and provided invaluable comment.

Gratitude goes to the MISTRA staff who contributed to the successful completion of this project: the fundraising, operations and project management teams; Terry Shakinovsky, who copyedited the book and managed the publication process; Professor Susan Booysen for her efforts to ensure that this publication meets the highest standards, and Joel Netshitenzhe for his thorough reading of the manuscript.

Final thanks go to the Jacana Media publishing team and to MISTRA's donors who make it possible to do this work.

PROJECT FUNDERS

Intellectual endeavours of this magnitude are not possible without financial resources. The Department of Science and Innovation (DSI) and Standard Bank deserve our special thanks for their support of this project.

MISTRA FUNDERS

MISTRA would also like to acknowledge the donors who were not directly involved with this particular research project but who support the Institute and make its work possible. They include:

- Albertinah Kekana
- Anglo American Platinum
- Anglo Coal
- Aspen Pharmacare
- Belelani Group
- Discovery Central Services
- Exxaro
- First Rand Foundation
- Goldman Sachs
- Harith General Partners
- Human Sciences Research Council (HSRC)
- Sishen Iron Ore
- National Institute for the Humanities and Social Sciences (NIHSS)
- Oppenheimer Memorial Trust (OMT)
- Pareto Ltd
- Phembani Group
- PEU
- Robinson Ramaite
- Safika
- Simeka
- Standard Bank
- Yellowwoods

Contributors

Zamanzima Mazibuko-Makena, co-editor of this volume, is the senior researcher in the Knowledge Economy and Scientific Advancement Faculty at The Mapungubwe Institute for Strategic Reflection (MISTRA) in Johannesburg, South Africa. She obtained her MSc (Med) in pharmaceutics cum laude from the University of the Witwatersrand. Mazibuko-Makena is the editor of and contributing author to the MISTRA volumes *Beyond Imagination: The Ethics and Applications of Nanotechnology and Bio-Economics in South Africa* and *Epidemics and the Health of African Nations*. Her research interests include healthcare systems in Africa, nanomedicine, systems of innovation, the low-carbon economy, and the beneficiation of strategic minerals in South Africa (particularly platinum group metals).

Erika Kraemer-Mbula, co-editor of this volume, is Professor of Economics at the College of Business and Economics, University of Johannesburg, South Africa and heads the DST/NRF/Newton Fund Trilateral Chair in Transformative Innovation, the Fourth Industrial Revolution and Sustainable Development. She holds a master's degree in science and technology policy from the Science and Policy Research Unit (University of Sussex), and a doctorate in development studies from the University of Oxford. Kraemer-Mbula specialises in science, technology and innovation policy analysis and innovation systems in connection to equitable and sustainable development. Prof Kraemer-Mbula is a steering committee member of the Open African Innovation Research (OpenAIR) network where she leads the theme on 'informal sector innovation', and is a Vice-president of the Global Network for Economics of Learning, Innovation, and Competence Building Systems (Globelics).

Michael Gastrow is the director of the Science in Society unit within the Impact Centre of the Human Sciences Research Council (HSRC), South Africa, and a Professor of Practice at the DST/NRF/Newton Fund Trilateral Research Chair in Transformative Innovation, the 4th Industrial Revolution and Sustainable Development at the University of Johannesburg. He holds a PhD in science communication from the University of Stellenbosch. He is a member of the Presidential Advisory Commission on the Fourth Industrial Revolution and is passionate about the role of science and technology in society, particularly in the ways they can foster constructive public discourse and inclusive human development.

Anelile Gibixego is an assistant researcher at the Knowledge Economy and Scientific Advancement Faculty at the Mapungubwe Institute for Strategic Reflection (MISTRA) in Johannesburg, South Africa. She holds a BSc from the University of KwaZulu-Natal and an honours degree from North West University in environmental and biological sciences. Gibixego has experience in microbiology, aquatic science and environmental management. Gibixego's research interests are in water resource management, river health and sustainable access to water resources and water governance. She forms part of projects in the Water Research Commission, in the low-carbon economy and the beneficiation of strategic minerals in South Africa.

Alison Gillwald is the executive director of Research ICT Africa (RIA). She serves as an adjunct-professor at the University of Cape Town's Nelson Mandela School of Public Governance and holds a PhD from the University of the Witwatersrand. Gillwald is a former regulator and was appointed to the founding Council of the South

African Telecommunications Regulatory Authority (SATRA) in 1997. She has advised the South African Presidency, the National Planning Commission, the Competition Commission and the Independent Communications Authority of South Africa, in addition to the African Union Commission, SADC, CRASA and the SADC Parliamentary Forum. She has been commissioned by the International Telecommunications Union (ITU), the World Bank and the African Development Bank to undertake research to inform policy across a number of Africa countries. She is the deputy chairperson of Giganet, the international academic internet governance conference, and was an inaugural associate editor of the ITU journal, *Discoveries*.

Edward Lorenz is Professor of Economics at Aalborg University, Denmark and Visiting Professor at the DST/NRF/Newton Fund Trilateral Chair in Transformative Innovation, the Fourth Industrial Revolution and Sustainable Development at the University of Johannesburg. His research focuses on the internationally comparative analysis of business organisation, employment relations and innovation systems. His work has been financed by various national and international organisations including the Directorate Research of the European Commission and the Organisation for Economic Co-operation and Development. He publishes in journals such as *Industrial and Corporate Change, Cambridge Journal of Economics* and *Research Policy*. He contributes regularly to United Nations events related to the SDGs, including the February 2019 workshop on 'Science Technology and Innovation for the SDGs' organised by UN DESA and ESCAP in Thailand and the 13 November meeting on 'Structural Transformation, Industry 4.0 and Inequality' organised by UNCTAD in Geneva.

Khwezi Mabasa is the senior researcher in the Faculty of Political Economy at the Mapungubwe Institute for Strategic Reflection and is a part-time lecturer in the Department of Political Sciences at the University of Pretoria (UP). Mabasa obtained his MA in political science (political economy) from the University of Pretoria. He is currently working on a PhD in development sociology at the

University of the Witwatersrand focusing on a gendered analysis of South Africa's agrarian question. He served as the National Social Policy Coordinator at the Congress of South African Trade Unions (COSATU), programme manager at Friedrich Ebert Stiftung (FES), and junior lecturer in the Department of Political Sciences (University of Pretoria). Mabasa's published work mainly focuses on heterodox political economy, political sociology and political philosophy.

Lerato Mahlangu is an intern at the Mapungubwe Institute for Strategic Reflection (MISTRA) in the Knowledge Economics and Scientific Advancement Faculty. She holds a BA degree in public governance and politics from the North West University and aspires to complete her undergraduate LLB degree.

Vukosi Marivate is the ABSA UP Chair of Data Science at the University of Pretoria and a visiting principal data scientist at the Council for Scientific and Industrial Research (SCIR). Marivate holds a PhD in computer science from Rutgers University, USA. Marivate works on developing Machine Learning/Artificial Intelligence methods to extract insights from data. A large part of his work over the last few years has been in the intersection of Machine Learning and Natural Language Processing. Marivate is interested in data science for social impact, using local challenges as a springboard for research. He is a founder of the Deep Learning Indaba, the largest Machine Learning/Artificial Intelligence workshop on the African continent.

Duduetsang Mokoele is a researcher in the Knowledge Economy and Scientific Advancement Faculty at the Mapungubwe Institute for Strategic Reflection (MISTRA). She completed her undergraduate degree in political studies and honours in international relations at the University of the Witwatersrand and her second honours degree in public policy and administration at the University of Cape Town in 2012. In 2011 she was the recipient of the Chief Justice Ismail Mohamed medal. Before joining MISTRA she was an intern at the Department of Science and Technology (now known as the Department of Science and Innovation) in the Global Project unit.

Nomaqhawe Moyo is the Information Technology Manager at the Mapungubwe Institute for Strategic Reflection (MISTRA). Moyo has an MPhil degree in information and knowledge systems from Stellenbosch University. She also holds an MSc and a BSc honours in library and information science from the National University of Science and Technology in Zimbabwe. She is an ITIL Expert (V3) certified by AXELOS. She has published on aspects of information science in Africa.

Mamokgethi Molopyane is a media commentator, researcher and writer. She currently works as a consultant in the public and private sector, drawing on her training in communications, human resource, and years of experience including from the 2018 US State Department Global Economic Cooperation Project. Previously, Molopyane worked in communications within organised labour and is currently studying sociology and international relations. In 2019, she presented a paper at the London School of Economics Millennium conference entitled 'Extraction, Expropriation, Erasure? Knowledge Production in International Relations'. Molopyane is a regular contributor to PowerFM and a columnist on Moneyweb. Her writing has appeared in *Business Day*, *finweek*, *The Citizen* and *Business Times (Sunday Times)*. Molopyane has a strong interest in the LGBT youth in rural and imine-hosting communities.

Caroline Ncube is a professor and the DSI/NRF SARChI Research Chair in Intellectual Property, Innovation and Development in the Department of Commercial Law at the University Of Cape Town (UCT). She holds PhD, LLM and LLB degrees. Her teaching includes the master's degree in intellectual property (MIP) jointly offered by the World Intellectual Property Organization (WIPO), the African Regional Intellectual Property Organization (ARIPO) and Africa University (AU). She is a member of various academic associations such as the International Association for the Advancement of Teaching and Research in Intellectual Property (ATRIP), the South African Association of Intellectual Property Law and Information Technology Law Teachers and Researchers (AIPLITL) and the Society of Law

Teachers of Southern Africa (SLTSA). She is affiliated to the University of Ottawa as an Associate Member, Centre for Law, Technology and Society.

Oghenekaro Nelson Odume is a senior researcher and director of the Unilever Centre for Environmental Water Quality within the Institute for Water Research at Rhodes University. He holds a BSc (Hons) degree from the Delta State University, Abraka, Nigeria and MSc and PhD degrees in water resource science from Rhodes University in South Africa. He has over 11 years of research experience in the field of water resources with interest in social-ecological systems, applied ecology, water quality, water governance and ethics, and ecosystem ecology. His research has received local and international awards, including being the first recipient of the Emerging River Leader Award by the International River Foundation, Australia and the Bronze Medal Award by the Southern Africa Society of Aquatic Scientists in recognition of his contribution to water quality management in South Africa. He was among the expert practitioners engaged to develop the occupational framework for the water sector by South Africa's Energy, Water Sector Training Authority (EWSETA).

Chidi Oguamanam holds full professor status in the Faculty of Law (Common Law Section), University of Ottawa, where he is affiliated with three centres of excellence: the Centre for Law, Technology and Society; the Centre for Environmental Law and Global Sustainability; and the Centre for Health Law Policy and Ethics. He holds numerous research fellowships and affiliations with leading global organisations and institutions, including the Centre for International Governance Innovation, Centre for International Sustainable Development at McGill University, and the IP Law Unit at University of Cape Town. Oguamanam leads and is associated with many research consortia, including the ABS Canada project and the Open African Innovation Research Partnership network (OpenAIR). He is the author of several books and publications that reflect a wide range of interdisciplinary research interests. He is named to the Royal Society of Canada College of New Scholars, Artists and Scientists. Recently, Oguamanam edited the Cambridge University Press open access book, *Genetic Resources, Justice and Reconciliation: Canada and Global Access and Benefit Sharing.*

Isaac Rutenberg is an academic and lawyer based in Nairobi, Kenya. He is currently the director of the Centre for Intellectual Property and Information Technology Law (CIPIT) at the Strathmore Law School, Strathmore University, where he is also a senior lecturer. He is also an associate member at the Centre for Law, Technology, and Society at the University of Ottawa. Rutenberg holds a JD (degree in law), a PhD in chemistry, a BSc in chemistry, and a BSc in mathematics/computer science. He is admitted to practice law in the state of California, patent law in the United States Patent and Trademark Office, and patent/trademark law in the Kenya Industrial Property Institute. He is a member of the Chartered Institute of Arbitrators (Kenya branch) and the American Chemical Society. He previously served as a member of the Board of Directors at the Kenya Copyright Board.

Marie Blanche Ting is an associate research fellow at the Science Policy Research Unit (SPRU), University of Sussex, UK. She has over 15 years of experience in the sustainable energy and development field. Recently she completed her PhD at SPRU, with a case study on South Africa's energy transitions (analysing the role of mining and energy, renewable energy and gas). In addition, she has also worked on the transformative role of science, technology and innovation (STI) in achieving the Sustainable Development Goals (SDGs). For this, she worked on STI policies in Ghana, Kenya, Senegal and South Africa. Her working experiences include the private sector, government, multilateral organisations, research councils and in academia. She holds two MA degrees, in climate change and development from the Institute of Development Studies (IDS), UK as a Mandela-Sussex Scholar, and applied science in chemical engineering at the University of Cape Town (UCT).

Mzukisi Qobo is a professor and Head of School at the Wits School of Governance, University of the Witwatersrand. He specialises in international political economy, global governance and foreign policy. Between 2007 and 2009 he was chief director responsible for trade policy at the Department of Trade and Industry. He serves on President Cyril Ramaphosa's Economic Advisory Council. He obtained his BA from the University of Cape Town, MA in international studies from Stellenbosch University and PhD from the University of Warwick, UK.

Acronyms and abbreviations

AAS	African Academy of Science
ABS	Access and Benefit Sharing
ACDC	African Centre for Disease Control
ADAM	Approach to Distribution Asset Management
AfDB	African Development Bank
AI	Artificial Intelligence
AIS	auto investment scheme
AI4D	Artificial Intelligence for Development
AIMS	African Institute of Mathematical Science
AOSTI	African Observatory for Science, Technology and Innovation
APDP	Automotive Production Development Programme
ASRIC	African Scientific Research and Innovation Council
AU	African Union
BPSA	Business Process Enabling South Africa
CAD	computer-aided design
CBD	Convention on Biological Diversity
CCMA	Commission for Conciliation, Mediation and Arbitration
CCTV	closed-circuit television
CEO	chief executive officer
CHROs	chief human resources officers
CHW	community healthcare workers
CNC	computer numeric control
CONMESA	Construction and Mining Equipment Suppliers' Association
COSATU	Congress of South African Trade Unions

CPA	Consolidated Plan of Action
CPS	Cash Paymaster Services
CSIR	Council for Scientific and Industrial Research
DABUS	Device for the Autonomous Bootstrapping of Unified Sentience
DBE	Department of Basic Education
DEA	Department of Environmental Affairs
DERs	distributed energy resources
DHET	Department of Higher Education and Training
DLI	Deep Learning Indaba
DLR	dynamic line rating
DMA	Directorate of Market Abuse
DMRE	Department of Minerals and Energy Resources
DoT	Department of Transport
DPE	Department of Public Enterprises
DSA	Data Science Africa
DSI	Digital Sequence Information
DSN	Data Science Nigeria
DST	Department of Science and Technology
EBP	Earth BioGenome Project
EC	Enforcement Committee
ECA	Economic Commission for Africa
EDCs	endocrine disrupting compounds
EDSS	environmental decision support systems
EHR	electronic health record
EMS	European Manufacturing Survey
ERP	enterprise resource planning
EPO	European Patent Office
ETDP	Education, Training and Development Practices
EVs	electric vehicles
FDI	foreign direct investment
FNB	First National Bank
FRAND	fair, reasonable and non-discriminatory
FSB	Financial Services Board
GAS I4P	Gas Independent Power Producers Procurement Programme
GATS	General Agreement on Trade in Services

GDP	Gross Domestic Product
GDRP	General Data Protection Regulation
GHG	greenhouse gases
GIS	geographical information systems
GSM	Global System for Mobile Communications
GSMA	Global System for Mobile Communications Association
GPS	Global Positioning System
GVC	global value chains
GW	Gigawatts
HFT	High Frequency Trading
HySA	Hydrogen South Africa
I4.0	'Industrie 4.0'
IBM	International Business Machines Corporation
ICANN	International Corporation for Assigned Names and Numbers
ICASA	Independent Communications Authority of South Africa
ICT	information and communication technology
IETF	Internet Engineering Task Force
IFR	International Federation of Robotics
IK	indigenous knowledge
IKS	indigenous knowledge system
ILO	International Labour Organisation
IoS	Internet of Services
IoT	Internet of Things
IP	intellectual property
IPLC	indigenous peoples and local communities
IPPs	independent power producers
IPRs	Intellectual Property Rights
IR	Information Retrieval
IRP	integrated resource plan
ISM	industrial, scientific and medical
ISPs	internet service providers
ISS	Institute for Security Studies
IR	information retrieval
IT	Information Technology
ITU	International Telecommunication Union

LTE	Long Term Evolution
LV	low voltage
MD	membrane distillation
MEC	member of the executive council
MEMSA	Mining Equipment Manufacturers of South Africa
MerSETA	Manufacturing, Engineering and Related Services SETA
MICT	Media, Information and Communication Technologies
MIDP	Motor Industry Development Programme
ML	Machine Learning
MLP	multi-level perspective
MNC	multinational corporation
MOOCs	massive online open courses
MPRDA	Mineral Petroleum Resource Development Act
MSA	Municipal Systems Act
MT	Machine Transport
MW	milliwatt
NAACAM	National Association of Automotive Component and Allied Manufacturers
NAAMSA	National Association of Automobile Manufacturers of South Africa
NDoP	National Department of Health
NDP	National Development Plan
NEMBA	National Environmental Management Biodiversity Act
NEPAD	New Partnership for Africa's Development
NERSA	National Energy regulator of South Africa
NGO	National Governmental Organisation
NGP	New Growth Plan
NHI	National Health Insurance
NIKMAS	National Indigenous Knowledge Management System
NIPS	National Integrated Power System
NIS	National Innovation System
NLP	natural language processing
NLU	natural language understanding
NPC	National Planning Commission
NRF	National Research Foundation
NRS	National Record System

NSI	National System of Innovation
LRA	Labour Relations Act
OCGTs	Open-cycle Gas Turbines
OECD	Organisation for Economic Co-operation and Development
OEM	original equipment manufacturer
O-RAN	Open Radio Access Network
OTT	Over the Top
P2P	Peer to Peer trading
PAIA	Promotion of Access to Information Act
PAIPO	Pan African Intellectual Property Organization
PAYG	pay-as-you-go
PBM	Participatory Business Model
PEM	polymer electrolyte membrane
PGM	Platinum Group Metals
PHC	primary healthcare
PI	production incentive
PLC	programmable logic controller
PQAAF	Pan African Quality Assurance and Accreditation Framework
PV	Photovoltaic
PWC	PriceWaterHouseCoopers
RAIL	Resources for Indigenous Languages
RBSA	Rooibos Benefit Sharing Agreement
R&D	research and development
RECs	Regional Economic Communities
REIPPP	Renewable Energy Independent Power Producer Procurement Programme
RPP	renewable power plant
SABC	South African Broadcasting Corporation
SADiLaR	South African Centre for Digital Language Resources
SAMERDI	South African Mining Extraction, Research, Development and Innovation
SAPP	Southern African Power Pool
SARChI	South African Research Chair in Industrial Development
SASBO	South African Society of Bank Officials

SASS	Stream Assessment Scoring System
SASSA	South African Social Security Agency
SCADA	supervisory control and data acquisition
SD	Surveillance Division
SDGs	sustainable development goals
SemEval	Semantic Evaluation
SEPs	standard-essential patents
SER	Standard Employment Relationship
SES	social-ecological system
SETA	Sector Education and Training Authority
SMME	small, medium and micro enterprises
SMQ MS	Smart Water Quality – Monitoring System
SONA	State of the Nation Address
SPRU	Science Policy Research Unit
ST	sustainability transitions
STI	science, technology and innovation
STISA_2024	Science, Technology and Innovation Strategy for Africa 2024
STEM	science, technology, engineering and mathematics
SWQMS	smart water quality monitoring system
TNC	Transnational Comparators
TCE	Traditional Cultural Expressions
THC	Traditional Healers Committee
TK	traditional knowledge
TOC	total organic carbon
UKIPO	United Kingdom Intellectual Property Office
UNCTAD	United Nations Conference on Trade and Development
USPTO	United States Patents and Trademarks Office
UJ	University of Johannesburg
V2G	Vehicle-to-Grid
VAA	volume assembly allowance
VPP	virtual power plant
VRE	variable renewable energy
WEF	World Economic Forum
WHO	World Health Organization
WIPO	World Intellectual Property Organization

WIPO-IGC	World Intellectual Property Organization Specialist-Intergovernmental Committee
WOAN	Wholesale Open Access Networks
WTO	World Trade Organization
WQSAM	Water Quality System Assessment Model
WWTW	Wastewater Treatment Works

Framing the Fourth Industrial Revolution in the context of Africa

ERIKA KRAEMER-MBULA AND
ZAMANZIMA MAZIBUKO-MAKENA

There is growing consensus about the current and potential impact that rapid advances in technologies such as Artificial Intelligence, robotics, the Internet of Things and biotechnology can have on our productive and social systems. These rapid developments were first described as the Fourth Industrial Revolution (4IR) by the World Economic Forum founder, Klaus Schwab, in 2016. According to this vision, the scale, scope and complexity of the 4IR are so fundamental and far-reaching that it is expected to bring about disruptive changes across all dimensions of human experience. One of the defining features of the 4IR concept is the convergence or integration of these technologies; in other words, the 'blurring of the boundaries' of physical, biological and digital systems (Schwab, 2016). Such advances are likely to present new opportunities and challenges across all economies and societies in the world.

The potential challenges and opportunities the 4IR presents to countries in Africa, a continent with unique contextual characteristics, have not been sufficiently addressed in research and policy (Ayentimi and Burgess, 2019). In particular, we are yet to take full stock of the impact of new technological advances in a continent whose economy is dominated by a large informal sector, poor public infrastructure, low technical skills and low education levels (Oluwatayo and Ojo, 2018), and where advanced technologies can be found in only a few sectors (Salahuddin and Gow, 2016). It is important to position the debates around the 4IR within the African context in order to guide the development of solutions that respond to Africa's specific needs, including those in healthcare (Mahomed, 2018), education (Xing and Marwala, 2017) and social empowerment (Van Rensburg et al., 2019). If discussion on the 4IR is focused on the African context, it can be determined how strategies formulated in developed economies are incompatible in different settings, such as South Africa.

This introductory chapter locates discussions around the emergence of the 4IR in the context of Africa, and South Africa in particular. It highlights the useful focus on technological change that the notion of 4IR has brought to the context of developing countries, as well as the discrepancies of a 4IR discourse based mainly on developed economies. We review here the notion of 4IR and concepts related to it, its development, and the main critiques of it in the literature. We then provide a brief historical account of the evolution of industrial revolutions and of how Africa is largely a passive participant in this history. Third, we position the 4IR in a developing context, paying particular attention to key developmental features that are characteristic of our continent, such as poverty, inequality and informality. Finally, we provide a synthesis of the arguments raised in the subsequent chapters of this book. This discussion is presented according to the organisation of this volume, and highlights emerging themes.

A CRITICAL REVIEW OF THE NOTION OF THE FOURTH INDUSTRIAL REVOLUTION

In 2016, Schwab described our society as entering a 'Fourth Industrial Revolution'. This terminology has since been used to frame and explore the various types of impacts that emerging technologies may bring to almost every dimension of human development, from evolving social norms and communication to economic development and international relations.

The literature on the 4IR has grown exponentially in recent years, as the notion has become widely adopted by scholars and practitioners worldwide. While a clear-cut and consistent definition of the term would be expected given such growth in the number of studies on the subject, recent research papers show inconsistency in the conceptualisation of the phenomenon. This inconsistency is largely related to the use of '4IR' as a synonym for similar concepts such as Industry 4.0, 'smart manufacturing' or 'digital transformation', among others. While the existence of multiple and often overlapping concepts brings nuance to academic debates, it also results in confusion when defining the scope and theoretical foundations of academic investigation concerning 4IR.

While the terminology of 4IR can be traced back to the World Economic Forum (WEF) and its publications in 2015 (Schwab, 2015) and 2016 (Schwab, 2016), the underlying vision emphasised by the notion of 4IR was preceded by earlier reports that identified an upcoming revolution in manufacturing and beyond (for example, Manyika, 2012; Bradley et al., 2013). Moreover, by that time, various government initiatives had already given different labels to a new manufacturing paradigm. For instance, 'Factories of the Future' was launched as a public–private partnership in 2008, constituting the European Union's main programme for realising the next industrial revolution via research and innovation in new production technologies and systems. Similarly, 'Industrie 4.0' (I4.0), launched in 2011 by the German government, is a national strategic initiative to strengthen Germany's competitive position internationally in manufacturing through increased digitisation and interconnection of products. In the USA, the Advanced Manufacturing Partnership was also launched in 2011, aiming to secure the country's leadership in emerging technologies and create high-quality manufacturing jobs (Kracke, 2012). Other related terminologies pre-dating the 4IR concept include 'Industrial Internet' (Evans and Annunziata, 2012), 'industrial revolution' (for example, Tien, 2012), and 'smart manufacturing' (for example, Radziwon et al., 2014).

When interpreting the notion of 4IR through the lens of innovation, three features can be identified: it is technologically driven; it requires new capabilities; and it drives further innovation. First, under 4IR, technological change becomes the main driver of transformative change, affecting all industries and parts of society. The notion of 4IR specifically focuses on the potential that a particular set of technologies (AI, robotics, 3D printing, machine learning and so on) have on efficiency. Technology has long been considered a source of economic progress, although simultaneously it has generated cultural anxiety throughout history. These concerns include the pressure that new technologies exert on employment and wages, sparking debates around job insecurity, growing inequality and mass 'technological' unemployment. But such technology-induced cultural anxiety goes far beyond concerns related to the future of work, as it also gives narrative to a range of sentiments linked to the fear of losing our human identity, alienation and social unrest, among others (Cave and Dihal, 2019).

Second, 4IR requires new capabilities. Although 4IR builds largely on digital technologies and their related infrastructure, the convergence of new technological fields requires a range of new skills and capabilities within firms as well as in institutions. New skills, such as those related to the development, use and oversight of cybersecurity, blockchain, machine learning, genome editing and many other fields, will have to become a central offering from educational institutions. The blurring lines between the physical, digital and biological spheres demand interdisciplinary skills and new approaches to learning. Indeed, educational responses to 4IR might require either new or restructured educational institutions so that they can train workers and youth in emerging interdisciplinary fields (Gleason, 2018). Within firms, management capabilities become essential for smart technology management, while at the institutional level, a weak capacity to coordinate and manage technological changes can become a serious constraint to navigating a dynamic and rapidly changing environment (Cirera and Maloney, 2017).

Third, 4IR is driven by technological innovation, but it is itself also a driver of further innovation. The application of a new wave of technological advances in interconnected robots, machine learning, Artificial Intelligence and other 4IR-related technologies, holds the potential to unlock further economic value by transforming production and so resulting in new products, services, production processes and business models. This is a distinguishing characteristic of industrial success achieved through the process of 'creative destruction' described by Schumpeter (1942) many decades ago. Firms or countries that are furthest from the technology frontier are said to benefit most from investments in innovation (Griffith et al., 2004). This should be a boon for developing countries. However, owing to the constrained levels of investments in innovation in South Africa, this potential benefit seems to be forfeited. Low investments in innovation, therefore, poses limitations to the capacity of 4IR to materialise in further innovation-driven development.

The unfolding of the 4IR involves a systemic change across many sectors, activities and aspects of human life. Studies informing the 4IR literature tend to focus on understanding the application of a specific disruptive technology and its effects on a firm or a sector, often delving into the specific characteristics of the individual case. Yet, historical examples indicate that technological disruptions do affect entire industries and also society. A range of disciplinary insights in the literature has helped us understand systemic changes, taking into consideration the complex interactions between technology and social relations. These insights include those from long wave theory, innovation systems, techno-economic paradigms and socio-technical transitions.

Nikolai Kondratiev (1892–1938) introduced the concept of long waves into economic theory, identifying ascending and descending phases in industrial activities, with major innovations clustered in the decade or two before the start of a new ascending phase. Later, Joseph Schumpeter (1883–1950) refined the idea of the long waves, relating them to processes of 'creative destruction', where during a period of recession, innovations converge in mutually enriching 'clusters' around a smaller number of disruptive innovations. These innovations then spread and cause major shifts in how goods are produced, marketed and consumed, and ultimately bring about economic recovery.

Drawing from this tradition, the 'neo-Schumpeterian' school of research – or evolutionary economics – has made important contributions to explaining the processes that lead to the emergence of these innovations: how they are produced, diffused and used in order to trigger systemic change. Some of the concepts developed by evolutionary economists include systems of innovation and techno-economic paradigms. A system of innovation, defined at national or regional level, is a network of institutions and actors in the public and private sectors who interact in producing, adapting, diffusing and using new knowledge and new technologies that contribute to social and economic development. The innovation systems framework was developed in the late 1980s and early 1990s (Freeman, 1987; Lundvall, 1992; Nelson, 1993; OECD, 1997) placing innovation as the main driving force in the evolution of social and economic systems. Emphasising the role of multiple stakeholders and the norms or institutions that guide their behaviour, this literature highlights the importance of networking, interactive learning and collaboration, providing an analytical framework to understand innovation practices as a multilateral and highly context-specific phenomenon. Innovation systems are not static; they are susceptible to constant change. As conditions change and knowledge accumulates, a system can evolve and thus the role of its components requires constant readjustment. It is, therefore, possible to influence the process of transforming an innovation system. This is not an easy task, and may imply the establishment of new sub-systems, new linkages, creating new policies or mechanisms to support innovation – or all of the above. The ability of a system to change and evolve depends to a large extent on the culture in which it is embedded, which can be either a favourable environment or, the opposite, one that is resistant to change. The insights from the innovation systems framework provide a useful platform to explore the production, demand, diffusion and use of knowledge in a 4IR society.

Evolutionary economists such as Carlota Perez, Luc Soete and Chris Freeman conceptualised the notion of techno-economic paradigms, where the convergence of several new technology systems gives rise to structural changes and the emergence of new social and institutional frameworks. This literature has identified at least five techno-economic paradigms since the mid-18th century: 1) the steam engine (1780–1830); 2) railways and steel (1830–1880); 3) electricity and chemicals (1880–1930); 4) automobiles and petrochemicals (1930–1970); and 5) information and communications technologies (1970–2010). According to this literature, the 4IR would still be in the realm of the fifth industrial revolution described, based on digital technologies. New techno-economic paradigms provide an opportunity for latecomers to leapfrog into new sectors. This is an important consideration to take into account when exploring the possibilities of 4IR in Africa.

The literature on socio-technical transitions also allows for exploring systemic changes, as they involve alterations in the overall configuration of various sub-systems (such as transport, energy or production systems), which entail technology, policy, markets, consumer practices, infrastructure, cultural meaning and scientific knowledge (Geels, 2004). This literature pays attention to the forces that reproduce and maintain the status quo, versus the actors that push for alternative practices. In this respect, the multi-level perspective (MLP) presented by Frank Geels (2002) describes three levels at which transformative change can happen: the niche level (micro level), regimes (meso level) and landscape (macro level). These three levels are interconnected as either changes in the landscape or a push for alternative practices in emerging 'niches' can cause a replacement of the current mainstream practice (regime). Technological transitions in general are complex, and usually take place incrementally over long periods of time. This view, in fact, questions whether a 'revolution' is such, in terms of abruptness, or in fact is rather an outcome of several adaptations and changes over time.

These theoretical contributions can be seen as 'lenses' through which to interpret the systemic changes brought by the 4IR. These perspectives can help us to identify the emergence and possible trajectories of alternative practices, underlying learning processes, relevant actors and dominant forces facilitating or hindering system-wide change.

The underlying features of the 4IR, therefore, are not entirely new and can be seen as complementary to earlier theoretical frameworks that have tried to unpack the complex dynamics of systemic change. However, the recent focus on the notion of 4IR has triggered a renewed impetus to discuss technological change, in particular disruptive technologies, their drivers and potential impacts. More importantly, such impetus has permeated the agendas of multiple stakeholder communities such as scholars, the private sector and policymakers, allowing space for valuable interactions and collaborations in the pursuit of collective goals.

Finally, it is important to bear in mind that the notion of the 4IR has not been undisputed. Several critiques deserve attention. First, the notion of 4IR has been connected to neoliberal rhetoric and a vision of capitalism that seeks the endless pursuit of the accumulation of capital (Avis, 2018). In this regard, the 4IR has been considered to represent the interests of the technocratic elites (elite politicians, academics and business leaders), disregarding the interests of vulnerable and marginalised populations, including those in precarious and low-paid jobs.

Second, the WEF's description of 4IR places a strong focus on a number of technological advances and their almost inevitable effects. However, such focus on technology has been perceived to downplay the complexity of social processes that are essential in shaping and adopting new technologies as well as new forms of social, political and economic organisation. In other words, the consideration of technologies as social artifacts implies that the same technological applications can be shaped and used differently as a result of differences in social, cultural and political environments (Collan and Michelsen, 2020).

Third, the orthodox 4IR conceptualisation is silent on its connection to broader sustainability issues. Schwab (2016) describes 23 technology shifts comprising the 4IR that span the digital, physical and biological worlds, and they are most powerful when they combine and reinforce one another. Although these shifts identify areas of disruption, they offer no assessment of the potential to deliver against sustainable development.

A BRIEF HISTORY OF INDUSTRIAL REVOLUTIONS AND 'THE ABSENCE' OF AFRICA

The world's advances in manufacturing processes and the modes of drivers of industry have been characterised by the term 'industrial revolution'. Traditionally, an industrial revolution has been defined by the degree of technological advancement that changed industry. However, the interplay between technology and society in all industrial revolutions cannot be denied. This interaction determines the uptake, direction, rate and patterns of innovation adoption, diffusion and use. This is a process involving not only technologies but also changes in consumer practices and needs; the skills and capabilities of all actors involved; infrastructures; governance; regulation; industry structure; ethical implications; and cultural meaning of the system.

Industrialisation is a marked shift to a more powered, more efficient means for machinery, mass production and factory (Allen, 2009). To fully appreciate the Fourth Industrial Revolution, it is important to understand the conceptualisations of the past three revolutions. The First Industrial Revolution, as termed by Thomas Aston (1948), was an epoch of technological change that began in Great Britain, spanning through the years 1760 and 1830 (Crafts, 1996). It introduced the steam engine into industry as well as the conveyer belt, leading to mechanised production that created significant changes in output. Communication was revolutionised through the invention of the telegraph and, for the first time, people could communicate almost instantaneously without being in the same place (King, 2019). This industrial revolution also saw the rise of banks and industrial financiers, as well as a factory system dependent on owners and managers (History, 2019). The conditions were ripe for urbanisation and were accompanied by a rapidly growing middle class (or middle strata) as well as scientific and technical advances. The First Industrial Revolution also resulted in often harsh employment and living conditions for the poor and working class (Humphries, 2010). Women were excluded from manufacturing positions leading to a male-dominated industry (Philbeck and Davis, 2019).

The Second Industrial Revolution occurred between the years of approximately 1870 and 1914 and made mass production possible. It was able to divide labour and maximise the means of production (Mokyr, 1998). Unlike the First Industrial Revolution, which resulted in industrial growth using coal, textiles, railroads and iron (Mohajan,

2019), the Second Industrial Revolution expanded the use of electricity, steel and petroleum as well as innovative models of public transportation established from the internal combustion engine (Mohajan, 2020). This was indicative of a system change that amalgamated around the new-found acceptance that science and technology contributed towards an improved standard of living. The Second Industrial Revolution shifted the focus more towards services than manufacturing with new service-sector jobs created far more rapidly than the old manufacturing jobs disappeared (Blinder, 2006). The Second Industrial Revolution broadened the activities and products of the first and introduced momentous changes in technical complexity, precision in manufacturing and technological infrastructure. Living standards and the purchasing power of money increased rapidly (Mokyr, 1998). However, these industrial revolutions also came with increased carbon dioxide emissions (Rifkin, 2012).

The Third Industrial Revolution commenced decades later, utilising automated production, information technology and mainframe computing (Li et al., 2017). Rifkin describes it as 'a new convergence of communication and energy'. It began in the 1960–1970s and is ongoing, with more developments in telecommunication, radio, 3D printing and the internet. It bloomed alongside the discovery of the double helix, the space race and the development of nuclear power. It shaped a post-war world that needed new economic structures and that had shifting conceptions of the human place in the cosmos, the natural world and the political order. It also connected the planet's societies through infrastructure and applications, creating new flows of information-sharing that continue to shape values, knowledge and culture (Philbeck and Davis, 2019). Rapid progress towards increasing computational power led to a more interconnected and complex world in many ways and is still driving change across sectors and regions at the beginning of the Fourth Industrial Revolution, just as the continuing spread of electricity access is still bringing the benefits of the Second Industrial Revolution to communities around the world. The Third Industrial Revolution required a new energy base that is more renewable and less carbon intensive (Jänicke and Jacob, 2009). Challenges of the Third Industrial Revolution include its requirement for a review of capital, professional skills and redistribution of wealth among sectors and regions (Jänicke and Jacob, 2009).

Similarly, the Fourth Industrial Revolution brings incredible opportunities for individuals, industries and nations. The Fourth Industrial Revolution is said to be an extension of the third, which encapsulates digital technology with physical and biological technology (Philbeck and Davis, 2019). The 4IR captures the idea of the confluence of new technologies and their cumulative impact on our world. Artificial Intelligence (AI) entails the capability of machines/robots to process information and to develop learning and logic that allows them to act independently. Thus AI can produce a medical diagnosis from an X-ray faster than a radiologist and does this with pinpoint accuracy; robots can manufacture cars faster and with more precision than assembly line workers; they can mine base metals like Platinum Group Metals and copper, crucial ingredients for renewable energy and carbon cleaning technologies. The argument for exploring the 4IR is compelling, particularly for the developing world. New technologies are advancing with exponential velocity, breadth and depth. Their systemic impact is likely to be profound. Klaus Schwab highlights how the 4IR is different from anything else: 'the new age is differentiated by the speed of technological breakthroughs, the pervasiveness of scope and the tremendous impact of new systems' (Schwab, 2017).

It is important in these discussions, however, to acknowledge that the previous industrial revolutions were not inclusive of African countries. The continent missed the First and Second Industrial Revolutions while it endured the invasion, occupation, division and colonisation of the African territory by European powers (Rodney, 1982). Elements of these revolutions were selectively introduced in the colonial territories in a subordinate relationship to serve the interests of the imperial metropolises. At the early stages of the Third Industrial Revolution, most countries in the continent were newly independent states, facing a range of social, economic and political challenges, in the midst of the difficult process of reconstructing their nations and repositioning themselves in the global scene (Mendes et al., 2014). As a result, to date, most countries in Africa display low levels of industrialisation. This has many implications for the state of development in Africa and how the continent engages with the advent of the 4IR. While the 4IR has the potential to drastically transform industries and improve the lives of many, this potential has not been adequately nor pragmatically discussed in relation to Africa. With Africa's insufficient public infrastructure, a large informal sector and low technical skills, what does the implementation of the 4IR look like in that context? Several crucial elements of the Second and Third Industrial Revolutions have not been perfected in Africa and the 4IR builds on these previous revolutions. The next section thus locates the 4IR in the African context, in order to move the discussion away from the context of developed economies.

POSITIONING THE 4IR WITHIN A DEVELOPING CONTEXT

While there is no doubt that the pathway for development in African countries requires a more inclusive and integrated approach that considers not only economic but social and environmental factors as well, the means to achieve this approach have not been agreed. Manufacturing has been a catalyst and a critical element in the development model in impoverished countries, particularly in Asia. However, emerging technologies are purported to eradicate the labour intensiveness of manufacturing and, while this leads to more efficiency, it takes away from the opportunity to leverage the substantial job creation that can be generated by manufacturing. For example, technological advancements in steelmaking in the US ensured that 82,800 people working in the industry in 2018 were able to produce 14 per cent more steel than approximately 399,000 workers managed to produce in 1980 (Levinson, 2019). The advent of the 4IR, and the advanced technological innovation it proposes, are raising questions about its potential to offer a developmental path for less-developed economies. The potential of innovation and technological change for economic growth and development has been asserted by several scholars over many years (Schumpeter, 1942; Cameron, 1996; Lundvall et al., 2009). Developed countries have used innovation (through previous industrial revolutions) to grow their economies and they continue to use emerging technologies such as those of the 4IR to further maintain their competitive advantage (Coleman, 2017).

For most countries in Africa, with their different levels of industrialisation, the realities of poverty, inequality, high unemployment rates, inadequate infrastructure and other socioeconomic challenges make innovation even more urgent. Innovation and learning capabilities are fundamental to solving, transforming and leveraging socioeconomic realities for growth and development. Historically, this recognition of the role of innovation led to a need to better understand its effects. The innovation systems framework allows for the efficient coordination of science, technology, engineering and innovation policies for managing economic development (Freeman, 1987; Freeman and Lundvall, 1988; Lundvall, 1988). It is, therefore, also useful in assisting with responding to the technological and structural change that the 4IR promises, and for managing its possible effects on economic development. Several scholars such as Adesida et al. (2016) and Muchie (2016), however, speak of the need for Africa to have an innovation system for integrated development in place of existing and prevailing development pathways to address the relentless problems of underdevelopment in Africa. Muchie (2016) proposes an 'African-centred innovation and development system for integrated and sustainable industrialisation' (Muchie, 2016: 16). This is a theme that runs throughout this book: although the 4IR has the potential to be transformative, imposing 4IR strategies that were developed in well-established economies will not necessarily translate to widespread 4IR benefits in developing countries.

Emerging technologies have been described as novel and knowledge-intensive, showing ground-breaking scientific innovation with the potential for widespread impact (Cozzens and Thakur, 2014). Although there is no unanimity on the definition of these technologies and there is uncertainty regarding their development process, they often receive heightened interest, which is noticeable in the number of publications in which they are covered, and their mention in several policies (Rotolo et al., 2015). New and emerging technologies require knowledge that can traverse industries and borders for the 'co-development of processes, products and engineering designs and/or acquiring them from other countries or organisations' (Perrot, 2018: 2).

Often, emerging technologies are cultivated and shaped by science and innovation policies in high-income, developed countries and are instituted in wealthy, well-educated population groups with the infrastructure required for advanced technologies (Cozzens and Thakur, 2014). Therefore, they rarely improve the lives of impoverished people, who need them the most. It is thus crucial for developing countries to have science and innovation strategies that take account of who the ultimate beneficiaries of new technologies are and that these technologies align with and directly serve the needs of the people they are intended to benefit (Cozzens, 2012; Perrot, 2018). Science and innovation policies that are aimed at economic development but are not cognisant of, and intentional about addressing, overall socioeconomic challenges may result in increased income inequalities even if overall incomes of the poor increase (Page, 2005; Perrot, 2018).

South Africa, as one of the most unequal societies in the world, is condusive for the most disadvantaged to miss out on the benefits of the 4IR and for inequality to deepen. Science and innovation policies from developed countries that are mechanically adopted by developing countries often fail to focus on local societal challenges or to seek meaningful benefit for the majority of the population, particularly the poor. The diffusion and widespread use of emerging technologies require behavioural or social changes and systemic transformation in order for their adoption to be successful. Access to new technologies is driven by social exclusion and intersecting inequalities, including gender, race, disability, social class, age, locality, and so on. African countries have largely been unsuccessful at innovation-based development due to several challenges, including dependence on raw materials and commodities, faltering leadership and incoherent innovation systems (Oyelaran-Oyeyinka, 2014). High levels of income inequality can be partially attributed to unequal access to education and unequal earning opportunities (AfDB, 2020). Systemic transformation and the ability to innovate around social challenges requires a well-functioning innovation system that enables the development of capabilities for high-skilled and high-productivity sectors. For the 4IR to be impactful and transformative on the African continent, intentional and targeted learning through the process of technical change is crucial. This requires broad-based systems of education and training able to build widespread capacity for continuous learning and adaptation (Lorenz et al., 2016).

Moreover, with Africa's extensive informality and grassroot innovations, discourse around the 4IR should also seek to address where and how the informal sector fits in. The disproportionate emphasis on research and development (R&D) that overlooks low-cost or low-tech innovations, which are usually aimed at alleviating socioeconomic challenges, should be reviewed. Soumonni (2016) argues that innovation in advanced technological fields should take place in lower- and lower-middle-income countries through co-evolution of both low-tech and high-tech innovation. Tondi (2019) presents a case for the inclusion of African indigenous knowledge systems (IKS) for socio-cultural change and in economic development strategies, including 4IR strategies. It is argued that the intentional inclusion of IKS has the potential to complement 4IR strategies while involving the users and custodians of IKS (Tondi, 2019).

SUMMARY OF THE BOOK CHAPTERS AND THEIR CONTRIBUTIONS

This book seeks to expand the 4IR debate beyond purely economic or technological perspectives to include broader social, cultural and policy perspectives. For countries in Africa, industrialisation has been a much slower process, one which, due to historical events, began years after more mature economies had already embarked on their journey. The current debates on innovation in the context of 4IR are of particular relevance to Africa, and it is essential to contextualise these debates to guide the development of tailored solutions that respond to Africa's pressing needs. There is the rhetoric around African countries having the opportunity to use the 4IR to leapfrog into accelerated development (Mathews and Lee, 2018). However, discourses on the 4IR identify contradictory implications and it is this contradiction that culminated in this book's exploration of the 4IR, particularly in the context of a society as unequal as South Africa.

The chapters in this volume are organised into three sections with the following thematic areas: emerging technologies and inclusive innovation; human capability formation; and the application of advanced technologies in sectoral developments. The themes and arguments examine the societal value of the 4IR and its broader social impact, while providing practical research outcomes such as policy advice and recommendations. Chapters draw largely on South African and some other African evidence in order to ground the book. A transdisciplinary approach is applied, traversing the limitations of orthodox disciplines to create a holistic perspective (Du Plessis et al., 2011) to better understand the 4IR and its significance in South Africa. The book incorporates perspectives from researchers in innovation studies; technological sciences; humanities; intellectual property law; labour and political economy, among others, to produce research that is inclusive and evidence based. This volume explores whether the 4IR can be just and sustainable and whether it can be used to achieve South Africa's socioeconomic objectives. More so, at the core of this project is the opportunity for South Africa – working with the rest of Africa – to shape and chart a path that benefits the continent's people.

Emerging technologies and inclusive innovation

The first section of this volume provides a critique of not only the conventional discourse around the 4IR but also of the way in which the development of emerging technologies, such as AI and machine learning, excludes certain population groups and the ethical considerations of 4IR technologies. Through the use of accessible language and the provision of definitions, chapters in this section also accommodate readers interested in aspects of the 4IR and its technologies but who are not experts in the field – in the true essence of inclusion and transdisciplinarity.

Gillwald, in the first chapter, poses the question of whether there should even be debate about the 4IR in South Africa, or if we should rather be developing a new digital deal. The chapter provides an alternative discourse to that presented in popular and technical literature, and argues against the uncritical adoption of the 4IR into development policies by governments and institutions. Gillwald refers to the WEF's notion of the 4IR as being largely based on technological determinism, paying insufficient attention to societies in which technologies are developed. Her chapter cautions against policy documents on 4IR solely based on the WEF's notion of the 4IR and drafted without appropriate local, independent expertise. As reasoned, these policy documents promise that 4IR technologies across industries in South Africa can contribute exponentially to economic growth, job creation and the empowerment of women and the youth with no evidence of how this will be done, nor of awareness of how structural inequality might restrict such positive outcomes.

Gillwald argues that even though developing countries need to prepare for emerging technologies, they still need to ensure that there is universal access to the infrastructure that enables advanced technologies. Implementing these advanced technologies over the existing structural inequalities in South Africa will only reproduce or exacerbate existing inequality in the country. To explore what needs to occur at different governance levels to enhance digital inclusion, Gillwald applies a conceptual framework constructed from the work of Best and Gheciu (2014) on the dynamic nature of what constitutes 'public', and from Frischmann (2005; 2012), on the need for public resources on the demand side to be valued, not just commercial supply-side valuation; and the notion of common-pool public resources. This framework is also used to examine which principles would reinforce the development of a transversal digital policy that will enable the country to create the conditions necessary to harness the benefits of advanced technological developments and mitigate the risks associated with them.

Central to this book's argument is the view that the 4IR cannot take place in Africa without the inclusive participation of Africans. This appears strongly in chapter 2 by Oguamanam, who discusses the exclusion of indigenous knowledge systems (IKS) and its custodians from the 4IR discourse and from 4IR strategies in Africa. His chapter argues for legal and policy protection of indigenous knowledge for inclusive economic growth, cultural survival and sustainability. Oguamanam illustrates how African science technology and innovation

(STI) strategies are not proactive, including those concerning African IKS. It is an unexplored chasm in the African STI ecosystem and because STI strategies are precursors to the uptake of the 4IR, strategies for 4IR are following the same path. The chapter examines this exclusion at a continental level through the African Union (AU) STI strategies as well as a study by the Economic Commission for Africa (ECA) on assessing STI readiness in Africa. In both cases, the chapter presents how the discourse is based on science, technology, and research and development (R&D) priorities, galvanised by the conventional science community with no intention to develop and include IKS institutions and stakeholders.

Using South Africa's experience, Oguamanam presents the case of the exclusion of IKS from the 4IR at a national level. Even with South Africa's progress in repositioning indigenous knowledge across diverse social, economic, legal and policy regimes, and despite references to IKS as part of the ecosystem of South Africa's STI, IKS has not been adequately integrated into the 4IR discourse. The chapter aims to explore the potential benefits and threats that the interface of the 4IR and indigenous knowledge poses for indigenous knowledge stakeholders. It offers insights into how indigenous knowledge and its custodians could be positioned to optimise the opportunities and mitigate the potential threats that may unravel with the advent of the 4IR.

Some of these threats are not easily identifiable, as certain aspects of the 4IR are already woven into everyday human activities. One example is Bostrom and Heinen's (1977) socio-technical framework, as applied by Mokoele, Moyo and Mahlangu in chapter 3, to explore the proliferation of algorithms in facilitating human decisions in our everyday lives. The authors argue that the decision-making by these algorithms is obscure and consequently unregulated. Algorithms are shaping the way we do things and the extent to which they are changing the human lived experience is a focus in the chapter. Communities are not always aware that algorithms are deciding on issues that have crucial implications for them, such as who gets into university, who gets insurance, what people pay for that insurance, and much more. Algorithms also have an ever-greater influence on our day-to-day life and work: they influence our political choices and votes; where and how we invest; the business models we create; and the content of our education and medical diagnostic systems, to name just a few. The chapter thus addresses how algorithms can be deployed to best serve humanity and the authors reason that the extent to which 4IR technologies develop alongside society will determine their success or failure in achieving the desired outcomes.

Algorithms are social constructs too, situated in a specific culture and reiterate societal values. This view allows for reflection on the way algorithms present ethical challenges and may have a negative impact on society. Authors in this chapter argue for policies that protect humanity in a world increasingly using algorithms, and that preserve human values at the heart of human decision-making. Furthermore, the authors show how adverse effects of technologies introduced into society can be avoided through mediation as early as the technology engineering phase and throughout the life cycle of that technology. This chapter further explores the implications of this (inherent) bias of algorithms for Africa.

The integral bias of algorithms is indeed not an error. As mentioned above, algorithms are social constructs; they rely on input from designers and data that reflects the unequal society in which we live. Who the designers are often determines who the algorithms are meant to benefit. Marivate in chapter 4 explores this issue, which links to the theme of inclusive innovation, through motivating for the use of local languages in AI and machine learning (ML) technologies. African languages are being left out of the development of these technologies, which means that Africans and their need for access are being excluded from AI and ML technologies. Marivate discusses the challenges faced by local languages across the continent in both development and in building ML/AI systems, using South African languages as a case study. Challenges that exacerbate low representation of African languages in ML/AI systems, such as the small amounts of data available for implementation into ML/AI systems, are discussed. Marivate uses an example of how developing a question and answer system that operates in, for instance, Setswana, isiNdebele or Xitsonga, would require large amounts of data to train such models, and that this data is largely not readily available. He indicates that most languages on the African continent are poorly resourced, therefore building natural language processing (NLP) systems for these languages is both a technical challenge as well as a broader societal challenge in ML systems. However, not undertaking this exercise would lead to further chasms in ML/AI systems which would amplify inequity due to unchallenged biases in the technology.

Using an adapted soft system perspective, Marivate presents these and other challenges and provides possible solutions in the nexus of machine learning, natural language processing and African languages. Gradual gains in African NLP are indicative of a trend that has to be harnessed in the next few years to not only preserve African languages but also to provide new research avenues and tools across the continent. The chapter argues that for African natural language processing to be established, different sectors of society need to collaborate to provide data, assist in citizen science, provide regulatory flexibility and clarity, and make local language a priority in building our 4IR strategies.

Human capability formation

An important component of a well-functioning innovation ecosystem is education and training for the development of human resources and R&D (Eggink, 2013). While section one of this volume deals with the concept of inclusive innovation, section two explores the changes required in education and skills development (including the need for multidisciplinary curricula) for Africans to participate in the 4IR in a meaningful way. It does so within the context of South Africa's unequal society. The section examines the future of work in a world that is increasingly reliant on advanced technologies and what this reliance means for a country already plagued with high rates of unemployment. The reality is that, even for people who manage to access higher education, unemployment is high. Broad human capabilities are essential for the development of innovation systems (Scerri, 2019); however, the current shortfall in human capability development in Africa hampers the prospects to shift on to a sustainable development path in the context of the 4IR. The fundamental policies required for the promotion of inclusive growth have not been universally determined. However, human capabilities (specifically providing good quality education) and the creation of jobs in high-productivity sectors have been shown to be crucial (AfDB, 2020) for growth. Such considerations link to the chapters described below on building human capabilities to assist with reducing inequality.

The first chapter of section two explores how the 4IR could shape concepts of education and skills development aimed at tackling South Africa's growth challenges. Gastrow examines how new approaches to education can play a role in addressing inequality, while considering how insufficient access to data and technology can deepen existing disparities. With an education system as divided as South Africa's, in which access to quality education still depends to a large extent on race, gender, class and location, the introduction of advanced technologies in education should not further entrench inequalities. Rather, it should be ensured that those in less-resourced schools (that is technologically under-developed and disconnected from the digital world) are not further disadvantaged. As learners move into the world of work, the 4IR has the potential to create new kinds of opportunities for black participation in the formal economy, including new avenues for growth in industrial development and the informal sector. Across the African continent, education and skills development are arguably the most important lever for positioning Africa to benefit from the 4IR. South Africa, therefore, bears a responsibility to the African collective to lead in the area of education and skills development in the 4IR.

Gastrow explores the types of changes that may be required if South Africa is to meet its goals of harnessing new technologies to meet its national development aspirations. The chapter focuses on three principles that can provide overarching guidance for policy makers in times of rapidly accelerating technological change: first, the 4IR should not leave black South Africans behind; second, education in the 4IR must aim to reduce structural inequality; and finally, adaptability – the overall reconceptualisation of education must grapple with the issue of change.

In chapter 6 Lorenz and Kraemer-Mbula discuss the impact of the adoption of 4IR technologies on two advanced-manufacturing sectors, namely the automotive and mining-equipment sectors. The chapter focuses on the impacts on employment, skills development and training – central concerns in South Africa due to the high levels of unemployment and skills shortages in the country. The chapter highlights the importance of paying attention to the specificities of how technology adoption played out in different firms and sectors, as evidence shows significant differences in impact across firms and sectors. According to the authors, social and contextual considerations are essential when examining the effects of 4IR technologies on employment in South Africa.

To complement the topic of skills development, the next chapter provides an analysis of the changing nature of work, influenced by the introduction of automation and advanced technologies. Molopyane in chapter 7 focuses on how the 4IR might affect the future of work in South Africa by examining the impact of robotic technology and automation on work in different countries, including China, Germany, Japan and the US. Through an analysis of the diffusion of robotics in varied countries, the chapter provides insight into the possible future of work and the widespread, collective effect of current and 4IR technologies on policies. Molopyane explores how technology can both displace and be in harmony with workers, and how this differs from country to country.

The chapter zooms in on the impact of robotics and digital transformation on work and employment in banking and mining in South Africa. The two industries are large creators of employment but have seen an increased shedding of jobs owing to, among significant factors, a rise in the use of technology. Furthermore, the chapter discusses the reskilling of mineworkers in other mining-intensive countries, undertaken to adapt to the advent of new technologies. Molopyane raises questions about what reskilling in South Africa would entail, considering the high levels of youth unemployment in the country, including unemployed graduates on the one hand, and the older mineworkers with low education levels on the other. Ultimately, Molopyane observes that the current discourse on 4IR does not take into account the role of stakeholders in shaping how technology is deployed in the economy and how it will impact on society.

Advanced technologies are also changing the structure of work in the services sector. Apps such as AirBnB and Uber are disrupting the hospitality and transport sectors respectively and there has been uncertainty about which laws these digital businesses should adhere to. In chapter 8, Mabasa and Qobo draw on heterodox political economy theory to understand ways in which new technologies and platforms have disruptive effects on the structure of work in the services sector, and the relationship between the state, labour and capital in the regulation of economic activity. In this way, they contend with the underlying socioeconomic contradictions and policy contestations developing in the transition to a digital economy. Using Uber operations in South Africa as a case study, the chapter examines how the structures of the services economy are being challenged and altered by disruptive businesses and how employment relations for the taxi drivers, who are 'driving partners' of Uber, are affected. The discourse on the shared political economy model is also used for examining the Uber case study.

Mabasa and Qobo argue that as South Africa transitions rapidly into integrating its real economy into a dynamic digital economy, there is an urgent need to explore new regulation pathways appropriate to this change, particularly for the protection of workers from exploitation. Furthermore, regulations should take account of social dynamics in order to offer cushioning for the social costs of AI and not have digital markets that focus solely on profit maximisation. There is a need to strike a balance between digital markets and AI-driven technologies providing innovation and competitive advantage, and regulation for social protection. The digital platforms are mostly accessible to those with secure incomes in the middle-class and upper layers of society, and who have access to banking services and technology, whereas the majority of the population lack connectivity or access to ICT products and services.

Application of advanced technologies in sectoral developments

The last section of this volume, section three, examines some of the most relevant sectors (advanced manufacturing, health, environmental sustainability, energy) in South Africa, where 4IR technologies can or are already being employed. This section explores past failures in implementing advanced technologies in some of these sectors as well as the potential benefits and threats of introducing 4IR technologies, in the context of the themes of inclusive innovation in a developing country. Section three also examines the extent to which intellectual property law in South Africa is prepared for the 4IR, which is an extension of the discussion on the regulatory regime required to manage the 4IR.

In chapter 9, Mazibuko-Makena raises questions about the multifaceted medical, economic and social implications of implementing 4IR technologies in healthcare in an unequal, resource-strained society with a weakened healthcare system. Her chapter assesses the conditions that 4IR technologies have to contend with in the South African healthcare system and examines the key challenges and opportunities for the 4IR in this sector.

Using the conceptual framework of co-creation, the chapter argues for inclusive innovation to ensure equitable access to 4IR technologies. Co-creation encourages the active participation of citizens in the development of technological and public service innovations meant for them, as in the discussion on including IKS and African languages in 4IR technologies. Furthermore, the chapter puts emphasis on ensuring that patients from all socioeconomic backgrounds are able to access new technologies and are educated on how to use them. Through a brief analysis of the electronic health records (EHR) implementation process, Mazibuko-Makena explores the barriers to adoption of technologies that the government has previously attempted to implement. Technology adoption is discussed as an intrinsically social and developmental process; therefore, involving the users of technology in the process of developing technologies is crucial.

Gibixego and Odume, in chapter 10, explore the extent to which the 4IR can limit or reverse the damage to key ecological boundaries and systems in South Africa while improving social-ecological interactions. They use the following five important features of South Africa's economies and ecologies to explore this: the quantity and quality of freshwater and health of the general ecosystem; land system change; mining; carbon emission and climate change; and biodiversity. The analysis is done using a systems view, with arguments embedded within a socio-ecological paradigm. The authors argue for the use of

4IR technologies to restore the environment, and to achieve social justice as well.

The chapter thus presents a case for a just and inclusive transition from the current system into a low-carbon economy that will require a paradigm shift in industrial policy. The authors indicate that for South Africa to be on an environmentally sustainable path, there should be an investment in developing 4IR technologies, together with the development of agile, responsive governance and frameworks. Previous industrial revolutions have led to advancements in world development, which have resulted in the overexploitation of natural resources. Consequently, the earth is approaching various ecological tipping points. Furthermore, there has been an exacerbation of inequality and poverty, as the environmental burden falls unevenly on the impoverished who least contributed to this deterioration. Therefore, regulations put in place for 4IR technologies should ensure that environmental risks associated with the 4IR, which may exacerbate existing environmental, social and economic inequalities, are avoided.

Chapter 11, by Ting, investigates the applicability of digital transformation in South Africa's electricity system. The author uses a transdisciplinary lens to reveal how, drawing on an energy justice framework to examine potential sociotechnical implications. Ting shows how the role of digital transformation in electricity systems may contribute to cost and energy savings, as well as new market opportunities, even though the same concerns that emerge in the book for other sectors apply, namely that benefits may be uneven, and in some cases reinforce inequality. The author argues for inclusive, participative and collaborative efforts with consumers. Findings in this chapter show that focusing narrowly on technology as a solution can be detrimental; there needs to be just, systemic change.

Nonetheless, with all the concerns and threats posed by advanced technologies, 4IR technologies are gradually being developed and introduced in South Africa and the rest of Africa. This requires a discussion of knowledge governance, specifically intellectual property (IP) law, to ensure the protection of technological inputs and outputs and to avoid the extensive appropriation and monetisation of technologies. Thus chapter 12 by Ncube and Rutenberg concludes this section by analysing global intellectual property rights (IPR) frameworks from the 1990s and early 2000s, and points out that the current IPR laws in most countries share far more similarities than differences. The authors show that the globalised standard framework for IPR evolved long before 4IR technologies, and that there is very little clarity among stakeholders as to how IPR should be applied and how it should further evolve in the face of emerging technologies. Judicial and administrative decisions have, of necessity, attempted to clarify some of the unexplored and controversial issues, but there remain many questions about the ultimate influence of intellectual property rights on the 4IR. Ncube and Rutenberg reveal additional complications found in many African countries that lie in the implementation of IPR systems. This chapter supplements the discussions on regulations relating to the introduction of 4IR technologies and on how governance often lags behind technology advancement.

In tailoring the discourse on the 4IR to the South African context and occasionally drawing from experiences in other African countries, the authors of this volume ensure the discussion is applicable to concrete realities. The 4IR needs to be clearly understood in this context to respond effectively to the challenges and opportunities it presents for both labour and citizens. In many ways, inclusive innovation is argued to be one of the interventions required to ensure equitable access to 4IR benefits. However, even with such an approach to innovation, a major constraint of the implementation of the 4IR lies in the lack of prerequisite conditions such as ICT infrastructure, skilled personnel and inadequate regulations. Complementary policies across all areas of human endeavour are, thus, required to ensure that the 4IR is tenable and that it does not exacerbate inequalities.

REFERENCES

Adesida, O., Karuri-Sebina, G. and Resende-Santos, J. 2016. *Innovation Africa: Emerging Hubs of Excellence*. UK: Emerald Group Publishing.

African Development Bank (AfDB). 2020. 'African economic outlook 2020: Developing Africa's workforce for the future'. https://www.afdb.org/en/documents/african-economic-outlook-2020, accessed 13 July 2020.

Allen, R.C. 2009. *The British Industrial Revolution in Global Perspective*.

Cambridge: Cambridge University Press.

Arocena, R. and Sutz, J. 2000. 'Looking at national systems of innovation from the South'. *Industry and Innovation*, 7(1), 55–75.

Aston, T.S. 1948. *The Industrial Revolution, 1760–1830*. Oxford: Oxford University Press.

Avis, J. 2018. 'Socio-technical imaginary of the fourth industrial revolution and its implications for vocational education and training: A literature review'. *Journal of Vocational Education & Training*, 70(3), 337–363.

Ayentimi, D.T. and Burgess, J. 2019. 'Is the fourth industrial revolution relevant to sub-Sahara Africa?'. *Technology Analysis & Strategic Management*, 31(6), 641–652.

Best, J. and Gheciu, A. (eds). 2014. *The Return of the Public in Global Governance*. Cambridge: Cambridge University Press.

Blinder, A.S. 2006. 'Offshoring: The next industrial revolution?'. *Foreign Affairs*, 85(2), 113–128.

Bostrom, R.P. and Heinen, J.S. 1977. 'MIS problems and failures: A sociotechnical perspective, part ii: the application of socio-technical theory'. *MIS Quarterly*, 1(3), 11–28.

Bradley, J., Barbier, J. and Handler, D. 2013. 'Embracing the internet of everything to capture your share of $14.4 trillion'. Cisco. https://www.cisco.com/c/dam/en_us/about/ac79/docs/innov/IoE_Economy.pdf, accessed 3 August 2020.

Cave, S. and Dihal, K. 2019. 'Hopes and fears for intelligent machines in fiction and reality'. *Nature Machine Intelligence*, 1(2), 74–78.

Cirera, X. and Maloney, W.F. 2017. 'The innovation paradox: Developing-country capabilities and the unrealized promise of technological catch-up. The World Bank. https://openknowledge.worldbank.org/handle/10986/28341, accessed 3 August 2020.

Coleman, G. 2017. 'Companies benefit from the fourth industrial revolution, but do countries?'. World Economic Forum. https://www.weforum.org/agenda/2017/06/fourth-industrial-revolution-country-competitiveness/, accessed 4 August 2020.

Collan, M. and Michelsen, K.E. 2020. *Technical, Economic and Societal Effects of Manufacturing 4.0*. New York: Springer.

Cozzens, S. 2012. 'The distinctive dynamics of nanotechnology in developing nations'. In: Aydogan-Duda, N. (ed.). *Making It to the Forefront*. New York: Springer, 125–138.

Cozzens, S. and Thakur, D. 2014. *Innovation and Inequality: Emerging Technologies in an Unequal World*. Cheltenham, UK: Edward Elgar.

Crafts, N.F. 1996. 'The first industrial revolution: A guided tour for growth economists'. *The American Economic Review*, 86(2), 197–201.

Du Plessis, H., Sehume, J. and Martin, L. 2011. 'The transdisciplinary research project: Background approaches and methodology'. In: Du Plessis,

H., Sehume, J. and Martin, L. (eds). *The Concept and Application of Transdisciplinary in Intellectual Discourse and Research*. Johannesburg: MISTRA, 17–60.

Eggink, M. 2013. 'The components of an innovation system: A conceptual innovation system framework.' *Journal of Innovation and Business Best Practices*.DOI:10.5171/2013.768378.

Evans, P.C. and Annunziata, M. 2012. 'Industrial internet: Pushing the boundaries'. *General Electric Reports*, 488–508.

Freeman, C. 1987. *Technology Policy and Economic Performance*. London: Pinter.

Freeman, C. and Lundvall, B. (eds). 1988. *Small Countries Facing Technological Revolution*. London: Pinter.

Frischmann, B.M. 2005. 'An economic theory of infrastructure and commons management'. *Minnesota Law Review*. https://scholarship.law.umn.edu/mlr/673, accessed 11 July 2020

Frischmann, B.M. 2012. *Infrastructure: The Social Value of Shared Resources*. Oxford: Oxford University Press.

Geels, F.W. 2002. 'Technological transitions as evolutionary reconfiguration processes: A multi-level perspective and a case-study'. *Research Policy*, 31(8–9), 1257–1274.

Geels, F.W. 2004. 'From sectoral systems of innovation to socio-technical systems: Insights about dynamics and change from sociology and institutional theory'. *Research Policy*, 33(6–7), 897–920.

Gleason, N.W. 2018. *Higher Education in the Era of the Fourth Industrial Revolution*. New York: Springer Nature.

Griffith, R., Redding, S. and Van Reenen, J. 2004. 'Mapping the two faces of R&D: Productivity growth in a panel of OECD industries'. *Review of Economics and Statistics*, 86(4), 883–895.

History.com. 2019. 'Industrial Revolution'. https://www.history.com/topics/industrial-revolution/industrial-revolution, accessed 2 July 2019.

Humphries, J. 2010. *Childhood and Child labour in the British Industrial Revolution*. Cambridge: Cambridge University Press.

Jänicke, M. and Jacob, K. 2009. *A Third Industrial Revolution? Solutions to the Crisis of Resource-Intensive Growth*. Berlin: Environmental Policy Research Centre.

Kracke, M. 2012. 'Overview of the advanced manufacturing partnership'. Conference paper. Conference and Exposition American Society for Metals.

Levinson, M. 2019. 'Job creation in the manufacturing revival'. Congressional Research Service. https://fas.org/sgp/crs/misc/R41898.pdf, accessed 4 August 2020.

Li, G., Hou, Y. and Wu, A. 2017. 'Fourth industrial revolution: Technological drivers, impacts and coping methods'. *Chinese Geographical Science*,

27, 626–637.

Lorenz, E., Lundvall, B.Å., Kraemer-Mbula, E. and Rasmussen, P. 2016. 'Work organisation, forms of employee learning and national systems of education and training'. *European Journal of Education*, 51(2), 154–175.

Lundvall, B.A. 1988. 'Innovation as an interactive process: From user–producer interaction to the national system of innovation'. In: Dosi, G., Freeman, C., Nelson, R. R., Silverberg, G. and Soete, L. (eds). *Technical Change and Economic Theory*. London: Pinter.

Lundvall, B.Å. (ed.). 1992. *National Systems of Innovation: Towards a Theory of Innovation and Interactive Learning*. London: Pinter.

Lundvall, B., Joseph, K.J., Chaminade, C. and Vang, J. (eds). 2009. *Handbook of Innovation Systems and Developing Countries: Building Domestic Capabilities in a Global Setting*. Cheltenham, UK: Edward Elgar.

Mahomed, S. 2018. 'Healthcare, artificial intelligence and the Fourth Industrial Revolution: Ethical, social and legal considerations'. *South African Journal of Bioethics and Law*, 11(2), 93–95.

Manyika, J. 2012. 'Manufacturing the future: The next era of global growth and innovation'. McKinsey Global Institute. https://www.mckinsey.com/~/media/McKinsey/Business%20Functions/Operations/Our%20Insights/The%20future%20of%20manufacturing/MGI_Manufacturing%20the%20future_Executive%20summary_Nov%202012.pdf, accessed 3 August 2020.

Mathews, J. and Lee, K. 2018. 'How emerging economies can take advantage of the Fourth Industrial Revolution'. World Economic Forum. https://www.weforum.org/agenda/2018/01/the-4th-industrial-revolution-is-a-window-of-opportunity-for-emerging-economies-to-advance-by-leapfrogging/, accessed 4 August 2020.

Mendes, A.P.F., Bertella, M.A. and Teixeira, R.F.A.P. 2014. 'Industrialization in Sub-Saharan Africa and import substitution policy.' *Brazilian Journal of Political Economy*, 34(1), 120–138.

Mokyr, J. 1998. 'The second industrial revolution, 1870–1914'. *Storia dell'economia Mondiale*, 219–245.

Muchie, M. 2016. 'Towards a unified theory of Pan-African innovation systems and integrated development'. In: Adesida, O., Karuri-Sebina, G. and Resende-Santos, J. (eds). *Innovation Africa*, UK: Emerald Group Publishing Limited, 13–35.

Murphy, T., Garg, S., Sniderman, B. and Buckley, N. 2019. 'Ethical technology use in the fourth industrial revolution: CEO leadership needed'. *Deloitte Insights*. https://www2.deloitte.com/content/dam/insights/us/articles/6275_Ethical-technology/DI_Ethical-technology-use-in-the-Fourth-Industrial-Revolution.pdf, accessed 11 July 2020.

Nelson, R.R. (ed.). 1993. *National Innovation Systems: A comparative Analysis*. Oxford: Oxford University Press on Demand.

OECD. 1997. *National Innovation Systems*. OECD. http://www.oecd.org/dataoecd/35/56/2101733.pdf, accessed on September 2020.

Oluwatayo, I.B. and Ojo, A.O. 2018. 'Walking through a tightrope: The challenge of economic growth and poverty in Africa'. *The Journal of Developing Areas,* 52(1), 59–69.

Oyelaran-Oyeyinka, B. 2014. 'The state and innovation policy in Africa'. *African Journal of Science, Technology, Innovation and Development,* 6(5), 481–496.

Page, J. 2005. 'Strategies for pro-poor growth: Pro-poor, pro-growth or both?'. African Development and Poverty Reduction: The Macro-Micro Linkage, the DPRU/TIPS Forum, hosted in association with Cornell University.

Perrot, R. 2018. 'The dynamics of new and emerging technologies in developing countries and the new role of the state: An introduction'. In: Mapungubwe Institute for Strategic Reflection (MISTRA). *Beyond Imagination: The Ethics and Applications of Nanotechnology and Bio-Economics in South Africa.* Mazibuko, Z. (ed). Johannesburg: MISTRA, 1–16.

Philbeck, T. and Davis, N. 2019. 'The fourth industrial revolution: Shaping a new era'. https://jia.sipa.columbia.edu/fourth-industrial-revolution-shaping-new-era, accessed 24 July 2020.

Radziwon, A., Bilberg, A., Bogers, M. and Madsen, E.S. 2014. 'The smart factory: Exploring adaptive and flexible manufacturing solutions'. *Procedia Engineering*, 69, 1184–1190.

Renjen, P. 2019. 'Success personified in the fourth industrial revolution'. *Deloitte Insights.* https://www2.deloitte.com/content/dam/insights/us/articles/GLOB1948_Success-personified-4th-ind-rev/DI_Success-personified-fourth-industrial-revolution.pdf, accessed 11 July 2020.

Rifkin, J. 2012. 'The third industrial revolution: How the internet, green electricity, and 3D printing are ushering in a sustainable era of distributed capitalism'. *The World Financial Review*. http://www.worldfinancialreview.com/the-third-industrial-revolution-how-the-internet-green-electricity-and-3-d-printing-are-ushering-in-a-sustainable-era-of-distributed-capitalism/, accessed 5 July 2019.

Rodney, W. 1982. *How Europe Underdeveloped Africa.* Washington: Howard University Press.

Rotolo, D., Hicks, D. and Martin, B.R. 2015. 'What is an emerging technology?'. *Research Policy,* 44, 1827–1843.

Salahuddin, M. and Gow, J. 2016. 'The effects of internet usage, financial development and trade openness on economic growth in South Africa: A time series analysis'. *Telematics and Informatics,* 33(4), 1141–1154.

Scerri, M. 2019. 'Human capabilities and the evolutionary prospects for systems of innovation in sub-Saharan Africa'. IERI Working paper, WP2019-002, April.

Schumpeter, J. 1942. 'Creative destruction'. *Capitalism, Socialism and Democracy*, 825, 82–85.

Schwab, K. 2015. 'The Fourth Industrial Revolution: What it means and how to respond'. *Foreign Affairs*. www.foreignaffairs.com/articles/2015-12-12/fourth-industrial-revolution, accessed 20 June 2020.

Schwab, K. 2016. *The Fourth Industrial Revolution*. New York: Crown Business.

Schwab, K. and Davis, N. 2018. *Shaping the Future of the Fourth Industrial Revolution: A Guide to Building a Better World*. London, UK: Portfolio Penguin.

Soumonni, O. 2016. 'Innovation in emerging technologies and socio-economic transformation in Africa: Fallacy or foresight?'. *Africa Growth Agenda*, 13(4), 18–22.

Tien, J.M. 2012. 'The next industrial revolution: Integrated services and goods'. *Journal of Systems Science and Systems Engineering*, 21(3), 257–296.

Tondi, P. 2019. 'The significance of indigenous knowledge systems (IKS) for Africa's socio-cultural and economic development in the dawn of the Fourth Industrial Revolution (4IR)'. *Journal of Gender, Information and Development in Africa (JGIDA)*, 8(1), 239–245.

Van Rensburg, N.J., Telukdarie, A. and Dhamija, P. 2019. 'Society 4.0 applied in Africa: Advancing the social impact of technology'. *Technology in Society*, 59, 101125.

Xing, B. and Marwala, T. 2017. 'Implications of the Fourth Industrial Age for higher education'. *The Thinker: For the Thought Leaders*, 73(3), 10–15.

Section One

*Emerging Technologies and
Inclusive Innovation*

.

ONE

A new digital deal rather than a Fourth Industrial Revolution policy?

ALISON GILLWALD[1]

INTRODUCTION

There has been relatively little critical engagement at the level of international policy with the World Economic Forum (WEF)'s notion of a Fourth Industrial Revolution (4IR), coined in 2016 alongside Klaus Schwab's publication of the same name. Despite its technological determinism and ahistoricism, the concept has gone viral in policy circles. Ivy League universities, international industry conferences,

---

1   The chapter draws on research undertaken for a number of different projects by Research ICT Africa (RIA) on interventions that could contribute to a new digital deal for South Africa. The research was led by the author and included high-level interviews with senior officials undertaken for a 2019 review of the National Development Plan (NDP) in relation to the 4IR. The author thanks RIA staff and associates Anri van der Spuy, Nils Bergland, Chris Geerdts, Shamira Ahmed and Fazila Farouk for their comments and collaboration on different aspects of the research in this paper and for the comments and review of this chapter. All views expressed and errors, however, remain those of the author.

multilateral agencies, governments, the media and the public have all been swept up in the frenzy around the potential for 4IR technologies – Artificial Intelligence (AI), robotics, drones, blockchain and others – to drive economic growth, and the dire fate of nations who fail to embrace these technological developments.

Of course, there has been some critical research (cf. Avis, 2018; Boyd and Holton 2018; Cammaerts and Mansell, 2020) and popular critiques in South Africa (cf. Gillwald, 2019; Badat, 2020). However, at an international level and in terms of national policy in South Africa, the 4IR discourse is widely accepted uncritically.

Since the African National Congress adopted 4IR as the central pillar of its communications policy at its 54th national elective conference in 2017, it has become the mantra of every official event from the State of the Nation address to the presidential inauguration – 4IR is a panacea for economic growth, job creation and the empowerment of women and the youth. Since then, WEF-inspired policies have been rolled out in the departments of Science and Innovation, Arts and Culture as well as Trade, Industry and Competition. With little significant public funding of independent digital policy research at public universities in South Africa, the government has established a WEF-affiliate Centre for the 4IR at the Council for Scientific and Industrial Research (CSIR) to support policy development (Mandaha, 2019).

The Department of Communications and Digital Technologies, which is also acting as secretariat for the Presidential Commission on the 4IR, released a policy document on 4IR in 2019. However, the department might have been better served by drawing on local, independent expertise for this document rather than on policy for which a private consultancy company, Accenture, holds the copyright. The report claims that 'using the methodology developed by the World Economic Forum in collaboration with Accenture', 4IR technologies across industries in South Africa can contribute five trillion rand worth of social and economic value and create four million jobs. However, it makes this claim without any reference to sources for the numbers in it or any transparency on the modelling used,

To put these claims in perspective, South Africa has produced between 270,000 and 300,000 jobs per annum over the past five years, with about 700,000 young people entering the job market every year. The Centre for Economic Development and Transformation estimates that absorbing these new entrants into the economy will require a growth rate of around 10 per cent (Mabaso, 2019 in Gillwald, 2019). With a nod to digital inclusion, the Accenture-copyrighted report does observe that to make this miracle happen, South Africa will need to meet several other conditions. These include improving connectivity; effective regulation; functioning markets optimised consumer welfare redressing poor education outcomes; and developing an appropriate digital skills base for the new economy. These conditions have been identified in policies and plans over the past decade, most recently in the national broadband plan, SA Connect (Electronic Communications Act, 2013), all of which the country has failed to meet. The report does not attempt to explain why this is so or what needs to be done to realise these preconditions for its forecasts to hold true.

POLICY AND GOVERNANCE RESEARCH PROBLEM

The introduction and proliferation of advanced digital technologies is likely to increase inequality rather than alleviate or reduce it unless policy interventions are specifically designed to redress unevenness in the opportunities to utilise these advanced technologies. This resultant 'digital inequality paradox' means that as more people are connected to – or become data subjects of – advanced technologies, the more inequality increases (Gillwald, 2018). Digital inequality not only exists between those online and offline (Gunkel 2003; Kenny 2003), but between those who have the technical and financial resources to use the internet optimally and those who cannot afford consistent, meaningful usage or lack basic digital literacy to use new applications and services.

Digital inequality paradox

This digital inequality paradox is arguably, after climate change, one of the most intractable policy problems facing governments today (De la Chapelle, 2019), particularly in highly unequal countries like South Africa. The introduction of advanced technologies will amplify current inequalities unless it forms part of a transversal national project that seeks to transform the economy and society, including developing some digital preconditions. These preconditions include enabling affordable access to and use of advanced technologies by individuals, the informal sector, business and the public sector. It also includes actively limiting harms and mitigating the risks associated with increased digitalisation and 'datafication' associated with hyperglobalisation and advanced capitalism – what Zuboff (2019) calls 'surveillance capitalism'.

This chapter moves from the policy critique that developing countries cannot divert all their resources, as more developed economies are doing, to focus narrowly on the potential and dangers of so-called 4IR technologies. While aiming for the benefits of AI, machine learning and data analytics in monitoring, planning and forecasting public-sector delivery, and in improving firm productivity, South Africa needs to develop a transversal digital policy that is far more comprehensive and inclusive than one focusing on 4IR. Policymakers need to recognise that while the technologies may be neutral, the uses to which they are put reflect particular interests; that the value for individuals tends to be unevenly optimised; and that the beneficial absorption of new technologies is determined by the country's level of human development. Prioritising 4IR policy over other far-from-completed earlier digital policy challenges will result in the continued preferencing of those who have the access and skills to use these advanced technologies to improve their life opportunities over those who do not.

Digitalisation and 'datafication'

This chapter proposes a policy framework that looks at the intensifying processes of digitalisation and datafication (and their economic, technological and innovation drivers) in an increasingly unevenly developing global economy, rather than the rapidly changing technologies themselves. Developing countries need to prepare for AI becoming the next general-purpose technology they need to continue to secure universal, affordable access to underlying information infrastructures, networks, and services if they want more equitable societies. It is these that underpin Over the Top (OTT) services, Internet of Things (IoT), social networking and e-commerce platforms, which already skew information flows. Developing countries need to address policy and institutional challenges at national and local levels, and the backlogs that have been created, in order to harness advanced technologies for development. This will require heightened institutional capacity that can respond with agility to increasingly dynamic and complex global governance contexts to secure economic opportunities and safeguard citizens' rights in a globalised gig economy.

Alternative policy and regulatory frameworks

Within this context, the chapter seeks to provide an alternative vision to that of the WEF's 4IR, which has been so widely adopted by development banks and multilateral institutions; it seeks a vision more suited to South Africa's political economy. Many of the country's past policy failures can be attributed to adopting so-called 'best practices' from more mature markets with more developed regulatory institutions. This chapter proposes an alternative framework to inform regulatory and governance challenges associated with the processes of digitalisation and datafication that characterise our increasingly globalised digital economy and that play out at the international, national and local level.

In doing so it draws on the United Nations Conference on Trade and Development (UNCTAD) Global Green New Deal 2019, which argues that 'to exit the age of anxiety that the global economic crisis ...' (compounded by the COVID-19 pandemic) has resulted in, will require '... a profound change in the thinking and policy mix that caused the economic crisis' and the negative impacts of hyperglobalisation. While the appropriate mixture of recovery, regulation and redistribution will vary across countries (with policy experimentalism of particular importance in the developing world), all policymakers can still usefully recall the original New Deal of the 1930s, which must now be translated globally to leverage the opportunities of today's inter-dependent world (UNCTAD, 2019: 3).

This new hyperglobalised environment (Marsden, 2001), undergirded by unevenly distributed global digital networks, brings with it a new set of policy challenges. In pursuit of the policy experimentalism so badly needed in developing countries to rectify the failures of past policy to reduce digital inequality, this chapter seeks to demonstrate how an alternative approach to resource allocation in just three policy areas, framed differently, could have the transformative effects anticipated in the South African constitution. It could also help the country to make progress towards the ICT targets directly associated with 6 of the 17 sustainable development goals (SDGs).

Research question

The overarching research question this raises is: with intensifying digitisation and datafication of an increasingly globalised economy, what policies and forms of governance are required to realise global public goods at the national level? What is required to redress digital inequality, harness the potential of new technologies for social and economic development and improve public sector efficiency and delivery while simultaneously protecting citizens from harms associated with being data subjects of big data collection and analytics? What is required to mitigate the risks of increasingly pervasive 'surveillance capitalism' (Zuboff, 2019)?

Conceptual framework

The conceptual framework is constructed from two different sources. The first is the work of Best and Gheciu (2014), who examine the changing nature of 'public'. The second is the work of Frischmann (2005; 2012) who writes about the need for demand-side value of public resources, not just the commercial supply-side valuation that has driven the liberalisation of markets since the 1980s World Trade Organization (WTO) General Agreement on Trade in Services (GATS). The concepts put forward in these two bodies of work are best understood through Kaul et al.'s (2003) extrapolation of the concept of public goods from the national to the international level to provide a basis for the increasing need for global governance of worldwide platforms and services.

The paper also draws on Best and Gheciu (2014) to explore 'disruption and transition' in shifting conceptions of what notions like 'public' and 'private' mean. Their conceptualisation of 'public' as a set of practices rather than a single institution contributes to the growing literature on the blurring of boundaries between public and private in governance discourse (Cammaerts and Mansell, 2020). Together with Hirschman's (1958) notion of private provisioning of social and public goods that does not exclude an enabling role for the state, these non-state forms of 'public' in global governance and in local delivery, provide a path out from the polarised views of the state as public and the market as private.

As something of a middle way, advocates of open access frequently call for the creation or the protection of a commons (Frischmann, 2005), which hypothetically serves as an alternative to state control and generally amounts to the rejection of resource management through the market. Drawing on Ostrom's (2010) theory of common-pool resources, Frischmann (2005; 2012) presents an economic case for why some classes of key resources need to be managed in a more accessible manner. While conventional economic analyses focus primarily on the supply-side value of infrastructure and the profit imperative in network investment and regulation, Frischmann (2012) explores demand-side considerations to analyse how infrastructure resources generate value for consumers.

Building on traditional economic concepts used in welfare analyses of infrastructure resources and societal demand, Frischmann (2005) puts forward a new theory of infrastructure from three key insights that emerge from adopting a demand-side, value-creation-focused analysis. The first is that infrastructure resources are fundamental and generate value when used as inputs into a wide range of productive processes. The second insight highlights that the outputs of infrastructure industries are generally public and 'non-market' goods that create positive multipliers in both economy and society. The third insight is that 'managing infrastructure resources in an openly accessible manner may be socially desirable when it facilitates … downstream activities' (Frischmann, 2005: 918 in Gillwald and Van der Spuy, 2019).

This demand-side valuation of public goods extends the notion of infrastructure resources to the wider information infrastructure or so-called infostructure (Melody, 2012). This structure includes the increasingly complex array of global public goods produced – and increasingly commercially valued and commodified – and the high levels of inequality that result from the intensifying process of digitalisation and datafication. The multifaceted notion of 'public' is also used to expand the conceptualisation of public provisioning, usefully expounded by Kaul et al. (2003). They identify the internet and knowledge as dimensions of global goods and governance. Such global public goods emerge to the extent that all countries help to produce them. Kaul et al. (2003) appeal for new forms of international cooperation and institutions that will support the development of global public goods and ensure greater digital inclusion.

Conceptually this analysis draws together notions of public, private and civil society in the governance and delivery of public goods, in varied forms. Increasingly, global cooperation will be required for delivery of these goods to be realised at national levels.

This conceptual framework is used to analyse three policy and regulatory cases at the local, national and international levels of governance that are often raised in the 4IR discourse. From an empirical point of view, this paper does not deal comprehensively with each case, given the limitations of a book chapter. Rather, the chapter draws on an analysis of relevant documents, laws and regulations pertaining to these three mini-cases. It triangulates the findings from that analysis with high-level interviews with senior officials in different institutions over a number of years, but particularly for the digital futures review undertaken for the National Development Plan (NDP) in the context of the 4IR in 2019. From this, the chapter seeks to demonstrate where regulation and governance of digital resources as public goods, at the local, national and international level, and through a demand-side valuation, could have transformative outcomes.

The first case reviews frequency spectrum regulation as a central tool for facilitating affordable, universal access to the internet. Under pressure from powerful global mobile interests, the spectrum policy agendas have been driven by 5G in the context of network readiness for the 4IR, rather than addressing the longstanding backlog in releasing spectrum[2] or ensuring access for the almost four billion

---

2 'Spectrum' refers to the full range of radio waves used for varied communication purposes, including radio broadcast, bluetooth, wifi and smartphones

people worldwide who are not online (ITU, 2019). The second case looks at data governance, an issue that is central to the development of a framework that reduces harms and mitigates the risks associated with the deployment of big data analytics, AI and Machine Learning. Dominated by five global tech monopolies, Google, Amazon, Facebook, Apple and Microsoft, their business is built on the gathering of data on the behaviour of millions of individuals around the world through 'free platforms'. These platforms are monetised through selling the data gathered to advertisers, developers or political strategists. Zuboff (2019) meticulously describes the development of 'behavioural surplus' which underpins what has famously become known as 'surveillance capitalism' in her book by the same name. The third case looks at 'smart cities' that are now widely proposed at the local and national level as part of wider 4IR narratives. 'Smart townships' are presented as an alternative national strategy to deal with the extreme digital inequality in urban contexts but which could also be extended to 'smart villages' using demand-side valuation of resources rather than the supply-side valuation that has come to drive the allocation of digital resources at the local level with private delivery of public goods.

NATIONAL RESOURCE ALLOCATION: DEMAND-VALUATION OF SPECTRUM AS A PUBLIC GOOD

In the context of digital policy, regulatory and institutional failure, this section assesses the issue of forthcoming spectrum assignments in relation to the public objectives of universal affordable access to digital services for users, both private and commercial. With citizens predominantly dependent on mobile communications, 85 per cent of which are prepaid mobile services, this is another area of digital readiness in which there are severe regulatory bottlenecks. In the context of the long overdue forthcoming auction by the Independent Communications Authority of South Africa (ICASA) of high demand spectrum for 4G or LTE services, the need for the release of 5G spectrum to support South Africa's preparedness for the 4IR has been strongly emphasised.

Release of spectrum is essential for the optimal evolution of next-generation technologies, even though operators deploy technology to overcome supply-side challenges and circumvent regulatory obstacles. The lack of success of any auction in Africa is cited by open access advocates to argue against the 'best practice' orthodoxy on spectrum assignment (Song, 2017). However, there are few other ways of valuing and assigning spectrum competitively whilst ensuring optimal efficiency (including regulating spectrum trading to enable the correction of any auction outcome errors). Auctions can also be qualified to compel bidders to provide services to unserved parts of the country first or to share spectrum with secondary users.

There are other more cautionary lessons to be drawn from more than two decades of supply-side valuation of spectrum in auctions. Theoretically, the more public-service obligations placed on a licence, the lower the price. Very high reserve prices for auctions, driven by revenue-hungry treasuries throughout the world, have exacerbated the artificial scarcity of spectrum. High prices resulting from artificial scarcity both delay the rollout of next-generation services, as operators recoup their costs of acquiring spectrum, and drive up consumer prices for new services (Melody, 2001). This has had extremely negative policy outcomes for downstream services and for consumer welfare, both of which are prime objectives of competition and sector regulation (Gillwald, 2005).

The extractive rents (prices) being commanded by ineffectively regulated operators and by indebted, developing-country governments (licence, auction fees) through commercial supply-side valuation of spectrum represents a key bottleneck in affordably meeting public demand in developing countries. Arguably worse than high spectrum prices, however, is not releasing spectrum, which is essential to the evolutions of global systems for mobile communications (GSM) networks and for access to innovative technologies in all available bands. However, granting it on the basis of supply-side valuation only (auction) without the caveats (demand-side value) identified above could intensify inequality and miss an opportunity to open the market to smaller players.

Although previous dominant supply-valuation of spectrum in

South Africa has resulted in significant infrastructure investments connecting millions of people to communications services, it has also been highly uneven. It has left large numbers of South Africans unable to enjoy the economic and social benefits of the internet. This problem became more urgent with the onset of COVID-19: as a relatively small number of South Africans have the resources to move their work, schooling and retail activities online with COVID-19, the vast majority are unable to enjoy what has become a minimal requirement for participation in the modern economy and society.

Wholesale Open Access Networks (WOAN)

In the absence of the ICASA completing a statutory market review to determine market power, an open-access network was conceived by the Department of Communications as a solution to the dominance of the mobile operators (Interview: Mjwara, 2019). It was argued that this would enable more competitors to come into the market, affordably accessing its common carrier objectives. At the time, proponents of an open access network believed that, in the context of a developmental state, the network should be a state-owned entity. Ownership of the wireless mobile network would have extended the national broadband infrastructure of the state, which was envisaged with the now largely defunct state-owned fibre company, Infraco (Gillwald, 2007).

But is the preferential treatment of the WOAN required by the ministerial directive justified by its ability to rectify inequities in the market? What is clear is that the policy intervention initially tried to deal with the regulatory failure around dominance in the market, despite the arguable political and financial capture of the policy instrument along the way. However, instead of enabling greater participation in the mobile market and more diverse spectrum holdings by historically disadvantaged individuals, policy makers have again resorted to supply-side measures that are likely to serve vested interests (Interviews: Mjwara, 2019; Booi, 2019; Silber, 2017; Ngwepe, 2019). Mandated favourable spectrum allocation may well be set up for failure – including the arguably anti-competitive or irrational compulsory use of the WOAN for 30 per cent of their spectrum use by dominant operators (McCleod, 2019; Vermeulen, 2019).

The WOAN is set up to compete against highly successful, dominant players, as was Infraco, in the fibre market (cf. Gillwald, 2005; Gillwald et al., 2018). Setting aside a significant amount of spectrum through demand-side valuation on the other hand has the potential to greatly increase spectrum access to smaller internet service providers and individual users in ways that would alleviate digital inequality. Considering the failure of certain public policy reforms in the telecommunications sector – leading to a lack of equitable access to the internet in South Africa – spectrum should be understood as a central resource in the mobile market that could serve as the primary vehicle for Internet take-up and as an input into downstream activities.

Complementary forms of supply- and demand-side management are necessary to meet constraints in developing countries. Building on Hirschman's (1958) notion of private provisioning of social and public goods, the state should attract or 'crowd-in' productive private investments where it lacks the resources or institutional endowments to provide them. As Frischmann (2005) notes, the state still has a responsibility to coordinate and regulate the information infrastructure to ensure widespread access to it by citizens, despite the fact that much of this infrastructure is provided by the private sector. The institutional arrangement between the state and its specialised agencies, industry and civil society needs to produce the technical capacity to ensure the level of cross-sector public and private coordination required for such a national project.

With COVID-19 came an increased demand for bandwidth. The South African government's response was to release temporary spectrum, in advance of a planned spectrum auction. However, this sudden need represented a missed opportunity for South Africa: it could have been a chance to correct its course to meet the national objective of affordable digital access for all to the full range of communication services. Particularly because part of the rationale was to reduce digital inequality, greater consideration should have been given to safeguarding the social value of spectrum as a public good. Some key resources – like spectrum – need demand-side valuation to recognise the full value of their utility. This could have been recognised by expanding the 'spectrum common' through the opening up of bigger tracts of spectrum for public access, such as is currently being done with public wireless computer networks (WiFi) in the unlicensed industrial, scientific and medical (ISM) bands intended for research and experimentation.

Wi-Fi operating on the licence-exempt ISM bands, for which it was not really intended, has already demonstrated the power of the commons[3] and its potential as an access and backhaul technology. The nature of Wi-Fi means there are low market barriers for manufacturing and deploying this technology, which has allowed people to build broadband networks and connect places deemed 'uneconomic' by operators in a manner unforeseen by policymakers (cf. Song, 2019). This has resulted in a proliferation of independent, non-profit, community-led initiatives, as well as commercial wireless internet service providers (ISPs), to be able to meet some of the pent-up small- and micro-scale demand.

GLOBAL GOVERNANCE OF GLOBAL PUBLIC GOODS: INTERNET AND DATA AS GLOBAL PUBLIC GOODS

This conceptualisation of critical resources can be extended to data as a key resource for consumption and production in the data-driven economy. Demand-side valuation, whether of spectrum or of data, enables public interest governance of a resource as a non-rivalrous, low-excludability public good that can be accessed for public planning, entrepreneurship and democratic accountability. There is increased recognition of the need for greater governance of increasingly global public goods with the digitisation of communications technologies – together with the reinforcing liberalisation of markets – driving the rise of global markets and players (UN HLPDC, 2019 in Gillwald and Van der Spuy, 2019).

The rise of the internet as a global public good necessary for global trade, and financial and information flows, requires new forms of

---

3 'The commons' refers to the cultural and natural resources accessible to all, including air, water and earth. 'Digital commons' involve communal ownership and distribution of information resources and technology.

global cooperation to govern the data and content being generated. To be effective, global governance instruments need to be adopted and implemented at national level. The following sections provide some examples of governance of these complex and adaptive global public goods, and the formal and informal ways in which this plays out at regional and national levels.

Global governance, specifically of the internet, is a complex, multi-institutional process that operates outside of the traditional institutions responsible for the international coordination of sovereign member states in the areas of telecommunications, trade and security. Institutions such as the International Telecommunication Union (ITU) and the World Trade Organization (WTO) that have played a critical role in the opening up of markets, and the setting of international standards to enable global markets, have already modified their structures to include the private sector interests driving market expansion.

New organisations dedicated to internet governance are not member state based or organised like traditional multilateral organisations. Generally, they adopt a 'multi-stakeholder' approach to governance, though the interests of government, the private sector and civil society are seldom the same. On a technical level, this includes policy platforms like the Internet Engineering Task Force (IETF) and the International Corporation for Assigned Names and Numbers (ICANN). There are agencies such as the WTO and the World Intellectual Property Organisation (WIPO) working towards the development of formal agreements between member states to enable free trade and the protection of intellectual property respectively. However, in addition, global interests such as the World Economic Forum, the Global System for Mobile Communications Association (GSMA), and the big platforms and applications – such as Facebook, Amazon and Google – dedicate significant resources to lobbying aimed at limiting formal regulation of their businesses. Included in such lobbying is the offer of self-regulated solutions to the need for identifying new harms and mitigating risks, and enforcing these solutions when the businesses do not have a physical presence within a government's jurisdiction.

Even with these modifications and the belated inclusion of civil society representation, national sovereign governments are arguably not equipped to deal with the governance of the privately delivered, global public goods spawned from the global expansion of the internet, along with the global monopoly platforms and applications they have given rise to. Challenges of global governance lie in the increasing complexity and adaptiveness of the global communications system over which nation states, and particularly developing countries, appear to have little control. Multiple, competing interests coalesce around these contemporary digital governance questions (Gillwald and Van der Spuy, 2019).

Like all public goods, in order to manage the negative externalities associated with the universality of the internet, someone has to provide the resources required to build and maintain the internet and mitigate the risks associated with it. For this reason, much funding for development of the internet and tools to govern it, where states are unable to provide these themselves, are invested by mature states' departments of international affairs or development and ministries, or the funding is provided by multilateral agencies or other third parties. While these efforts sometimes have positive outcomes, depending on where the funding comes from, they are also used to launder foreign countries' policies, to promote laws modelled in dominant regimes in developing countries by insisting on policy harmonisation or to shift policy discussions to more favourable forums (Hosein, 2006), often to serve dominant commercial interests.

As a result, global North institutional and governance frameworks are frequently exported and adopted uncritically for local use even in the absence of local research and policy formulation. Yet, for most African countries this formulation of policy is something that happens externally to them and from which they need to safeguard their national interests and sovereignty (Calandro et al., 2013). Dominant interests determine the agenda and influence decision-making, whether through international treatises or regional regulation at the economic community level. Implementation occurs formally, but also through networks of development banks and consultants who are active in developing countries, often paid for by industry associations or the platforms themselves (Haas, 1992).

Data governance and justice

Despite the low levels of internet penetration on the continent, Africans' digital identities and data are being used to feed into, improve and alter emerging technologies, particularly AI and associated Machine Learning. With data being collected by third parties, who are responsible for operating government systems in public–private partnerships or supplier relationships, even people who may not be online have become data subjects and vulnerable to exploitation. This was evident in the case of the South African Social Security Agency (SASSA) and private contractor Cash Paymaster Services (CPS). SASSA now has one of the largest databases in the country as a result of millions of beneficiaries having had their fingerprints, photographs and even their voices captured for an automated payment system that forms part of the contract with CPS. Following a Constitutional Court order instructing SASSA and CPS to pay pensions on time, CPS then had to be prevented by the courts from exploiting the SASSA public biometric databases. The Court ruled that CPS, together with its mother company, Net1 UEPS Technologies Inc (Net1) and working in partnership with Grindrod Bank and Mastercard, should be prohibited from marketing financial services to the SASSA beneficiaries (Razzano, 2017).

An alternative way of allocating the critical resource of data may be by shifting from a commercial, supply-side valuation of the data only (reflected in the profitability of global monopolies such as Google or Facebook) to a complementary demand-side valuation of data. This would allow for its use where it is most needed as an input in public (planning or service delivery) or private (start-up innovation) uses. Public allocation on this basis should allow for the resource to be more transparently and accountably managed and for the privacy, anonymity and identification of data subjects to be better safeguarded.

Indeed, the need to harness data commons for development purposes has gained traction over the past year in various intergovernmental processes. For example, in the outcome report of the UN Secretary-General's High-Level Panel on Digital Cooperation (UN HLPDC, 2019), a 'Digital Commons Architecture' is proposed in order to better synergise efforts by governments, civil society and the private sector 'to ensure that digital technologies promote the SDGs' and to 'address risks of social harm' (UN HLPDC, 2019; Gillwald and Van der Spuy, 2019).

With growing deployment of AI and biometrics in digital governance systems, citizens are increasingly data subjects even if they not connected to the internet. Greater biometric identification across the continent raises issues of data justice as people are compelled to provide personal data in order to access government services, or to participate in a system without sufficient protection, opening themselves to risks of their privacy being breached or being subject to illegal state surveillance. With such biometric systems increasingly implemented within the public–private partnerships discussed above, third-party commercial surveillance or misuse of data present risks without data protection frameworks and the widespread raising of awareness of peoples' rights. Policymakers, governments and development agencies have used data to classify and categorise people or – where data is unavailable – to omit them from data sets and the planning such data enables, such as for housing or health care. In low-income or remote environments, authorities' ability to gather data is often limited, meaning that some communities are at risk of suffering from data-driven discrimination due to digital invisibility (Taylor, 2017 in Gillwald and Van der Spuy, 2019).

These actions all influence the way people are seen and treated by the state and private sector – leading to significant 'ethical, political and practical implications' (Taylor, 2017). As far as the private sector is concerned, for instance, innovative global technology platforms have used such data to become some of the biggest corporations on earth. Tech giants like Google, Apple, Facebook and Amazon are neither bound by national boundaries nor by most national laws, and are therefore accountable to few (Johnson and Post, 1996). But policymakers, especially in Europe, are starting to push back against the corporations' behaviour through a combination of taxation and data and consumer protection laws, such as the global digital tax being proposed by the G20, the Organisation for Economic Co-operation and Development (OECD) and the General Data Protection Regulation (GDRP) of the European Union.

Attached to this is the fact these tech giants, including social media platforms of various kinds, collect and store personal data on individuals, including private information and behavioural trends. Who this data belongs to and how it is used and shared is a fundamental question of both IP and the right to privacy.

Internet governance, unlike earlier forms of telecommunications regulation, has to be coordinated at global level by the very nature of the globalised markets, services and products that operate across it. While African countries need to align their cybersecurity and data protection frameworks to international norms and treaties, they also need to ensure that laws and guidelines are implementable, given the institutional endowments of their respective nations. As much larger numbers of marginalised people come online, extra effort will be required to ensure people are able to exercise their rights to data justice in digital policy frameworks which respect online privacy and keep users safe and secure. African perspectives are necessary to inform global governance of what was recently described as the latest 'civilisational challenge' – that of organising billions of people around the world in a highly complex and adaptive system, through global collaboration on the basis of principles and norms that will accommodate the greatest possible diversity (De la Chapelle, 2019).

African governments have, however, been largely disengaged, either with the technical regulation of the internet through ICANN, or with the more informal, multi-stakeholder internet governance forums that discuss the future of the internet. There is very little reference in these forums to the challenges of global governance and the ability of developing countries to feed into this governance. Data drives the digital economy. Creating trusted environments for equitable digital participation and innovation at the national level – through the shaping of and alignment with global governance – is as important as infrastructure development and the anticipation of advanced technologies to safeguard the country's digital future.

LOCAL GOVERNANCE: SMART TOWNSHIPS

Issues of digital governance have not only shifted 'upwards' to the international level from the national sovereign state level where telecommunications have traditionally been regulated, but also downwards to local government, where global public goods like the internet are often realised. At the local government level smart cities have increasingly been presented, in the context of the 4IR, as being the solution for everything from climate change to city inefficiencies (Wahba, 2020). This is aligned with global trends over the past three decades towards techno-managerialism, which has seen public resources increasingly allocated on the basis of their supply-side value. At the municipal level, this is often reflected in smart-city public–private partnerships, particularly with technology suppliers – even multilateral coordination conceives of smart cities in these terms. Describing itself as the largest and most ambitious undertaking to advance the responsible and ethical use of smart city technologies on a global level, the G20 Global Smart Cities Alliance on Technology Governance's primary purpose is to establish global norms and policy standards for the use of connected devices in public spaces. This, they claim, often opens up markets to suppliers of these technologies who are hamstrung by red tape and less publicly minded bureaucracy and outright corruption. There is much talk of ethics, technology, efficiency and trust but very little about inequality, spatial dislocation, poverty and people.

The latest of these tech-invested initiatives is that of the Siemens Smart Cities. It indexes the readiness of cities, of which Johannesburg is one, for the 4IR. The data and indicators reflect the same problems for South Africa as in other indices such as the WEF Digital Readiness Index or the United Nation's ITU ICT Development Index. Johannesburg scores only 2 out of 10 in terms of wider readiness; the weaknesses identified include the city's non-digital infrastructure, including its poor cargo rail network and an unreliable energy supply. Digitisation of these infrastructures – through, for example, smart grids – is presented as one of the solutions to becoming 'smart' (Siemens, 2020). While these are clearly possible solutions, such technologically driven solutions fail to deal with the underlying causes of urban decay, poverty and inequality.

The ITU's definition tries to overcome some of the technology determinism of the Smart City by incorporating the potential social good aspects (FG-SSC, 2014):

A smart, sustainable city is an innovative city that uses information and communication technologies (ICTs) and other means to improve quality of life, efficiency of urban operation and services, and competitiveness, while ensuring that it meets the needs of present and future generations with respect to economic, social, environmental as well as cultural aspects.

In Africa, the 'smart city' agenda found a home in the Smart Africa project led by President Paul Kagame of Rwanda, the board of which includes presidents from a dozen other countries. Kigali, already a small island of development in a sea of underdevelopment, was to be the model city of this utopian vision for Africa. Politicians, government officials, development banks and futurists seemed blind to the irony of spending billions of dollars on building largely high-end office parks and residential areas which are providing enhanced infrastructure to those who already had it, rather than spending it on infrastructure to connect divided parts of the city, and the city to the hinterland.

Megacities failed – arguably, predictably – to deliver on their promise as a result of their inability to tackle underlying issues with a generally sporadic application of smart technologies to localised urban problems. As a result of this failure, the smart city discourse has shifted over the last decade to the building of 'grandiose new landmark metropolis'. These are new, built-from-scratch developments often on the peripheries of overpopulated cities such as Vision City in Rwanda, King City in Ghana, Eko Atlantic in Nigeria, Kenya's Konza and Cairo Vision 2050 in Egypt (Watson, 2014 in Ahmed et al., 2020).

Echoing this vision, President Cyril Ramaphosa in his 2020 State of the Nation Address (SONA), revealed plans for a smart city development in Lanseria, on the outskirts of Johannesburg, highlighting the transformative potential of ICT. This comes in the context of the Presidency's commitment to harnessing so-called 4IR technologies to make South Africa globally competitive and improve the lives of women and the youth (Ramaphosa, 2019). This R84-billion smart city development is intended to create a 'world class city' around Lanseria airport, an international hub consisting of upmarket commercial and residential property and creating 200,000 jobs. 'It will not only be smart and 5G ready but will be a leading benchmark for green infrastructure continentally and internationally' (Ramaphosa, 2020: 14).

Perhaps the inequalities highlighted in the country by COVID-19, including the inability of the majority of South Africans to move their lives online, and the areas of greatest need that will be revealed in post-COVID-19 economic reconstruction, will compel a reassessment of how 'transformative' such a project would be.

This is not to say that digital technologies and services cannot be deployed by local government to improve wellbeing, if this is done in such a way that it prioritises the application of technologies where they can most redress the extreme inequality of service between suburban, township and informal settings. Rather than seeking to attract the kind of foreign private-sector investment needed to build a new futurist city, a large amount of local capital could be crowded in to redress the digital and spatial inequality in current South African cities and towns. What if, instead of a R84-billion new city, those resources were diverted to supporting the development of smart townships and smart villages to redress digital inequality and enable greater social and economic inclusion? Such an integrated local government strategy, which would support the evenness of delivery not only within cities but between them and smaller municipalities, could be a critical element of an integrated digital new deal.

SMART TOWNSHIPS

Another example of the challenges associated with sustaining demand-side valuation of digital resources at the municipal level is the short-lived smart township initiative, run under the auspices of the now defunct Cape Digital Foundation. The initiative sought to shift the Western Cape Broadband Initiative, which had focused on connectivity, to building 'smart, digitally savvy business communities by providing digital training to township SME owners' that would be engaging with and enabling 'smart citizens' (Interview: Emma Kaye, 2020 in Ahmed et al., 2020). Imizamo Yethu was selected: a mixed formal and informal township that had developed alongside suburban areas. Over the past two decades, a large number of Malawian immigrants had settled in the community, becoming a significant part of it. It was recognised that with the infrastructure deficits that existed in informal settlements and townships, the smart township initiative would only be possible through partnerships that provided access to reliable, affordable internet and digital skills training. This training would entail daily use of the internet supported by relevant, hyper-local content – created 'by the people for the people' (Interview: Emma Kaye in Ahmed et al., 2020).

Connecting the unconnected by working with partners who could provide fast and affordable Wi-Fi into homes and public spaces was central to the project. It also included the provision of digital skills, supported through a number of training workshops for community members (Interview: Craffert, 2020).

The project enlisted the support of a local micro provider, Too Much Wi-Fi,[4] who were able to connect resellers to local fibre connections and mesh networks for an initial connection fee of R1,000 with R1,000 of free data. Residents were able to access high-quality bandwidth at a much lower cost than the prices offered by commercial operators. There was also an understanding of locally relevant content as critical to digital inclusion and residents were therefore able to download local programming developed by a locally trained producer. The programming included local hairdressers and hot food vendors, with strong Malawian community programming, including recipes and Malawian cooking content, which attracted a regular following of about 4,000 daily viewers (Interview: Xubuzana, April 2020).

The project managed to complete the training of a first round of entrepreneurs in basic IT skills before the funding of the Digital Foundation dried up and, with it, the miniscule funding for the initiatives. The Western Cape received much attention when the smart township initiative was being hailed by the national Minister of Communications, and at international 'smart' conferences, as providing an appropriate smart framing for African cities. However, it was at best a fringe project, initiated, implemented and driven by an individual unable to get it institutionalised within any of the provincial

---

4   Too Much Wifi: https://toomuchwifi.co.za/

departments (Interviews: Kaye, 2020; Dyers, 2020; Cloete, 2020).

As with some public Wi-Fi initiatives in the province, there was no demand-side valuation. In addition, once the minimal funding of the Digital Foundation dried up, there was no allocation of resources to sustain the gains that had been made towards supporting the basic infrastructure required for other potential smart initiatives. Public-resource allocation towards digital solutions in the township, such as the deployment of sensors to detect when public toilets require emptying, to the increased allocation of closed-circuit television (CCTV) to limit criminality and xenophobic attacks, provide examples of how such investments might have contributed to a fairer allocation of digital resources. The potential for digital initiatives can be seen in successful commercial micro-entrepreneurial ventures, which use IoT sensors, in informal settlements (Lewis, 2018).

The processes of digitalisation and datafication are central to the modernisation of cities, as are digital infrastructure and public data, discussed in the national and international sections above. As such, digital strategies at the municipal level must be integrated into wider national strategies and indeed global governance of public goods, as they are often the point at which global public goods such as internet, data and cybersecurity are realised at the national level. Also, due to the uneven development of the South African economy and enormous geospatial inequality, there is far less capacity, often none in municipalities outside of the main metropoles, to deploy ICTs. This deployment is required to enhance the efficiencies of municipalities' operations and to ensure that their citizens have public access to the internet to benefit from online services or create livelihoods.

CONCLUSION

In the first Nelson Mandela Lecture held online, as a result of the COVID-19 pandemic, United National Secretary General, Antonio Guterres (2020), said the world is witnessing two seismic shifts: the climate crisis and digital transformation, both of which will increase inequality even further. South Africa should be developing an entirely new digital deal rather than continuing on the digital policy trajectory that has brought us to the 4IR and which is associated with the problems of hyperglobalisation and increasing global inequality, which COVID-19 has so starkly highlighted. A profound change in the thinking and policy mix that caused the economic crisis, even before the devasting effect of the pandemic, is required. Developing countries planning to deploy digital solutions to mitigate the negative outcomes of future pandemics or to rebuild their economies will need to assess the social and economic outcomes of supply-side valuation for the allocation of resources. This valuation has characterised technological and market advances over the past three decades.

Rather than looking at the impacts of futuristic advanced technologies on the economy and society, this chapter has argued that what is needed is a transversal digital policy that is designed to deal with the complexity and dynamism of the global communication systems arising from intensified digitalisation and datafication. It has been argued that this will require an adaptive policy and governance framework, able to engage at the global, national and local level in the digital ecosystem. A comprehensive policy will necessitate a shift from the sectoral silos in which national digital policy is being formulated. This is what is required to achieve the necessary integration across the public sector – particularly international affairs and trade and revenue services – as well as coordination between the public and private sectors, and the predominant informal sector, to meet the needs of the South Africans currently denied digital access.

In the spirit of the New Green Deal, this chapter has tried to demonstrate the policy experimentalism needed, particularly in the developing world, in order to get the right mixture of recovery, regulation and redistribution, particularly post COVID-19. This is especially relevant in the digital realm that will need to extend beyond the national level to both local and global to leverage the opportunities of today's interdependent world for the delivery of global public goods such as the internet, public data, cybersecurity and innovation. With the increasing complexity and adaptiveness of the global communications system, both new and more traditional forms of governance are proving incapable of providing adequate tools for the governance of global public goods. 'Datafication' of economy and society provides new challenges for the governance of these global goods. Developing countries need to address these while continuing to redress digital inequalities and unevenness in access to and use of underlying infrastructures, services and applications.

This chapter used three cases associated with the discourse of 4IR – spectrum allocation; data governance associated with big data analytics and AI; and smart cities – to explore how enhanced transformation outcomes might be achieved. Time and space constraints prevented the exploration of a digital policy that would enable South Africa to create the conditions necessary to harness the benefits of advanced technological developments and mitigate the risks associated with them. In this way, the chapter examined the processes of digitalisation and datafication, at national and international levels, to provide examples of where the inclusion and prioritisation of demand-side valuation in the allocation of resources could have the transformative outcomes required to address digital inequality and enable social and economic inclusion. The analysis looked at spectrum as an impure public good at national level. It considered the rent extraction associated with supply-side valuation in the allocation of this resource, both by the state through auctioning and licence fees, and by dominant players in the prices charged in both wholesale and retail markets. It explored how the supply-side incentives can be tempered with demand-side valuation of spectrum through setting aside more spectrum for a 'digital commons'. With 5G spectrum now also up for assignment, policymakers need to ensure that 5G technology, which operates well within a spectrum-sharing environment with expectations of traffic off-loads to WiFi, is harnessed for public purposes and not just niche commercial applications.

In the next case, the international level, the chapter examined the challenges of global governance of data emanating from digital platforms. It explored ways in which demand-side valuation of data in global and national governance could combat the global supply-side valuation evidenced in the platform economy, and in what Zuboff (2019) has called 'surveillance capitalism', to constrain potential harms of privacy breaches and algorithmic bias. This chapter highlighted the need to create more trusted environments as greater numbers of more vulnerable people come online and how this can only be achieved through global collaboration, given the challenges of enforcement.

Finally, the case of the smart city, promoted globally and in South Africa in the context of the 4IR, was used to explore how urban and local policies deploying smart or digital solutions could reduce the inequality resulting from the current distribution of smart solutions at the municipal level if allocated on a demand-side basis. It demonstrated that demand-side allocation of resources for public Wi-Fi, or the prioritisation of broadband, smart sensors, or device deployment in townships before suburban and city centres, could reduce the inequalities currently being compounded in smart cities. It proposed that rather than create a new smart city, with digital inequalities associated with them, productive private-sector investment instead be crowded into the delivery of digital public goods through an integrated smart township or smart village national project. Such a project could contribute to the structural transformation of the country at the heart of government strategies for post-COVID 19 economic reconstruction, and for a more equitable South Africa.

REFERENCES

Ahmed, S. and Gillwald, A. 2020. 'Smart townships will build smarter cities.' Research ICT Africa, Policy Brief number 3. https://researchictafrica.net/publication/smart-townships-will-build-smarter-cities/, accessed 20 September 2020.

Avis, J. 2018. 'Socio-technical imaginary of the fourth industrial revolution and its implications for vocational education and training: a literature review'. *Journal of Vocational Education & Training*, 70(3), 337–363.

Badat, S. 1 June 2020. 'The 4IR superhighway: A dangerously utopian future'. *Daily Maverick*. https://www.dailymaverick.co.za/opinionista/2020-06-01-the-4ir-super-highway-a-dangerously-technocratic-utopia/, accessed 25 June 2020.

Best, J. and Gheciu, A. (eds). 2014. *The Return of the Public in Global Governance*. Cambridge: Cambridge University Press.

Boyd, R. and Holton, R.J. 2018. 'Technology, innovation, employment and power: Does robotics and artificial intelligence really mean social transformation?'. *Journal of Sociology*, 54(3), 331–345.

Calandro, E., Gillwald, A. and Zingales, N. 2013. 'Mapping multi-stakeholderism in internet governance: Implications for Africa'.

Research ICT Africa. https://www.researchictafrica.net/publications/ Evidence_for_ICT_Policy_Action/Discussion_paper_-_Mapping_ Multistakeholderism_in_Internet_Governance_-_Implications_for_ Africa.pdf, accessed 15 July 2020.

Cammaerts, B. and Mansell, R. 2020. 'Digital platform policy and regulation: Toward a radical democratic turn'. *International Journal of Communication*, 14(20), 135–154.

De La Chapelle, B. 2019. 'Internet Jurisdiction Policy Network'. Closing address in Berlin, Germany, June. https://www.internetjurisdiction.net/ uploads/pdfs/Berlin-Roadmap-and-Secretariat-Summary-3rd-Global-Conference-of-the-Internet-Jurisdiction-Policy-Network.pdf, accessed 15 July 2020.

Electronic Communications Act No. 35 of 2005. 6 December 2013. https:// www.gov.za/sites/default/files/gcis_document/201409/37119gon953.pdf, accessed 12 May 2020.

European Union. 2016. 'General Data ProtectionRegulation (GDPR) 2016/679'. European Union. https://www.google.com/url?sa=t&rct=j& q=&esrc=s&source=web&cd=&cad=rja&uact=8&ved=2ahUKEwiu3d XnuNvqAhUsVRUIHexbA6oQFjAAegQIAxAB&url=https%3A%2 F%2Fgdpr-info.eu%2F&usg=AOvVaw1akHzzz224Oq1yU0pd6qSw, accessed 25 July 2020.

Frischmann, B.M. 2005. 'An economic theory of infrastructure and commons management'. *Minnesota Law Review*. https://scholarship.law.umn.edu/ mlr/673, accessed 25 June 2020.

Frischmann, B.M. 2012. *Infrastructure: The social Value of Shared Resources*. Oxford: Oxford University Press.

Gillwald, A. 2005. 'Stimulating investment in network extension: The case of South Africa'. In: Mahan A.K. and Melody W.H. (eds). *Stimulating Investment in Network Development: Roles for Regulators, World Dialogue on Regulation*. Denmark: Regulate Online. https://www. infodev.org/infodev-files/resource/InfodevDocuments_12.pdf, accessed 12 May 2020.

Gillwald, A. 2007. 'Straddled between two stools: Broadband developments in South Africa'. *Southern African Journal of Information and Communication*, (8), 53–77.

Gillwald, A. 2018. 'Understanding the paradox of digital inequality as more Africans come online'. Public Lecture: Nelson Mandela School of Public Governance. https://researchictafrica.net/2018/05/02/public-lecture-understanding-the-paradox-of-digital-inequality-as-more-africans-come-on-line/, accessed 12 May 2020.

Gillwald, A., Mothobi, O and Rademan, B. 2018. 'After Access – State of ICT in South Africa'. Policy Paper no. 5, Series 5, Research ICT Africa, Cape Town. https://researchictafrica.net/after-access-south-africa-state-of-ict-2017-south-africa-report_04/, accessed 27 November 2020.

Gillwald, A. 2019. 'South Africa must harness technology in a way that helps fix its problems'. *The Conversation*, 3 October. https://theconversation.com/south-africa-must-harness-technology-in-a-way-that-helps-fix-its-problems-121191, accessed 27 November 2020.

Gillwald, A. and Van der Spuy, A. 2019. 'National delivery of global public goods, data governance from a developing country perspective'. Giganet. https://www.giga-net.org/2019symposiumPapers/34_Gillwald_VanderSpuy_Global-Governance.pdf, accessed 14 April 2020.

Global Smart City Alliance. 2019. https://globalsmartcitiesalliance.org/, accessed 25 July 2020.

Gunkel, D. 2003. 'Second thoughts: Toward a critique of the digital divide'. *New Media and Society*, 5(4), 499–522.

Haas, P. 1992. 'Introduction: Epistemic communities and international policy coordination'. *International Organization*, 46(1), 1–35. https://www.jstor.org/stable/2706951, accessed 12 May 2020.

Hirschman, A.O. 1958. *The Strategy of Economic Development*. New Haven, CT: Yale University Press.

Hosein, G. 2006. 'Policy laundering, and other policy dynamics'. In: Halpin, E., Trevorrow, P., Webb, D. and Wright, S. (eds). *Cyberwar, Netwar and the Revolution in Military Affairs*. London: Palgrave Macmillan, 228–241.

International Telecommunication Union (ITU). 2014. 'Technical report on smart sustainable cities: An analysis of definitions'. United Nations, International Telecommunication Union (ITU-T), Focus Group on Smart Sustainable Cities (FG-SSC).

Johnson, D.R. and Post, D. 1996. 'Law and borders: The rise of law in cyberspace'. *Stanford Law Review*, 1(1), 1367–1402.

Kaye, E. 17 April 2019. 'Cape Digital Foundation: Smart Townships and Digital Innovation'. *eResearch Africa 2019*. http://www.eresearch.ac.za/presentations-2019-day-one, accessed 14 April 2020.

Kaul, I., Grunberg, I. and Stern, M. (eds). 2003. *Global Public Goods: International Cooperation in the 21st Century*. London: Oxford University Press.

Kenny, C. 2003. 'Development's false divide'. *Foreign Policy*. doi: 10.2307/3183524.

Lewis, A. 2018. 'Everyone should be able to feel safe in their own home'. Accion. https://www.accion.org/insurtech-fire-protection-south-africa, accessed 2 August 2020.

Mabaso, S. 16 June 2019. 'Powerful'. *Sunday Independent*. https://www.pressreader.com/south-africa/the-sunday independent/20190616/281517932636476, accessed 25 July 2020.

Mandaha, D. 15 April 2019. 'South Africa and World Economic Forum announce intention to establish 4IR Affiliate Centre'. The Council for

Scientific and Industrial Research (CSIR). https://www.csir.co.za/south-africa-and-world-ec.onomic-forum-announce-intention-establish-4ir-affiliate-centre, accessed 25 July 2020

Marsden, C. 2 October 2001. 'Towards the hyperglobalisation of the individual: How the ubiquitous internet will make the international political economy increasingly dynamically unstable'. https://papers.ssrn.com/sol3/papers.cfm?abstract_id=1578203, accessed 14 April 2020.

McCleod, D. 27 July 2019. 'Spectrum policy released – the good, the bad and the (not so) ugly'. Tech Central. https://techcentral.co.za/spectrum-policy-released-the-good-the-bad-and-the-not-so-ugly/91361/, accessed 24 July 2020.

Melody, W. 2001. 'Spectrum auctions and efficient resource allocation: Learning from the 3G experience in Europe'. *Info: The Journal of Policy, Regulation and Strategy for Telecommunications*, (3)1, 5–10. https://www.ingentaconnect.com/content/mcb/272/2001/00000003/00000001/art00002, accessed 25 July 2020.

Ostrom, E. 2010. 'Beyond markets and states: polycentric governance of complex economic systems'. *American Economic Review*, 100(3), 641–672.

Ramaphosa, C. 2019. 'President appoints Commission on the Fourth Industrial Revolution'. The Presidency of the Republic of South Africa. http://www.thepresidency.gov.za/press-statements/president-appoints-commission-fourth-industrial-revolution, accessed 10 June 2019.

Razzano, G. 24 April 2017. 'SASSA Grants: The small information win hiding in the grant crisis'. *Daily Maverick*. https://www.dailymaverick.co.za/opinionista/2017-04-24-sassa-grants-the-small-information-win-hiding-in-the-grant-crisis/#gsc.tab=0, accessed 24 July 2020.

Schwab, K. 14 January 2016. 'The Fourth Industrial Revolution: What it is, how to respond'. *World Economic Forum*. https://www.weforum.org/agenda/2016/01/the-fourth-industrial-revolution-what-it-means-and-how-to-respond/, accessed 9 March 2020.

Siemens. 2020. 'Atlas of Digitalisation for Smart Cities'. https://atlas.dc.siemens.com/cities/methodology/?stc=wwcg222103&s_kwcid=AL!462020, accessed 24 July 2020.

Song, S. 21 April 2017. 'Failure of spectrum auctions in Africa'. https://manypossibilities.net/2017/04/the-failure-of-spectrum-auctions-in-africa/, accessed 25 July 2020.

Song, S. 22 January 2019. 'African telecommunications infrastructure in 2018'. *Many Possibilities*. https://manypossibilities.net/2019/01/african-telecommunications-infrastructure-in-2018/, accessed 25 July 2020.

State of the Nation Address. 13 February 2020. https://www.gov.za/speeches/president-cyril-ramaphosa-2020-state-nation-address-13-feb-2020-0000, accessed 25 July 2020.

Taylor, L. 2017. 'What is data justice? The case for connecting digital rights and freedoms globally'. *Big Data & Society*, 4(2): doi: 10.1177/2053951717736335.

United Nations. 2019. 'Financing a Global New Green Deal, Trade and Development Report 2019'. *United Nations Conference on Trade and Development* (UNCTAD). https://unctad.org/en/PublicationsLibrary/ tdr2019_en.pdf, accessed 24 June 2020.

United Nations Conference on Trade and Development (UNCTD). 2019. 'Trade and Development Report 2019: Financing a Global Green New Deal'. United Nations.

United Nations Human Rights Council (UNHRC). 2016. 'The promotion, protection and enjoyment of human rights on the Internet'. Human Rights Council, 32nd session (A/HRC/32/L.20). https://digitallibrary.un.org/ record/845728?ln=en, accessed 10 June 2019.

United Nations' Secretary-General's High-level Panel on Digital Cooperation (UNHLPDC). 2019. 'The age of digital interdependence'. https://digitalcooperation.org/wp-content/uploads/2019/06/ DigitalCooperation-report-web-FINAL-1.pdf, accessed by 24 June 2020

Vermeulen, J. 4 November 2019. 'How South Africa's national cellphone network will work'. *My Broadband*. https://mybroadband.co.za/news/ cellular/326113-how-south-africas-national-cellphone-network-will- work.html, accessed 24 July 2020.

Wahba, S. 14 February 2020. 'Here's how technology is tackling inclusion issues in smart cities'. *World Economic Forum*. https://www.weforum. org/agenda/2020/02/smart-cities-inclusive-future-technology-growth/, accessed 25 July 2020.

Watson, V. 2014. 'African urban fantasies: dreams or nightmares?' *Environment and Urbanization*, 26(1), 213–229.

Zuboff, S. 2019. *The Age of Surveillance Capitalism: The Fight for a Human Future at the New Frontier of Power.* Profile Books: Public Affairs.

The Guardian. https://www.theguardian.com/books/2019/feb/02/age-of- survelliance-capitalism-shoshana-zuboff-review, accessed 10 June 2020.

Research interviews
Carlos Rey Moreno, Zenzeleni Community Network, 30 January 2019.
Christopher Geerdts, former chair: Wireless Access Providers Association (WAPA), 25 March 2019.
Derek Kotze, Chief Executive Officer: MLabs, 17 April 2019.
Emma Kaye, Chief Executive Officer: Cape Digital Foundation/Smart Townships, 15 March 2020.
Evan Jones, Business Process Enabling South Africa (BPESA), 26 April 2019.
Joe Mjwara, then Director General of Communications, 24 April 2019.
Leona Craffert, Lab: University of the Western Cape, 7 April 2020.
Graham de Vries, Executive: Legal and Regulatory Affairs, MTN, 14

April 2019.

Kobus Roux, Meraka Institute, Council for Scientific and Industrial Research (CSIR), 15 April 2019.

Lucas Gumbi, Business Development, CSIR, responsible for World Economic Forum (WEF) Internet for All Infrastructure Mapping, 15 March 2019.

Marc Cloete, Western Cape Premier's Office, 2020.

Mike Silber, Executive Legal & Regulatory: Liquid Telecom, 23 March 2019.

Mlamli Booi, CEO: Sentech, 4 May 2019.

Naledi Pandor, Former Minister of Higher Education, 4 April 2019.

Olivia Dyers, Director: Department of Tourism and Development, Western Cape.

Owen Xubuzana. Producer: IYTV. 24 April 2020.

Phatang Nkhereanye, Head: Legal and regulatory, Broadband Infraco, 19 March 2019.

Phil Mjwara, Director General: Department of Science and Technology, 15 April 2019.

Robert Nkuna, Director General: Department of Communications, 25 April 2019.

Robert Urquart, Research Head, Harambee, youth accelerator, 26 April 2019.

Setumo Mohapi, Chief Executive Officer: State IT Agency, 30 April 2019.

Sipho Maseko, CEO: Telkom, 15 April 2019.

Willington Ngwepe, Chief Executive Officer, ICASA, 4 April 2019.

TWO

From science, technology and innovation to Fourth Industrial Revolution strategies in Africa: The case for indigenous knowledge systems

CHIDI OGUAMANAM

INTRODUCTION

We are in the human or Anthropocene epoch – the 'first time in the history of the world that human activities are the primary force in shaping all life sustaining systems on earth' (Schwab, 2016: 111). As a tool and as a process, industrialisation is perhaps the most obvious evidence of humankind's transformative imprint on earth. Conventional assessment of the trajectory of that imprint seems to agree on three phases of industrial revolution – the First, Second and Third Industrial Revolutions. Whether or not the Fourth Industrial Revolution (4IR) has crystallised or whether contemporary industrial, scientific and technological transformations are the continuation of the Third Industrial Revolution remains an open question (Mariani and

Borghi, 2019).[1]

Africa has consistently been conscious of science and technology as tools of socio-cultural, economic and industrial transformations. At continental, regional and national levels, Africa has elaborate policies on science and technology, in some cases dating back to the immediate postcolonial era. Lately, the policies or strategies have morphed into those of science, technology and innovation (STI). At various national levels, they are complemented by existing or periodically designed industrial policies, vision statements and miscellaneous strategy documents. A recent focus on information and communication technologies (ICT), which catalyse interest in the 4IR, has resulted in ICT becoming a prominent sectoral adjunct to the African STI policy ecosystem. A 2019 release by the African Union (AU) noted that '[a]t least 25 African countries were reported to have STI strategies … although there are gaps in STI policies, countries have been very active in developing ICT policies. As of 2016, at least 45 African countries had ICT policy frameworks that were in general being effectively implemented' (AU, 2019: 10–11).

The 4IR is a reality in which the previous industrial epochs appear to have converged to a variable degree, driving society – via what we can view as a kind of 'technology steroid' – across digital, physical and biological domains (Schwab, 2016; 2018). Countries, academic institutions, researchers, corporations, civil society, international development entities and global governance institutions have embraced the idea of a 4IR or 'Industry 4.0' (Madsen, 2019). On the African continent, South Africa is among the leading countries in enunciating a policy on 4IR in furtherance of its overarching STI strategy. At continental and national levels, African STI strategies are precursors to the emerging interest in 4IR. But those strategies have not been proactive about including African indigenous knowledge systems (IKS) and associated stakeholders. That is one of the unexplored gaps in the African STI ecosystem now being carried over to the policy space emerging for the 4IR. This chapter explores South Africa's experience against the backdrop of the broader AU-led continental strategy on

---

1   This research provides a robust literature review on 4IR/Industry 4.0.

STI, with the spotlight on how this strategy was implemented in Nigeria and Kenya.

The chapter uses the South African experience on 4IR policy to elaborate on both real and potential risks of the exclusion of IKS from 4IR strategy, with a few sector-specific illustrations. The chapter calls attention to the implications or ramifications of the IKS gap for Africa's optimal participation in the 4IR.

STI POLICY LANDSCAPE IN AFRICA – A FORETASTE OF EMERGENT 4IR STRATEGY

The starting point for exploring the policy landscape is the adoption by the AU in 2014 of a ten-year 'Science, Technology and Innovation Strategy for Africa 2024', also known as STISA-2024. The vision statement laid out in this document is to 'accelerate Africa's transition to an innovation-led, knowledge-based economy' (AUC, 2014: 11; AU, 2019: 23). STISA-2024's priority is on the demands of science, technology and innovation in six areas linked to the AU Agenda 2063. These are to be implemented according to ten-yearly phased incremental strategies. The six priority areas are: eradication of hunger and achieving food security; prevention and control of diseases; communication (physical and intellectual mobility); protection of our space; harmonious and peaceful co-existence; and wealth creation. In addition to STISA-2024 being incorporated into the national development plans of AU member states, it is embraced by the continent's Regional Economic Communities (RECs). Currently, it constitutes an integral part of the institutional policy ecosystem of the AU and its financial, development, and policy organs and partnerships. Put succinctly, 'STISA-2024 places STI at the epicentre of Africa's social and economic development within the long-term AU Agenda 2063' (AUC, 2014: 15).

Like earlier efforts on STI in Africa, notably the 2005 Consolidated Plan of Action (CPA),[2] STISA-2024 is largely inspired by the often-

---

2   CPA is an implementing instrument of all the decisions of the Assembly of AU Heads of States and Governments on science, technology and innovation before STISA.

quoted speech of Ghana's President Kwame Nkrumah at the 1963 inaugural summit of the AU where he said (AUC, 2014: 5):

> We shall accumulate machinery and establish steel works, iron foundries and factories; we shall link various states of our continent with communications; we shall astound the world with our hydroelectric power; we shall drain marshes and swamps, clear infested areas, feed the undernourished, and rid our people of parasites and disease. It is within the possibility of science and technology to make even the Sahara bloom into a vast field with verdant vegetation for agricultural and industrial development.

Missing in Nkrumah's forthright vision is a reference to the relevance of African indigenous knowledge systems in the discourse about science, technology and innovation in Africa. However, the focus of Cluster 1 of the Consolidated Plan of Action (CPA) by AU heads of state and government on STI is biodiversity, biotechnology and indigenous knowledge. Cluster 1 makes reference to 'Securing and using Africa's Indigenous Knowledge Base' (AUC, 2014: 14). Also, on paper, the STISA-2024, a successor to the CPA, contains occasional buzz words and references to the promotion of 'research, invention and innovation in traditional medicine and strengthening local health ecosystems, taking into account the socio-cultural and environmental situation of the people' (AUC, 2014: 22).[3] Overall, STISA-2024 recognises the cross-sectoral and multidisciplinary nature of STI and their role in Africa's socio-cultural and economic development aspirations. But the document falls short of recognising the equally cross-sectoral and multidisciplinary nature of African IKS in that same equation.

A situational analysis of STI in Africa as a prelude to the STISA-2024 official document alludes to the relevance of African indigenous knowledge through a dismissive reference to the opinion of civil

---

3   This appears under STISA-2024 priority 2 titled 'prevention and control of disease'.

society advocate organisations. Strangely, it noted (AUC, 2014: 18):

> Civil society organizations and Think-Tanks are championing the use of African Indigenous Knowledge to support sustained economic growth, and inform public attitudes and understanding of the relevance and importance of STI. While they contribute to STI policy debate in areas including biosafety, climate change, biodiversity and environment regulation and ICT, most contributions are not supported by evidence.

This observation is characteristic of the tangential treatment of African indigenous knowledge and African indigenous knowledge systems – terms which, for convenience, are deliberately used interchangeably here – in STI.

Perhaps not many indicators are more symbolic of the omission of IKS in Africa's STI strategies than the institutional building, programming, language and flagship projects of the AU's Consolidated Plan of Action (CPA) through to the STISA-2014. Programmes and institutions associated with STI are based on formal university and institutional research and development (R&D) complexes and models. STI initiatives from both sources encourage collaborations across international, regional and national institutional platforms. Prominent actors include but are not limited to the African Development Bank (AfDB), NEPAD[4] Centres of Excellence, the Pan African University, the World Bank African Centres of Excellence in Higher Education, the African Observatory for Science, Technology and Innovation (AOSTI), Pan African Quality Assurance and Accreditation Framework (PQAAF), the African Centre for Disease Control (ACDC), the Pan African Intellectual Property Organization (PAIPO), and the African Scientific Research and Innovation Council (ASRIC). The last institution represents the scientific community in setting agendas across research and flagship programmes for African STI and by extension African participation in the knowledge economy.

4　New Partnership for Africa's Development, a development arm of the AU Agenda 2063.

None of these entities represents IKS interest. Neither is in any position to be sensitive to an indigenous knowledge-inclusive approach to STI. Emphases are on providing an enabling environment – including intellectual property and other regulatory interventions, and infrastructure – for building and developing a strong science, technology and innovation culture through enhanced technical and professional competences in science, engineering, mathematics and related disciplines. Yet the STISA-2024 recognises the relevance of STI to 'community driven solutions that leverage the knowledge of African share values' (AUC, 2014: 23) in addition to a multidisciplinary approach that incorporates social sciences and humanities. It is recognised that given the inherent cross-sectoral nature of both STI and IKS, the two do not interface or converge in all sectors at even levels or degrees. However, most of the STISA-2024 language and focus is on science, technology, and R&D priorities driven by the hard-core scientific community and all its conventional institutions with no deliberate attempt to develop and include institutions and stakeholders that cater to IKS.[5] That approach is reflected in the assessment of the profile of, and readiness for, STI in Nigeria and Kenya, which are indicative of the rest of the continent.

MEASURING STI READINESS IN AFRICA: THE CASE OF NIGERIA AND KENYA

In 2018, the Economic Commission for Africa (ECA) released the report on its study of the framework for assessing STI readiness in Africa. Nigeria and Kenya are two pilot countries for the report. Without going into the details, the study anchored the articulation of STI policies in developing countries, including Nigeria and Kenya, on their National Systems of Innovation (NSI) interchangeably referred to as National Innovation System (NIS) (ECA, 2018: 31). It noted that each NSI is contingent upon a country's peculiarities and history. However, that observation is without prejudice to the shared centrality

---

5   In a 2020 study for the Open African Innovation Research (Open AIR), Francis Kariuki explores the idea of indigenous knowledge holders' institutions. See reference to the study in the bibliographic list to this chapter.

of IKS in many African countries, including Nigeria and Kenya. Of interest are the nine criteria for the review of STI set out in the study. Save for one, those criteria do not consider the role of IKS and its stakeholders. According to ECA, in assessing STI preparedness, questions must be posed relating to the institutions responsible for STI policies; the engagement of critical stakeholders or potential investors in STI; tensions over local and international aspects of STI conflicts; the application of policies relating to the use of indigenous knowledge; the sectoral integration of STI; science-industry-university linkages and promotion of science education; the applications of green technologies; and the promotion of technological learning (ECA, 2018: 39). Clearly, the range of these criteria constrain the relevance of IKS as a cross-sectoral and multidisciplinary matter that is also relevant to STI, and in need of mainstreaming in other criteria. For example, even though broad, the study's discussion of institutions and stakeholders in STI and NSI contexts does not accommodate IKS institutions and stakeholders. Save for the reference to policies relating to indigenous knowledge, the same is true of all other criteria. Worse than the nine criteria enumerated above, the study's list of selected key indicators for measuring STI (ECA, 2018: 17) completely omitted any direct consideration or accommodation for African IKS.[6]

Nigeria and Kenya have identical trajectories and frameworks for their respective STI strategies. Both countries have contemporary STI and adjunct national industrial strategies or policies, as well as national-level vision documents on development. They are participants in the continental AU and at the level of regional economic communities. Both countries are committed to AU Agenda 2063 and STISA-2024. In addition, they also have a mosaic of policies, instruments and regulations around ICTs at national and sub-national levels. It seems needless to indicate that any national strategy on 4IR in these two countries, and indeed in most African countries, would build on the strides made in STI. For our present purpose, as adapted into the ECA study, Nigeria's STI objectives include the intention to '[e]ncourage

---

6   The list is mainly OECD-centric and includes the following: patents, bibliometrics, R&D intensity, hi-tech exports, royalties, hi-tech industries, tech achievement index to mention the few (ECA, 2018: 17).

and promote the creation of innovative enterprises making use of Nigerian indigenous knowledge and technology to market goods and services' (ECA, 2018: 95). Similarly, but more elaborately, Kenya's STI policy objectives include (ECA, 2018: 139):

> … supporting the application of traditional knowledge in the formal and informal sectors of the economy to enhance livelihoods and promote the use of full potential of science, technology and innovation to protect, preserve, evaluate and update, add value to and utilize extensive indigenous resources and traditional knowledge available in various Kenyan communities.

The above recognition is evident in Nigeria and Kenya, where there are minimal considerations of indigenous and traditional knowledge in STI policies. At national and continental levels in Africa, therefore, these policies are not in sync with the infrastructure and programming within STI institutions. Rather, on the continent, the focus seems to be exclusively around skills development and higher education in science, technology, innovation; on STI-driven research, manufacturing, and value-add, including the establishment of hi-technology hubs. Nonetheless, the AU has endorsed a 2018 review of African STI policies by the African Academy of Sciences (AAS). Referring to the report, the AU observed that (AU, 2019: 58)

> where there are STI policies (in Africa) they often focus solely on business and industrial development and that 'social and environmental goals are not adequately integrated into national STI policy framework of African countries'. The danger is that by focusing on industry and market-oriented development alone, the opportunity for STI to positively impact on non-commercial sustainable development is missed … There is also value in interfacing [STI policies] more broadly with social and environmental development policies.

Inspired by the AAS review and approaches from other places, notably

India, the AU clearly supports an STI approach that caters to and emboldens 'location-based research' and a combination of 'bottom-up' and 'top-down' and context-specific solutions (AU, 2019). All of these can benefit from deliberate and practical incorporation of African IKS and stakeholders with matching institutional support for STI strategies. This deliberate policy approach is necessary to plug the apparent gap in the exclusion of IKS from STI, and to ensure that the gulf is not replicated as Africa contemplates its strategy for the 4IR.

BEYOND STI: SOUTH AFRICA'S 4IR STRATEGY

South Africa has perhaps the most advanced physical and institutional infrastructures for industrial production and R&D in science, technology and innovation on the African continent. Human capital remains the country's most pressing challenge to competing globally, with human resource shortfalls in indigenous science, technology, engineering and mathematics (STEM) and digital skills. This is especially the case in the context of the 4IR. South Africa's 1996 Science and Technology White Paper, launched at the end of apartheid, was the premise of the country's National System of Innovation (NSI). Recent changes in the technological sphere, geopolitical shifts, and perhaps most importantly the advent of the 4IR were some of the motivations for the review of the South Africa's NSI, an exercise that culminated in the 2019 release of the White Paper on STI. Even though the STI White Paper references efforts to integrate and manage South African indigenous knowledge systems into the STI strategy, there can be little dispute about the peripheral location of IKS and the focus of STI on conventional disciplines and investment in the development of institutional, programming and infrastructural resources. Unlike Kenya and Nigeria, South Africa's NSI and 2019 White Paper on STI are designed with evident consciousness of 'the Fourth Industrial Revolution (4IR) and its attendant risks and opportunities' (DST, 2019: 3). It is not surprising that South Africa is one of the earliest and the most proactive on the continent to develop a focused strategic response and preparedness for the 4IR, a move that has ramifications for the rest of Africa.

In 2019, South Africa inaugurated a 'Presidential Commission on the Fourth Industrial Revolution' also referred to as 'the Commission'. In its earlier incarnation, the Commission was proposed as the Digital Industrial Revolution Commission. Among other objectives, the Commission aims at generating a 'single plan or blueprint which brings together all key role players into a single focus'. In addition to its advisory role, in substance (DTPS,[7] 2018: 6):

> The Commission will coordinate the development of South Africa's national response action plan to deal with the 4IR … identify policies, strategies and plans that are needed to position South Africa as a leading country in the evolution of and development of the 4IR.

Membership of the Commission is drawn from across gender, racial and sectoral categories. Without any IKS stakeholder, there is representation in 17 sectors: energy; ICT/telecom; audio-visual; agriculture; banking; finance; fintech; SMME; labour; business consultancy; education; astronomy; local government; science and technology; creative industry; education and skill development; software and engineering. Some sectors have more than one representative in the Commission.[8]

The Commission submitted a draft diagnostic report in November 2019, which included aspects of its mandate to explore how South Africa can leverage the 4IR to enhance its global competitiveness in the above-enumerated sectors through R&D, skills development, establishing institutional frameworks, fostering inclusiveness and entrepreneurship. In executing its mandate, the Commission is expected to adopt an approach that allows for the impact of all government interventions in the 4IR to be measurable. According to the President of the Republic, Cyril Ramaphosa, South Africa's national strategy for harnessing the 4IR is premised on three pillars: 1) the ability to

---

7   In 2019, the Department of Telecommunications and Postal Services (DTPS) was merged with the Department of Communications to form the Department of Communications and Digital Technologies (DCDT).
8   For example, ICT/telecoms has 10 representations.

respond with agility and adapt to the pace of change; 2) a focus on key sectors with high growth potential;[9] and 3) preparing citizens for technological change and shielding them from the adverse impacts of this change (Ramaphosa, 2020). As part of the implementation of South Africa's 4IR strategy, the Commission has various work streams on specific issues,[10] including law and policy. Recently, in addition to the 4IR, IKS has become an important site of law and policy reform in South Africa.

SOUTH AFRICA'S PROGRAMME ON INDIGENOUS KNOWLEDGE SYSTEMS (IKS)

South Africa has been proactive about revamping laws and initiating progressive policies of direct relevance to innovation and the digital economy (DST, 2019). Examples include three initiatives dealing with intellectual property, indigenous knowledge and related matters. They are the Intellectual Property Laws Amendment Act, 2013; the Copyright Amendment Bill, 2018 (pending since 2010) and the Protection, Promotion, Development and Management of Indigenous Knowledge Act, 2019 (IK Act). Of specific interest to the present chapter is the priority South Africa attaches to IKS. The IK Act is a significant step in implementing the IKS agenda. As far back as 2004, under the auspices of the Department of Science and Technology (DST), the Cabinet adopted a comprehensive IKS Policy, which is currently in its 16th year of implementation. The IKS Policy resulted from consultations and collaboration among diverse stakeholders, multiple government departments, NGOs, science councils, tertiary education institutions and, of course, specific knowledge holders (DST, 2004). The IKS Policy now serves as a catalyst and template for law making; for new departmental initiatives regarding indigenous (also traditional) knowledge, and for complementing existing programmes, laws and regulations. However, it is missing in South Africa's 4IR strategy.

---

9   These include agriculture, mining, manufacturing, ICT, electronics.

10 Others are infrastructure and resources, research, technology and innovation, human capital, industrialization.

IKS MISSING IN THE 4IR STRATEGY

Despite references to IKS as part of the ecosystem of South Africa's STI and NSI, it does not seem that IKS have been adequately integrated, let alone mainstreamed, into the new 4IR strategy. As perhaps a reflection of the gap in the AU's STI approach, recent interest on the continent in 4IR fails to make a link to African IKS. That is also true of a major continental study on unlocking the potential of 4IR in Africa commissioned by the African Development Bank (AfDB, 2019). In making what it calls 'the business case' for the adoption of 4IR in Africa, the study made no reference to African IKS across any of the 4IR technological paradigms it explored. This gap is also evident in the South African 4IR Commission. As became clear in the lead up to the Commission, its sectoral representations, composition and *modus operandi* outlined above, there is little direct mention of IKS and associated stakeholders. The latter are not part of the sectoral composition of the 4IR Commission even as IKS is not part of the design of the 4IR strategy.[11] Yet, according to IKS Policy, 'despite the clear association with heritage and cultural tradition, indigenous knowledge is very much at the cutting edge' (DST, 2004: 4) of innovation in a range of fields.

In international law, IKS – also referred to mostly as traditional knowledge (TK) – are no longer seen as part of a knowledge system associated with a stagnant culture and frozen in time. The dynamic nature of IKS is associated with innovation, practices and ways of life and aspects of living cultures, among other things.[12] Echoing similar sentiments, the DST's IKS Policy is projected to assist South Africa with the accelerated development of 'novel and more powerful technologies, market new (IK-based) products and services and stay ahead of the pack' (DST, 2004). South Africa recognises that the IKS Policy was propounded at a time of rapid change in innovation and

---

11 The IK Unit is now a full-fledged unit in the DST.

12 For example, Article 8(j) of the Convention on Biological Diversity, the 2007 United Nations Declaration on the Rights of Indigenous Peoples (among many other instruments) make mention of indigenous knowledge as part of innovation and practices.

technology. The policy is part of a conscious effort to integrate and empower IKS in the emergent economic and technological order. The most current iteration of that order is the 4IR.

So far, pursuant to the IKS policy, there has been increasing integration of IKS communities, practitioners and stakeholders into the digital sequence information (DSI) programmes and into various forms of partnerships with public and private sectors, and research councils, notably the Council for Scientific and Industrial Research (CSIR). The latter started as a science council parastatal created by the apartheid government in 1945. It was, and is, dedicated to conducting multidisciplinary research relating to the development of science and technology. Under that mandate, post-apartheid, it has witnessed transformations in its role, and in partnerships with indigenous knowledge holders.

Before the inauguration of the IKS Policy, the CSIR had been in partnership with the Traditional Healers Committee (THC). The THC voluntarily shared information relating to about 20 indigenous genetic resources, which were of medicinal and other economic value, in accordance with 'a royalty-bearing benefit-sharing agreement' (Bagley, 2018a: 14).

That arrangement has since resulted in new patents, development of new products and various forms of international collaborations with significant economic benefits for South Africa and indigenous knowledge holders and communities. The CSIR has continued to play a leading role in this and its experiences remain relevant in other partnerships involving indigenous communities and knowledge holders under the new law and policy framework. More importantly, the IKS Policy underscores the interdisciplinarity of indigenous knowledge and by extension the interdepartmental scope required of any related governance and administration. This insight is prescient for any policy on 4IR. In the words of the IKS Policy (DST, 2004: 20):

> IK is an important area of focus for a number of governmental departments. Some key departments are: Agriculture, Arts and Culture, Science and Technology, Education, Environmental Affairs and Tourism, Health; Trade and Industry; Provincial and

Local Government; Land Affairs; Water Affairs and Forestry; Sports and Recreation and Foreign Affairs. The role of DST has been to coordinate the different departments to ensure there is a coherent approach to matters relating to IKS.

Uniqueness of IKS in South Africa

IKS assumes a special significance for South Africa for several reasons; and two of those will suffice for now. The first is the breadth and scale of the country's invaluable indigenous resources and cultural heritage. These are related to the depth of South Africa's ethnic diversity, which correlates with the status of the country as one of the world's mega biodiversity hotspots and one of the richest indigenous knowledge regions of the world, taking into consideration the affinity between indigenous knowledge, ethnic diversity and biological diversity/ resources (Oguamanam, 2006).

The second relates to the historical and cultural context of apartheid in which indigenous knowledge systems and knowledge holders were repressed. The suppression of IKS under apartheid resulted in the marginalisation or exclusion of the vast majority of the population and the consequential setback to South Africa's economy (DST, 2004: 10), its people's cultural heritage and identity. Consequently, South Africa understands the potential of a revamped IKS to recognise, empower and affirm indigenous people's cultural identity; grow an inclusive economy; eradicate poverty and contribute substantially to improvements in the living conditions of the majority of its citizens, as well as advance sustainable development goals.

Given the transformational potential of the 4IR, governments' failure to directly reflect on its ramifications or interface with IKS risks inadvertently reversing the progress already made on the IKS front in South Africa and across the continent. Attempts will be made to address this lack, within the limitations of this chapter. The author will explore how the 4IR, in all its inherent contradictions, could nonetheless be harnessed to position South African – and indeed African – IKS and its stakeholders to take advantage of the opportunities presented by the 4IR. As a starting point, the next section provides an abridged sense of the scope and scale of IK and IKS in South Africa and internationally.

INDIGENOUS KNOWLEDGE AND INDIGENOUS KNOWLEDGE SYSTEMS: SCOPE AND SCALE

IK and IKS in the South African context

Unravelling indigenous knowledge and the undergirding worldviews and ways of knowing – the organic experiences and operations of which constitute IKS – is outside the framework of the present chapter. A pragmatic approach is to highlight how the IKS Policy itself has construed IK, supplemented with insights from the work of the World Intellectual Property Organization (WIPO), which has a specialist Intergovernmental Committee on Intellectual Property and Genetic Resources, Traditional Knowledge and Folklore (It also covers traditional cultural expressions) (WIPO-IGC). The WIPO-IGC runs arguably the most authoritative ongoing international process on TK.[13] It constantly grapples with the nature, scale and scope of IK, which it refers to as TK. The two terms are used interchangeably in this chapter.

The South African IKS Policy adopts a critical outlook on IK. Curiously, it refers to IK as counterpoise to a dubious category – an 'international knowledge system' or 'Western knowledge system' (2004: 10):

> [I]ndigenous knowledge manifests itself in areas ranging from cultural and religious ceremonies to agricultural practices and health interventions ... [it] is generically used synonymously with traditional and local knowledge developed by and within distinctive indigenous communities from the international knowledge system generated through universities, government research centres and private industry, sometimes incorrectly called the Western knowledge system.

---

13 WIPO-IGC uses the term TK. That is the basis for subsequent use of the term in this chapter, even though for all practical and analytical purposes, TK and IK are used throughout the chapter interchangeably.

The IK Act defines IK in an open-ended way as:

[K]knowledge which has been developed within an indigenous community and has been assimilated into the cultural and social identity of that community, and includes – a) knowledge of a functional nature; b) knowledge of natural resources; and c) indigenous cultural expressions.

In a similarly loose fashion, it defines indigenous cultural expressions as:

[E]xpressions that have a cultural content that developed within indigenous communities and have assimilated into their cultural and social identity, including but not limited to – a) phonetic or verbal expressions; b) musical or sound expressions; c) expressions by action; and d) action tangible expression.

IK at the international level

At the international level, indigenous knowledge is a complex subject, traversing intellectual property; trade and development; biodiversity conservation and the environment; cultural property and cultural heritage and human rights to mention a few areas (McManis, 2007; Helfer and Austin, 2011). As the most authoritative forum on IK (Robinson et al., 2017), the WIPO-IGC has devoted nearly two decades to elaborating and untangling the philosophical, jurisprudential and conceptual conundrums relating to IK. That much is implied in its long name which evinces the intersection of IK (which includes, as we have seen, indigenous or traditional cultural expressions) with intellectual property (the legal mechanism for governance of knowledge) and, most importantly, with genetic resources.

WIPO-IGC's mandate is to negotiate (an) international legal instrument(s) for effective protection of TK. The mandate is premised on the understanding that conventional intellectual property lacks the capacity to protect IK effectively. Nor is it capable of empowering TK holders or custodians to benefit equitably from the utilisation of their knowledge and innovations, which are constantly exploited through other economic frameworks.

An important aspect of WIPO-IGC is its proactive association of TK and traditional cultural expressions (TCE) with genetic resources. WIPO-IGC's text-based negotiations adopt a tripartite pathway and it has currently three draft texts on genetic resources, traditional knowledge and traditional cultural expressions. The WIPO-IGC definition of TK and TCE uses identical language. For convenience, here is the Committee's definition of TK (WIPO, 2019: 5):

[K]nowledge originating from indigenous [peoples], local communities and/or [other beneficiaries] that may be dynamic and evolving and is the result of intellectual activity, experiences, spiritual means, or insights in or from a traditional context, which may be connected to land and environment, including know-how, skills, innovations, practices, teaching, or learning.

As outlined above, TK/IK is construed nationally and internationally in general terms with an open-ended sense of its scope. What constitutes TK and the contexts for its production do not lend themselves to exhaustive enumeration except to say that TK transcends economic, social and cultural as well as other practices and experiences, especially those within intangible domains like the spiritual dimensions of TK. For example, at the level of expression, TK manifests in all sites of creativity including but not limited to language, dialect, phonology, arts, craft, carvings, designs, songs, dance, choreography, poems, storytelling, incantations, recitations, rituals, symbolisms, education, ceremonies – all of which represent complex and practically demonstrable bundles of knowledge production and ways of life. In their production, processing, marketing, application, evolution, diffusion and transformations, these aspects of IK are amenable to the applications of 4IR technologies such as Artificial Intelligence and Machine Learning, 3D printing, big-data, data science and Internet of Things, robotics, drones, etc.

Even outside of the framework of 'expression', TK encompasses knowledge production and practices rooted in a relationship with land, flora and fauna, the entire ecosystem and an ecological order within specific indigenous cosmogonies of nature and environmental stewardship. These are navigated within and through complex relationships, among indigenous peoples themselves, with other humans and with non-human life forms, the living and the dead, the seen and the unseen, the present and the future, including, of course, the future generation.

The centrality of genetic resources in TK/IK

It is evident from the WIPO-IGC that TK is associated with genetic resources, which constitute a major and pivotal site for its production. TK involves practices and innovations in agricultural, farming, breeding, medicinal, phytomedicinal, pharmaceutical, food, beverages, nutrition and aspects of therapeutic knowledge systems and knowledge production, all of which, as noted above, are accessible to the defining technologies of the 4IR. But to underscore further genetic resources as a core site of practically all aspects of TK, even expressive aspects of TK (a.k.a. traditional cultural expression) are associated with the use of genetic resources (Oguamanam, 2018). For example, traditional arts and crafts, painting and carvings, bead-making, weaving and other functional aspects of expressive culture not only reflect cultural symbolisms; they also inspire them. In their original state, most of them are produced with the use of genetic materials endemic to a given indigenous society. The majority of the functional aspects of TK are equally as expressive as other intangible aspects.

There are rich examples that illustrate how the holistic nature of TK is evident across the uses of endemic genetic resources; expressions of cultural symbolisms; the communication and negotiation of gender and of sexual, ancestral, environmental and historical dynamics, as well as to produce other aspects of tangible and creative culture. These are captured to some degree across the following experience of TK and TCEs: among Zulu women bead-makers (South Africa); Massai cultural beadworks (Tanzania); Kente fabric and designs (Akan, Ghana); Adire fabrics and designs (Nigeria). Outside of Africa, the famed aboriginal bark paintings (Australia); and totem poles of the Indigenous Peoples of the Pacific Northwest (United States and Canada) represent similar trends.

TK/IK: INTELLECTUAL PROPERTY, BIOPIRACY, ACCESS AND BENEFIT SHARING

As evident below, virtually all of the above-enumerated sites in which TK and TCEs are produced are also susceptible to the application of one or multiple categories of intellectual property rights (IPRs): patents, copyrights, trademarks, designs, geographical indications, etc. However, other aspects of TK-TCEs, especially intangible ones, fall outside of the design of conventional IPRs. These aspects include embodied spirituality and culturally rooted symbolisms, as well as complex communal dynamics that mediate the process of IK production and customary rules of ownership, custodianship or possession. Coupled with the market's economic fetishisation of intellectual property, these aspects compound the gaps between intellectual property and IKS. This state of affairs results in plural and often competing policy directions regarding IK. One such direction follows the pathway for conservation and safeguarding of aspects of IK construed as intangible cultural heritage (Kono, 2007). At the international level, UNESCO champions this.[14]

Another pathway is one that uncovers intellectual property's role as a tool for appropriation and unrequited transfer of IK, especially those dealing with genetic resources, outside the control of indigenous communities and knowledge holders. This phenomenon, which exposes the provenance of IK and associated genetic resources to contestation, is known as biopiracy (Mgbeoji, 2006; Robinson, 2010; 2015). As demonstrated later, biopiracy is further implicated in 4IR technologies and warrants the search for effective protection of IK.

National reforms in South Africa on IK follow two principle but complementary pathways. The first is protecting traditional knowledge through the IP system pursuant to the Intellectual Property Laws Amendment Act, 2013 and the Copyright Amendment Bill. In the bigger scheme, the Act and the Bill represent the repositioning of the country for the 4IR. The second is through a *sui generis* approach that

---

14 UNESCO has jurisdiction over a number of instruments dealing with preservation of culture and heritage. A quick example is the 2003 Convention for the Safeguarding of the Intangible Cultural Heritage.

attempts to plug the gaps between IK and conventional IPRs. This objective recognises unique facets of IK, including its non-intellectual property aspects, and is pursued through the IK Act, 2019.

Perhaps most importantly, with regard to biopiracy, South Africa is one of the most proactive countries on the continent. The country has a national access and benefit sharing (ABS) framework, aimed at empowering IK and its holders through environmental and related legislation consistent with two flagship international frameworks for dealing with biopiracy, namely the Convention on Biological Diversity (CBD) and the Nagoya Protocol ABS modules.

Biopiracy: IK and access and benefit sharing case studies in South Africa

In Africa, and even globally, South Africa is a source of important, controversial cases about biopiracy. Through its proactive legal and regulatory regimes, South Africa has become a source of inspiration for how IK holders and communities can be empowered and repositioned through progressive application of ABS and other reward measures that strengthen IKS and their practitioners. Recently, South African law and policy on biopiracy has been put to the test with laws supervising and mediating the interests of indigenous communities alongside those of R&D entities, transnational corporations[15] and other value chain intermediaries. These laws include IK practices and the uses of the following genetic resources for therapy and as medicines, antioxidants, sweeteners, beverages and health foods: hoodia plant (*Hoodia gordonii*), lemon bush (*Lippia javanica*), African ginger (*Siphonochilus aethiopicus*), molomo monate (*Sclerochiton ilicifolius*) and, most notably, rooibos (*Aspalathus linearis*) (Bagley, 2018a; Schroeder et al., 2020).

The use of rooibos as a beverage with medicinal properties is owed to the San and Khoi peoples of southern Africa. These ancient civilisations are reputed to hold rich traditional and economically valuable insights into indigenous southern African flora. Such knowledge is the source

---

15 CSIR was the major R&D entity in most of these cases while some of the multinationals included Nestlé and Pfizer.

of the famed use of rooibos as a primary tea, rich in antioxidants among other properties. Since 2014, rooibos became the first non-alcoholic product in South Africa to enjoy a geographical identification status, covering the Cederberg Mountain region where the beverage is grown. The San and Khoi IK of rooibos is associated with the prevalence of rooibos-related products spanning 'cosmetics, novel foods, slimming products, extracts and flavours' (Schroeder et al., 2020: 4). Rooibos is an engine of indigenous and rural economic development in South Africa employing 5,000 and producing about 15,000 tonnes of processed tea leaves annually, generating income estimated at half a billion rand (Schroeder et al, 2020: 3). As of 2020, there are 'in excess of 140 patents pending' relating to the 'biological and health properties' of rooibos (Schroeder et al., 2020: 3). On the back of previously unsuccessful attempts, after almost 20 years, the San and Khoi signed a ground-breaking benefit-sharing agreement with the rooibos-based industrial complex in the Greater Cederberg Biodiversity Corridor. Notwithstanding racial asymmetry in ownership of land in the region, and concerns over the exclusion of other historic stakeholders, the agreement is the most significant of its kind under the CBD and Nagoya Protocol and congruent with South Africa's unique IKS legal framework. Reflecting on the Rooibos Benefit Sharing Agreement (RBSA), Schroeder et al. (2020: 15) observe:

> With no precedent in the benefit-sharing world, the RBSA stands as a concrete example of the 'art of the possible.' It serves to confirm that such agreements can be concluded in support of indigenous communities, industry, and governments implementing the CBD and Nagoya Protocol. It offers lessons and inspirations to indigenous communities, and industry sectors within both Africa and other regions around the world.

The above observation underscores the practical strides taken in South Africa to position IK as a matter that spans several departments. This is evident in the focus of IKS Policy within the National Environmental Management and Biodiversity Agency (NEMBA), which is coordinated by both the DST and the Department of

Environmental Affairs (DEA). The next section presents a random sample of sectoral sites where the challenge that 4IR poses to IK becomes evident, thus reinforcing the importance of integrating IKS and its stakeholders in 4IR strategies.

SAMPLES OF SECTORAL SITES WHERE 4IR TECHNOLOGIES INTERFACE WITH IKS

The audio-visual context

In addition to the biological-resource sphere, South Africa has continued to exert robust influence in other aspects of IK production. This influence includes areas where ownership and control of IK are mediated by new technological transformations from both the Third and Fourth Industrial Revolutions and are compounded further by unfair intellectual property laws, especially copyrights – pre- and post-digital era – which are outside the scope of this chapter. For example, the amazing posthumous story of Zulu cultural icon and indigenous music innovator, Solomon Popoli Linda of the *Mbube* legend. Between 1939 and 2019, Solomon Linda's *Mbube* experience unravelled how new technologies, including digital technologies, advance the fluidity and malleability of creative innovations across multiple audio-visual platforms while at the same time facilitating cultural migration, adaptation and cultural appropriation through network effects. Solomon Linda's 1939 album, *Mbube*,[16] transitioned to American's Pete Seegars *Wimoweh* (1952), then to George David Weiss's *The Lion Sleeps Tonight* (1961) and its spin offs and associations with Walt Disney's *The Lion King* (1994) movie[17] and François Verster's *The Lion's Trail* documentary (2002) and, lately, to Netflix's documentary, *The Lion's Share* (2019).

South Africa is rich in an expressive repertoire of traditional music genres. Spanning conventional and uniquely traditional musical styles and variegated audio-visual works, the majority of these musical forms are globally acclaimed, and subject to international collaborations and

---

16 Solomon's was inspired by Mbube-style, which is an authentic musical genre in its own right.

17 Including Disney's remake of *The Lion King* in 2019.

partnerships. One symbolic depiction of this trend is Joseph Shabalala's (Ladysmith Black Mambazo) famous partnership with America's Paul Simon in *Graceland*. Within the digital domain of the 4IR, the South African traditional creative industry, and virtually all aspects of IK, are increasingly exposed. This is because they attract global patronage in contexts in which technological mediation – including 3D – renders the issues of provenance, ownership, originality and benefit sharing complex at best, but often contested, and undetermined.

3D printing in context

As a practical example, in the 4IR, 3D printing and the broader framework for additive manufacturing holds significant potential for TK and TCEs. In South Africa and most of Africa, 3D printing has been presented on a highly positive note (Makoni, 2017; AfDB, 2019). The emphases are on its capacity to democratise manufacturing, and its applications and potential for sourcing industrial parts for agriculture, automobile, aviation, renewable energy, health, etc. (AfDB, 2019; Makoni, 2017). Without question, 3D printing and 3D printers as a process and a resource are very handy in the reproduction and restoration of traditional arts, crafts, designs, etc. of all sources and configurations. The technology, however, requires expertise in several fields, including software engineering, coding, digitisation, design and postproduction (Li, 2017). Information or data is digitised via a computer-aided design (CAD) file which is the only form in which 3D printers can reproduce an object in 3D. IK holders lack the ability to leverage and optimise this technology as they do not have access to specialist educational and digital skills. This demonstrable gap in the application of 3D technology is one that is reflected, in varying degrees, in other defining technologies of the 4IR. It exposes both the vulnerability of IK to internal and external cultural appropriation and the weakness of IK stakeholders' ability to optimally participate in the 4IR. It emphasises the imperative for capacity building. This is feasible where polices around STI and 4IR both take account of the possibilities for increased convergence between these sectors and IKS and associated stakeholders.

Unsurprisingly, in South Africa and other parts of Africa, 3D technology is progressively applied in the arts and creative arenas mainly by a tech-savvy, urban elite (Miller, 2016). The technology yields a radical disruption of intellectual property (Kenney, 2011; Cronin, 2015; Li, 2017) – patents, designs and copyrights – regimes that, historically, have been problematic for IKS. For example, using software to compress TK or TCEs of any kind into CAD creates copyrightable material. Its reproduction or transformation into a functional 3D object by a 3D printer relates to patents while the product's association with a brand, original manufacturer, provenance, or tradename entails trademarks. When applied to TK and TCEs, 3D and additive manufacturing technologies reduce them to their intellectual or aesthetic essence or expression (Cronin, 2015), and their functionality as may be applicable. The inability of 3D-produced TK or TCEs to extend the latter's internalised symbolism, spiritual essence and aura demonstrates the indispensability of provenance, and authenticity in the valuation and validation of TK and TCEs. Consequently, the issue of disclosure of source and origin by third parties dealing with TK and TCEs is now fully extended to all IP regimes, including design law (Bagley, 2017; 2018b), and constitutes part of access and benefit sharing (ABS) explored in the next section.

In essence, 3D printing may have all the potentially empowering elements of openness and democratisation of production. Concerning IK holders, however, it raises the issue of inequity in accessing undergirding technologies and skills. In addition, it speaks to IP's fraught relationship with IKS and remains a source of serious concern regarding cultural appropriation. As a critical part of 4IR, the relationship between 3D technology and IKS requires a more thoughtful policy approach to complement and balance the technology's generally positive outlook in Africa, and to ensure that IK stakeholders are positioned to participate in the technology. This is also true of the applications of all 4IR technologies at the interface of IKS.

The biological domain – digital sequence information

The biological domain is perhaps the most intense site of contestation, one that merits prioritisation in the 4IR's interface with IKS. As already demonstrated, from the experience of South Africa, intense

IK production in the realm of genetic resources traverses medicine, therapeutics, pharmaceuticals, agriculture, health food, beverages, cosmetics and other fields. These have proven to be growth sectors and, as such, they are the logical targets of the South African 4IR policy. They have the capacity to boost inclusive economic development, involving a significant portion of South African indigenous peoples.

In the 4IR, the uses and applications of genetic resources transcend genetic engineering – a form of modification of gene functions that found traction in the 1980s. Under the 4IR, technology has advanced to the 'ability to edit biology' (gene or genome editing) across all cell types, resulting in the production of genetically modified plants and animals and in the modification of the cells of mature organisms, not excluding humans (Schwab, 2016: 22). All of this is done with an unprecedented degree of efficiency, and precision in pliability across biological and taxonomic boundaries. Gene editing is part of complex and advanced multidisciplinary (such as genomics, transcriptomics, proteomics, metabolomics) platforms, including notably synthetic biology. Through this biology, scientists are able to design, or fabricate, new or non-existing biological components or systems, even via 3D. They can also redesign existing biological parts or systems with the objective of modifying the behaviour of single organisms, or of integrating the behaviour of several others into a cohesive and single biological unit.

The new biological R&D complex, including synthetic biology, relies on the sequencing of information or data on genetic resources, using digital tools. This form of digitally enhanced virtualisation contrasts with dealing directly with genetic resources in their physical or natural states. While there has yet to be agreement over the appropriate characterisation of this phenomenon, the official designation of it in the CBD process is 'digital sequence information' (DSI)[18] on genetic resources. Without getting into the complexity of the debate over DSI, it designates the aggregation of information on genetic resources in digital forms, in contrast with their tangible, physical and material forms. Digitally sequenced information can be

---

18 Also called digital DNA.

stored in online databases and made accessible to researchers. Through digital sequencing, entire genomes or genes can be constructed from scratch, allowing for customisation and storage of vital information or data which can then be ordered or accessed from commercial (or open access) laboratories (Oguamanam and Jain, 2017: 106). Thus, it is cheaper and more efficient to source genetic information as virtual and intangible assets, bypassing reliance on their physical or intangible forms in nature. Genetic engineering, gene editing, synthetic biology and DSI are evidence of the confluence of the digital and biological aspects of 4IR. Physical hardware or devices (physical dimension) are deployed to convert or translate the knowledge downstream to deliver real live therapeutic, medicinal, agricultural, nutritional and environmental innovations and interventions.

While these new applications of genetic resources have monumental potential for improvements in the quality of lives, they also raise major legal, regulatory, ethical and practical concerns with regard to IK and its custodians. Six of the major concerns are discussed below.

First, the conceptual framework for understanding IK relating to genetic resources is squarely based on direct physical dealings with those resources. In most indigenous communities, this is tied to their relationship with the land and constitutes aspects of environmental stewardship.

Second, as a consequence of the first, the international and national legal framework on access and benefit sharing is conceptually based on physical access to genetic resources and associated IK and on a sense of their provenance. Therefore, the tendency to bypass or de-link corporal genetic resources and associated IK from their provenance in indigenous communities has the potential to undermine the ability of IK and its holders to participate in the opportunities presented by the biological domain of the 4IR. Already, debates are rife over whether the CBD and its Nagoya Protocol pre-empted DSI and the synthetic biological process as a whole.[19]

Third, in relation to practical, scientific and governance matters,

---

19 Through the activities of Ad Hoc Technical Groups on Synthetic Biology and Digital Sequence Information at the CBD.

data is a pivotal aspect of omics, gene editing, synthetic biology and digital sequencing. These are sources of large data sets – so-called big data. Big data is facilitated by computational infrastructure amenable to analytical velocity, now within the 5G purview (Eldred et al., 2019) and optimisable via internet architecture and design frameworks, requiring large storage capacity, for example, via the cloud. Building on the reality of the third, the fourth point is that IK stakeholders, including countries like South Africa, do not have the financial muscle or the expertise to level the playing field in the infrastructural and architectural (Kwet, 2019) asymmetry that characterise the operations of the biological component of the 4IR. As we have noted, the STI-driven R&D, programming, institutional and infrastructural outlook in the AU and in individual African countries are not designed for the contingencies that arise at the inevitable interface of new technologies with IKS – contingencies now magnified in complex 4IR scenarios.

Fifth, as part of its IK Policy, South Africa's elaborate legislative and administrative framework on IKS includes a National Record System (NRS) for IK under the superintendence of the DST. The NRS uses multimedia platforms for recording and documenting IK, especially those relating to traditional medicine and indigenous foods across diverse IK categories. NRS is designed as a defensive anti-biopiracy strategy to ward off appropriation of IK in their undocumented forms through the IP system. It is part of a broader National Indigenous Knowledge Management System (NIKMAS). In essence, NIKMAS is 'a digital repository for ... collected [indigenous] knowledge' (Bagley, 2018a: 17), secured and catalogued in searchable database form under a rigorous access-and-use protocol. The ultimate objectives of the NRS are to enhance the preservation, promotion and protection of IK for developmental outcomes. The valorisation of the TK database in this context is part of indigenous peoples' emergent interest in data sovereignty (Oguamanam, 2019; Dagne, nd) which poses a significant challenge for inclusivity and equity in the 4IR.

The importance of data resonates in 4IR dynamics as well as in the context of IKS. The supporting architecture and infrastructure for big data stokes asymmetry over the ability of IK holders to optimise the potential for their knowledge. Conversely, IK databases symbolise a practical aspect of the attempt by IK stakeholders to leverage and assert ownership and control over their own knowledge to advance their development aspirations (Oguamanam, 2019). As a matter of importance, data governance constitutes a pivotal subject for a potential rapprochement between 4IR and IKS yet to be explored in South Africa, and indeed Africa, through various STI, NSI and attention to 4IR.

CONCLUSION

Historically, there have been strong interests and efforts in Africa at continental, regional and national levels in the elaboration of policies and strategies around STI. As a contemporary matter, ICT is an important sectoral site for the evolution of STI policies on the continent. The implementation of those policies has resulted in elaborate investment in institutional and infrastructural resources in most African countries. As evident from Nigeria and Kenya, African STI policies make marginal references to considerations of African IKS. However, frameworks for the adoption, implementation and promotion of STI via programming, infrastructure, investment and the mobilisation of disciplinary stakeholders have clearly excluded IKS and its stakeholders. In part, this trend is evident in the dedicated focus on STI-driven institutions, education, training programmes and disciplines within conventional STI matrixes. With the advent of the present 4IR, it is expected that any dedicated African strategy at continental and certainly at national level would build on existing STI architecture. As is to be expected, such an approach is fraught with the risk of repeating the gap around IKS in STI and NSI. This is already evident from the experience of South Africa, which is the African country with the foremost official strategy on 4IR. At an empirical level, some sites of innovation, such as audio-visual contexts; 3D printing; aspects of data-driven, hi-tech developments in the biological sphere such as digital sequence information on genetic resources, have implications for the interface between 4IR technologies and IKS. Those contexts raise questions over intellectual property, biopiracy and issues around the role of IKS and its provenance as well as the interest of its custodians. As African countries mull over strategies for their effective and beneficial participation in the 4IR, there is a need to ensure that 4IR-related policies will proactively interrogate how best to accommodate IKS and stakeholders in these technological transitions as a matter of priority.

REFERENCES

African Development Bank (AfDB). 2019. 'Potential of fourth industrial revolution in Africa'. *African Development Bank Group*. https://4irpotential.africa/wp-content/uploads/2019/10/AFDB_4IRreport_Main.pdf, accessed 10 May 2020.

African Union (AU). 2019. 'Contextualizing STISA-2024: Africa's STI Implementation Report 2014–2019'. African Union. https://au.int/sites/default/files/newsevents/workingdocuments/37841-wd-stisa-2024_report_en.pdf, accessed 28 July 2020.

African Union Commission (AUC). 2014. 'Science, technology and innovation strategy for Africa 2024'. *African Union*. https://au.int/sites/default/files/documents/37448-doc-stisa-2024_english.pdf, accessed 28 July 2020.

Bagley, M.A. 2017. 'Illegal designs? Enhancing cultural and genetic resource protection through design law'. *Centre for International Governance Innovation*. https://www.cigionline.org/publications/illegal-designs-enhancing-cultural-and-genetic-resource-protection-through-design-law, accessed 10 May 2020.

Bagley, M.A. 2018a. 'Toward an effective indigenous knowledge protection regime: Case study of South Africa'. *Centre for International Governance Innovation*. https://www.cigionline.org/publications/toward-effective-indigenous-knowledge-protection-regime-case-study-south-africa, accessed 28 July 2020.

Bagley, M.A. 2018b. 'Ask me no questions: The struggle for disclosure of cultural and genetic resource utilization in design law'. *Vanderbilt Journal of Entertainment & Technology Law,* 20(4), 101–151.

Convention on Biological Diversity, 5 June 1992, 1760 UNTS 79, 31 ILM 818 (entered into force 29, December 1993).

Copyright Amendment Bill 2017. B 13 – 2017. 5 July 2016. https://www.parliament.gov.za/storage/app/media/uploaded-files/Copyright%20Amendment%20Bill%20Draft.pdf, accessed 10 May 2020.

Cronin, C. 2015. '3D printing: Cultural property as intellectual property'. *The Columbia Journal of Law & the Arts*, 39, 1.

Dagne, T. (Undated). 'Embracing the data revolution for development: Data justice framework for farm data in the African context'. Unpublished research paper for *Open African Innovation Research*

Department of Science and Technology. 2004. 'Indigenous Knowledge Systems Policy'. https://www.dst.gov.za/images/pdfs/IKS_Policy%20PDF.pdf, accessed 17 April 2020.

Department of Science and Technology. 2019. 'Science, Technology and Innovation Policy'. https://www.dst.gov.za/images/2019/FINAL-White-Paper-to-Cabinet_11-March-2019.pdf, accessed 10 May 2020.

Department of Telecommunications and Postal Services (DTPS). 2018. 'Invitation to nominate candidates for the presidential commission on fourth industrial revolution'. https://www.gov.za/sites/default/files/gcis_document/201812/42078gen764.pdf, accessed 10 May 2020.

Economic Commission for Africa (ECA). 2018. 'Country STI profiles: A framework for assessing science, technology and innovation readiness of African countries'. United Nations Economic Commissions for Africa. https://www.uneca.org/sites/default/files/PublicationFiles/sti_report_en_revised.pdf, accessed 28 July 2020.

Eldred, C., Kenny, M., Kushida, K.E., Murray, J. and Zysman, J. 2019. '5G: Revolution or hype?'. *SSRN*. https://papers.ssrn.com/sol3/papers.cfm?abstract_id=3443740, accessed 13 March 2020

Helfer, L.R. and Austin, G.W. 2011. *Human Rights and Intellectual Property: Mapping the Global Interface*. New York: Cambridge University Press.

Intellectual Property Laws Amendment Act No. 28 of 2013. 10 December2013. https://www.gov.za/documents/intellectual-property-laws-amendment-act-0, accessed 17 March 2020.

Kariuki, F. 2020. 'Harnessing traditional knowledge holders' institutions in realising Sustainable Development Goals in Kenya'. Unpublished Research Report for Open African Innovation (Open AIR) Network.

Kenney, C.C. 2011. 'Reframing indigenous cultural artefacts disputes: An intellectual property based approach'. *Cardozo Arts and Entertainment Law Journal*, 28, 501–552.

Kono, T. (ed). 2007. *Intangible Cultural Heritage and Intellectual Property: Communities, Cultural Diversity and Sustainable Development*. Antwerp: Intersentia.

Kwet, M. 2019. 'Digital colonialism: US empire and the new imperialism in the Global South'. *Race & Class*, 60(4), 3–26. https://doi.org/10.1177/0306396818823172.

Li, J. 19 July 2017. 'To print or not to print: Innovation and intellectual property issues in 3D printing'. *IP-Watch*. https://www.ip-watch.org/2017/07/19/print-not-print-innovation-ip-issues-3d-printing/, accessed 28 July 2020.

Madse, D.Ø. 2019. 'The emergence and rise of industry 4.0 viewed from the lens of management fashion theory'. *Administrative Sciences*, 9(3), 71.

Makoni, M. 13 March 2017. '3D printing in Africa: Huge benefit or big IP threat?' *IP-Watch*. https://www.ip-watch.org/2017/03/13/3d-printing-africa-huge-benefit-big-ip-threat/, accessed 28 July 2020.

Mariani, M. and Borghi, M. 2019. 'Industry 4.0: A bibliometric review of its managerial intellectual structure and potential evolution in the services industries'. *Technological Forecasting and Social Change,* 149. https://www.sciencedirect.com/science/article/pii/S0040162519311345, accessed 13 March 2020.

McManis, C.R. (ed). 2007. *Biodiversity and the Law: Intellectual Property, Biotechnology and Traditional Knowledge.* New York: Earthscan.

Mgbeoji, I. 2006. *Global Biopiracy: Patents, Plants and Indigenous Knowledge.* Vancouver: UBC Press.

Miller, A. 18 July 2016. 'South Africa's art machine features 100% 3D printed artwork'. *3D Printing Industry.* https://3dprintingindustry.com/news/south-africas-art-machina-features-100-3d-printed-artwork-87383/, accessed 28 July 2020.

National Environmental Management Biodiversity Act. 2004. Regulations on Bioprospecting Access and Benefit Sharing. No. R 138 of 2008. 8 February 2008, https://www.environment.gov.za/sites/default/files/legislations/nemba_regulations_g30739rg8831gon138_0.pdf, accessed 17 April 2020.

National Environmental Management Biodiversity (NEMA) Act No. 10 of 2004. 7 June 2004. https://www.gov.za/documents/national-environmental-management-biodiversity-act-0, accessed 13 March 2020.

Oguamanam, C. 2006. *International Law and Indigenous Knowledge: Intellectual Property, Plant Biodiversity and Traditional Medicine.* Toronto: University of Toronto Press.

Oguamanam, C. and Jain, V. 2017. 'Access and benefit sharing, Canadian and Aboriginal research ethics policy after the Nagoya Protocol: Digital DNA and transformation in biotechnology'. *Journal of Environmental Law & Practice,* 31(1), 79–112.

Oguamanam, C. 2018. 'Wandering footloose: Traditional knowledge and the public domain revisited'. *Journal of World Intellectual Property,* 21(5–6), 306–325.

Oguamanam, C. 2019. 'Indigenous data sovereignty: Retooling indigenous resurgence for development'. *Centre for International Governance Innovation.* https://www.cigionline.org/publications/indigenous-data-sovereignty-retooling-indigenous-resurgence-development, accessed 28 July 2020.

Protection, Promotion, Development and Management of Indigenous Knowledge Systems Act No. 6 of 2019. 19 August 2019. https://www.gov.za/documents/protection-promotion-development-and-management-indigenous-knowledge-act-6-2019-19-aug, accessed 10 May 2020.

Ramaphosa, C. 10 January 2020. 'A national strategy for harnessing the fourth industrial revolution – the case of South Africa'. *Brookings.* https://www.brookings.edu/blog/africa-in-focus/2020/01/10/a-national-strategy-for-harnessing-the-fourth-industrial-revolution-the-case-of-south-africa/, accessed 28 July 2020.

Robinson, D.F. 2010. *Confronting Biopiracy: Challenges, Cases and International Debates*. London: Earthscan.

Robinson, D.F. 2015. *Biodiversity, Access and Benefit-Sharing: Global Case Studies*. Oxon: Routledge.

Robinson, D.F., Abdel-Latif, A. and Roffe, P. (eds.) 2017. *Protecting Traditional Knowledge*. Oxon: Routledge.

Schroeder, D., Chennells, R., Louw, C., Snyders, L. and Dodges, T. 2020. 'The Rooibos benefit-sharing agreement – breaking new grounds with respect, honesty, fairness, and care'. *Cambridge Quarterly of Healthcare Ethics,* 29(2), 285–301

Schwab, K. 2016. *The Fourth Industrial Revolution*. New York: Currency.

Secretariat of the Convention on Biological Diversity. 2011. 'Nagoya Protocol on access to genetic resources and the fair and equitable sharing of benefits arising from their utilization to the convention on biological diversity'. https://www.cbd.int/abs/doc/protocol/nagoya-protocol-en.pdf.

World Intellectual Property Organisation (WIPO). 2019. 'The protection of traditional knowledge: Draft articles'. *WIPO/GRTKF/IC/40/4 Rev 2,* https://www.wipo.int/meetings/en/doc_details.jsp?doc_id=433260, accessed 13 March 2020.

THREE

When algorithms meet humanity

Duduetsang Mokoele, Nomaqhawe
Moyo and Lerato Mahlangu

'What is changing in our young, fast-growing digital civilisation is that we can delegate decisions in our individual, family or social lives to technology. Human existence can be subcontracted to software ... We've already started putting aside our feelings, intuitions and dreams in favour of choices calculated by an algorithm powered by data.' (Cathelat, 2018: 132)

INTRODUCTION

In April 2011, Peter Lawrence's book entitled *The Making of a Fly* initially cost just over a gargantuan US$1,200,000 and a few days later it peaked at US$23,698,655.93 (plus US$3.99 shipping) on Amazon. This was monumental in academic circles, and he must have been the envy of his peers. This immediately begs the question: what determined this hefty price? The answer: the pricing algorithms had gone haywire. For this chapter, two important points emerge: First, it took several days before the algorithmic malfunction was picked up and rectified. It required reflection on the place of human intervention vis-à-vis the rise of algorithms. Second, this is one of the most obvious ways in which algorithmic activity is apparent in our lives; their presence in the shadows was revealed. Which raises a second question: in what other silent but more profound ways (barely noticeable to humans) are algorithms shaping our lives? This chapter will discuss some ways in which algorithms are changing human lived experiences regarding agency. As a corollary, the key question to be addressed is: How can algorithms be deployed to best serve humanity? The authors believe that the success of the Fourth Industrial Revolution (4IR) will be measured by the extent to which technologies such as algorithms develop in unison with societies (Philbeck et al., 2018: 4). It is with this in mind that the chapter will focus on the hot button issues that keep governments, business, innovators, policymakers and society at large up at night, namely the effects of technology on human agency and capabilities.

ALGORITHMS: APPROACHES AND PARADIGMS

Where can algorithms be found and how often do we interact with them? The answer is everywhere, and all the time. These well-informed, invincible algorithms have touched almost all aspects of human society, making it necessary to create policies that protect the principle of maintaining humanity in this age, and that preserve human values at the heart of human decision-making.

Algorithms power the internet and search engines, and have afforded us the opportunity to awaken things that were once asleep such as our refrigerators, televisions and children's toys, making their potential power virtually infinite. Smartphones are nothing but algorithms. Computer and video games are algorithms telling a story. YouTube and Netflix recommendations on what to watch next would cease to function without algorithms. Algorithms handle many financial decisions and transactions worldwide. Communities are not always aware that algorithms are deciding on issues that have crucial implications for them, for example, who gets into university; who gets insurance; what people pay for that insurance, and much more. Algorithms also have an ever-greater influence on our day-to-day life and work: they influence our political choices and votes; where and how we invest; the business models we create; and the content of our education and medical diagnostic systems, to name just a few areas (O'Neil, 2016: 10).

Algorithms are defined as a finite, well-defined series of steps (input) in a computational procedure to solve a specific problem; the solution delivered would be the output (Cormen et al., 2009: 5). Beyond computer science and mathematical understanding, we use them all the time, for example, a manual to assemble a crib, video game instructions, following a recipe, and so on. These daily and relatable experiences can be thought of as algorithms. Like computer science applications, the intention is to solve a specific problem and yield predictable outcomes. If you programme the algorithm to identify pictures of koala bears the output should match exactly that and not show pictures of giraffes. The process runs on efficiency and predictability, which means it can be repeated with the same result. This is an important point: as illustrated in the introductory paragraph, and later in the chapter, this is not always the case, and programmers of algorithms, even with good intentions, may trigger unintended consequences.

There is an infinite number of algorithms just as there are limitless problems that require solutions from them, and there is no agreement on how to group them. It is helpful to think about their functions, which fall into four categories.

Classification

Algorithms learn many details about you, such as gender, age, where you live and your interests, to name a few examples. Women often give accounts of how, when they reached the legal age for marriage, they would be flooded with advertisements for diamond rings on social media; at child-bearing age, they would see baby-related merchandise; and when nearing retirement age, information about retirement homes would spike (Fry, 2018: 11). Advertisers opt for these kinds of algorithms.

Association

Algorithms are sometimes tasked with finding connections and relationships between things or people. These algorithms are usually preferred in dating sites such as Bumble, and career networking platforms like LinkedIn. Often when shopping online and you are about check out your cart, the online store's algorithm will ask if you would like to add specific items to your cart before checking out.

The algorithm looks for patterns amongst customers by assessing the item(s) in your cart and what other customers have added to accompany the contents of your cart. Once, a shopper on Amazon was about to purchase a baseball bat when the user was asked: 'Perhaps you'll be interested in this balaclava?' (Fry, 2018: 12).

Prioritisation

The aim here is to compile an ordered list of possible choices based on your preferences (Fry, 2018: 11). For example, YouTube will recommend which videos you may be interested in watching next. A global positioning system (GPS) such as Waze will suggest the fastest route, and a search engine such as Google ranks the search results in order of pages by running a mathematical procedure which predicts the pages that you are looking for (Fry, 2018: 11).

Filtering

The main objective here is to isolate what is most relevant or essential. Twitter filters which tweets, news or trending topics make it to your timeline based on your activity (Fry, 2018: 12). The same principle is applied on social media platforms such as Facebook and its company, Instagram. Some other algorithms in this category have to mute noise and focus on a signal, such as the speech recognition algorithms found in Siri (namely, Alexa and Cortana) that have to first eliminate background noise and then decrypt what is said (Fry, 2018: 12).

As algorithms evolve and are required to solve increasingly complex problems, most of the algorithms can be a combination of the above actions. They follow either one of the following two approaches:

Rule-based algorithms

Humans are at the centre of this paradigm because the programmer inputs instructions that are clear and finite from beginning to end (Fry, 2018: 13). This does not mean that such algorithms are simple or unsophisticated – powerful programmes can also be rule-based. The advantage of the rule-based paradigm is that algorithms produced within it are easily comprehensible, and humans can follow their logic (Fry, 2018: 13). The disadvantage of these algorithms is that they can only solve problems that humans know how to write instructions for (Fry, 2018: 13).

Machine learning-based algorithms

This category falls under Artificial Intelligence (AI). Machines learn similar to the way that people and other living creatures learn. A machine is given data with a clear objective: feedback as to whether it is on the right or wrong track and, most importantly, it is left to figure out the solution on its own (Fry, 2018: 14). The advantage of these algorithms is that they are not confined to human knowledge; they are dynamic and adaptable and are efficient problem solvers (Fry, 2018: 14). The machine-learning algorithms' strength is also their weakness: because they have room to learn on their own, it is not always possible for even the smartest programmers to know how exactly they arrived at the output (Fry, 2018: 14). Although the goal may be achieved, its logic and what happens on the inside as it works cannot be explained. This predicament is known as the black box (Fry, 2018: 14). The way in which a machine sees the world may not be strange or outrageous to the machine. A hypothetical example: AI-powered algorithms may be tasked with finding a cure for cancer. The machine's solution may be to kill all those suffering from the illness. The goal would be achieved, but humans weigh ethical considerations and appreciate the broader consequences of actions in a way that algorithms do not. We can opt for a less macabre scenario: the algorithm's objective may be to identify images of octopus (Fry, 2018: 14). When a minute detail is altered, such as a one-pixel change to an image of an octopus, the machine may identify it as a lighthouse which will be baffling to humans (Fry, 2018: 14).

Algorithmic accountability and transparency

Terms such as 'algorithmic accountability' and 'algorithmic trans-parency' are gaining traction as the sociotechnical interface raises fascinating, yet, challenging issues. The terms are often used interchangeably, but they mean different things. Algorithmic transparency refers to the principle that the people who regulate, use and are affected by algorithms have a right to know which factors are used to determine the decisions made by those algorithms (Diakopoulos and Koliska, 2017: 812). This is at the centre of modern privacy law. Failure to do that risks online violation of human rights such as privacy, safety and freedom (UNESCO, 2019).

Algorithmic accountability is holding institutions and individuals who deploy algorithmics responsible for their actions (Diakopoulos and Koliska, 2017: 812). Finding, agreeing on and implementing good accountability measures is complex. It requires taking into account overlapping technical, legal, institutional and political considerations. This becomes murkier in the context of machine-learning algorithms rather than rule-based systems. The former is given the freedom to learn and adapt as they operate with little human inference. As explained in the earlier sections of the chapter, the logic of the machine can remain a mystery even for the programmers themselves and the smartest humans on the planet. On the other hand, rule-based algorithms are human centred and easier to regulate and hold accountable. We are turning to machine-learning algorithms more frequently because they handle complexity better than rule-based algorithms; hence they are deemed appropriate for solving many of our problems.

One of the most popular recommendations in literature on algorithms is that one of the most effective ways to realise algorithmic accountability is by leaving a latch open and allowing for 'human-in-loop' mechanisms which give humans the ability to have oversight of the algorithm as it works (Cormen et al., 2009: 56). We argue that this is a deficient accountability mechanism. Machine-learning algorithms essentially speak a language that humans cannot understand, and their logic is hard to follow. It would not make much difference if humans could have a look inside of the black box, as what they found would still be unintelligible. It would almost be like encountering another black box inside a black box. The better approach would be setting transparency and accountability measures at the procurement stage of the machine-learning systems, which is when specifications and parameters are decided.

Humankind has only just started to realise how 4IR technologies are posing fundamental challenges to how we think about the world and the possibilities for detrimental effects (Matthews et al., 2019: 54). The fact is that these technologies are widespread and have begun unravelling gains made towards social cohesion (Matthews et al., 2019: 58). It can be argued that they can be a fracturing force which deepens inequality and transforms everything, from the structures of an economy to humanity itself (Matthews et al., 2019: 59). It can no longer be taken for granted that what is good for the economy and technological development will automatically be in the best interest of societal progress. There are many cases across the globe where material conditions have, on average, improved but inequality and an undesirable quality of life for many has intensified (Matthews et al., 2019: 53).

There are two pervasive views on technology policy in the private and public sectors. First, technologies are embedded and almost automatically lead to more significant opportunities (Philbeck et al., 2018: 4). The second view is that because of the sheer pace of technological development, it is mostly beyond people's control to shape the forces imbued in the technology (Philbeck et al., 2018: 4). Neither of these views is entirely accurate. They mask the complexity of society's relationship with technology.

Failure to develop a better comprehension of technologies such as algorithms, and of the moral implications of the role that they play in society, hampers the quality of decisions made by policymakers and regulators about how technology is developed and used (Rainie & Anderson, 2017). A better approach would be to acknowledge technologies as resources that interpret, transmute and help to make meaning of the world. Algorithms are not tools that are separate from humans; they are social constructs, situated in a specific culture, and they echo societal values (Rainie and Anderson, 2017). This view allows for reflection on the manner in which algorithms present ethical challenges and may have a negative impact on society. It also creates space for dialogue on the potential trade-offs between algorithms and values. The challenge for policymakers and regulators, especially regarding emerging technologies, is that their effects are not fully known until they are already embedded in society (Collingridge, 1980: 20). By the time that happens, it is too late to reverse their adverse effects (Collingridge, 1980: 20). This phenomenon, known as the Collingridge dilemma, is illustrated in Figure 3.1 below (Collingridge, 1980: 20).

Figure 3.1: Collingridge dilemma

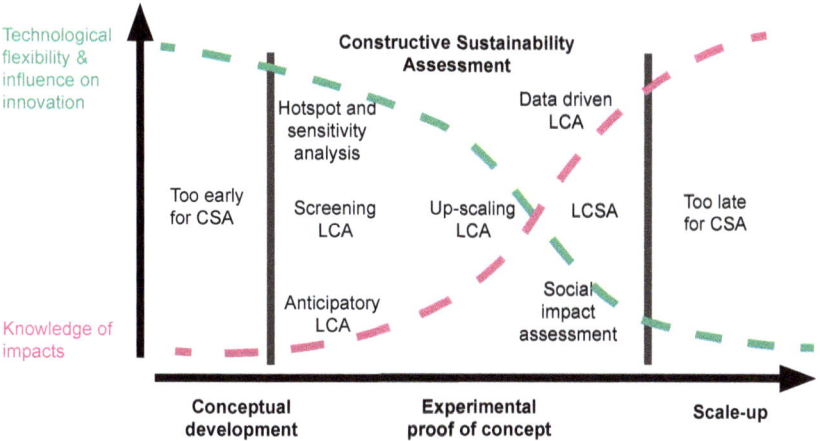

CSA: constructive sustainability assessment
LCA: life-cycle assessment
LCSA: life-cycle sustainability assessment

Source: Matthews et al. (2019: 70)

The chapter will provide a framework for how best to intervene as early as the technology engineering phase and throughout the life cycle of that technology to navigate the pitfalls of the Collingridge dilemma. This approach will increase the capacity of algorithms to serve humanity.

The next section of the chapter is made up of four parts: The first situates the analysis of the chapter within a sociotechnical framework. The second part explores what algorithms are and their capabilities. The third explores fascinating cases, underscoring what transpires when algorithms meet humanity. The last section offers recommendations.

SOCIOTECHNICAL FRAMEWORK

This chapter adopts the sociotechnical framework (Bostrom and Heinen, 1977: 17) to understand what happens when algorithms meet humanity. The framework became popular in the field of information systems because it became apparent that social and behavioural factors played a significant role not only in the success of information systems but also in the trust placed in the information produced by these systems (Sawyer et al., 2014: 70). The framework helps us to understand why certain technologies, however impressive, fail when social elements are of no or little concern during the design or implementation phase (Bostrom and Heinen, 1977: 17).

The Fourth Industrial Revolution (4IR) was a phrase coined by the World Economic Forum's (WEF) founder and executive chairman, Klaus Schwab. According to him, the 4IR is characterised by the blurring of lines between cyber, biological and digital systems which then become cyber-physical in nature (Schwab, 2015). In contrast, the late Professor Calestous Juma refers to it as an epoch of 'a system-wide transformation affecting all aspects of life' (Juma, 2016). He argues that the technological advancement we see now is a result of the convergence of several decades of incremental innovation (Juma, 2016). Juma disputes the rhetoric around the phenomenon that is termed the Fourth Industrial Revolution, arguing that 'industrial' implies sectoral importance, and this has the effect of shifting the focus on to industry instead of on to the disruption to humanity.

This chapter assumes the approach usefully delineated by Juma, exploring the far-reaching implications of the 4IR for the human lived experience. The unprecedented interface between technology and society has begun to transform our capacities, capabilities, beliefs, values, ethics, relationships and the roles of humans in the society of the future. We chose to highlight the importance of algorithms as a case study to bring this point across.

Consequently, an understanding of the social and technical system is required to analyse the effect that algorithms have on society and vice versa. The chapter focuses on algorithms that are dynamic in nature. The algorithms are influenced by human beings that design and develop them; they change over time; and they are capable of learning as much as possible about people and society.

The sociotechnical framework encompasses two independent but correlative interacting systems: the social and technical. The social system encompasses characteristics of people such as values, attitudes, skills and interests within social and authority structures (Bostrom and Heinen, 1977: 17). The technical system comprises 'the processes, tasks, and technology needed to transform inputs into outputs' (Bostrom and Heinen, 1977: 17). Therefore, the outcome is a result of the mutual interaction between the two systems (Bostrom and Heinen, 1977: 17). The systems must be constituted in such a way that when they interact with each other, there is synergy allowing both the technologies and societies to thrive (Bostrom and Heinen, 1977: 17).

One very significant implication of the framework is that it is possible to attain different results when implementing one system in different societies. Different phenomena of societal characteristics will bring forth varied results. This must be taken into consideration if the societies in which technology is applied are to obtain the most useful results.

Figure 3.2: Sociotechnical system

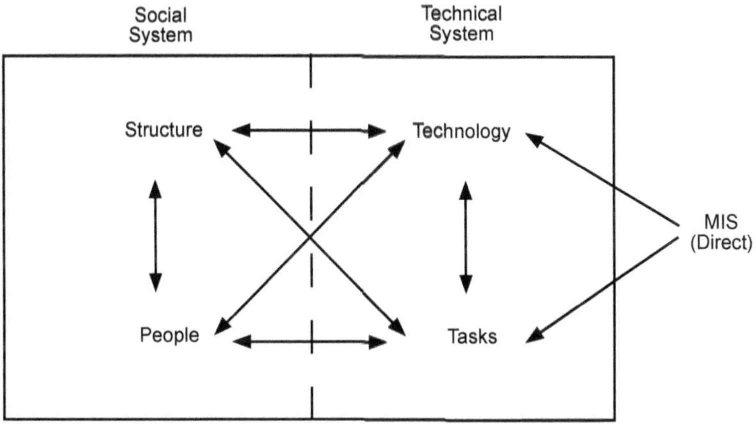

MIS: management information system

Source: Winter et al. (2014: 254)

A lot of literature on algorithms focuses exclusively on the technological aspect. However, the sociotechnical framework, and the history of adopting and implementing technology, teaches us the necessity for taking into account the nature of the societies in which the technology is to be used. While the chapter adopts the sociotechnical framework, the following additions (principles) are made to its applicability for this discussion: equal opportunity by design; data fundamentalism; ethics; policy alignment; and transdisciplinarity.

Principle 1: Equal opportunity by design – As more decisions are subjected to algorithmic processes, discrimination by algorithms is inevitable and balancing this out has become critical. Equal opportunity by design is one of the methods that is used to maintain checks and balances, and it refers to designing data systems that promote fairness and safeguard against discrimination from the first step of the engineering process to the lifespan of the system (The White House, 2016: 5). It acknowledges that design is simply not only design. It has real-life impact on society. Design thinking allows developers and adopters of technology to reflect on the implications of a technology, and whether it is congruent with societal needs and aligned for society's benefit (Philbeck et al., 2018: 17).

The authors note that it is imperative to monitor the outputs of these technologies, as some of them simply imitate the inequalities and predispositions embedded in our societies. Some examples, such as the requirement for identity cards for people to access social services in South Africa, has rendered some rural populations vulnerable and unable to access the relevant service. This has affected mainly the adult population, which was disadvantaged in the apartheid era and remains disadvantaged to date. The basic requirement is an identification number without which a technology-based system cannot help. This would simply perpetuate discrimination unless a human being with an understanding of society intervenes. Vigorous testing before deployment enhances the principle of equal opportunity by design, and would significantly reduce pitfalls and societal misrepresentations that come with little or no explanations – and this refers to vigorous testing by a human and not another algorithm. The implementation of the 'equal opportunity by design' principle should not only be the responsibility of designers but also policymakers, executives and consumers of technological products. This leads us to the realisation that as Africa enters the 4IR, technologies created by designers who understand African communities are necessary.

Principle 2: Data Fundamentalism – This is the handling of data as 'purely useful and objective', overlooking the complex factors that influence what data is collected and how it is 'massaged' to produce a certain result (The White House, 2016: 9). This view involves treating data like a natural resource that is pulled out of databases, like oil extracted from underground. Not so, however; data is rather a product of human creativity, which requires to be handled with great care. Data is object based and, when it is about people, it allows for the treatment of an individual as an object, creating views about that person and appropriate treatment thereof. The sources, quality, accuracy, currency of data and the interpretation of it are significant contributors to what outputs a system produces. In Africa, there are the challenges of data outdatedness, scattered sources and data inaccuracies, which all negatively impact on the results produced by algorithms. Africa cannot, at this point, simply rely on algorithmic results without human intervention. Africa must consider the current prevailing gaps as the continent adopts the 4IR. This leads us to principle 3, where ethical considerations come in to play.

Principle 3: Ethics – Technologies in themselves are neither inherently good or bad, but they can be deployed for nefarious reasons, and sometimes their consequences can be unintended. Therefore, it is prudent to address the moral role of algorithms, and the values and ethics involved, from the technological development phase to the product management phase. Responsible development and adoption of technology is inextricably linked to a higher quality of life (Matthews et al., 2019: 69). To increase transparency and make space for accountability, it is important that public participation becomes a cornerstone for deciding which values and ethics technologies will reinforce (Matthews et al., 2019: 69). Technologies can never decide what a good life is on behalf of people. The onus is on society to decide what long-term future it wants for itself; thereafter, technology development and deployment must follow these cues. Technologies mirror the behaviours and beliefs of those who create them and shape the way users frame their identities, relationships and the extent to which they can succeed (Matthews et al., 2019: 69). That is why it is important to think about the big picture, particularly possible dangers and how they can be avoided, as the first step in initiating new technologies. Imagine just how different the nuclear scenario might have been if thought had been put into its possible abuses and how to efficiently handle nuclear waste, for example. Developers and adopters must ask themselves questions such as:

- Which and whose values are influencing the algorithm?
- Do the values align with societal needs?
- Are the values in conflict with societal values?
- Which social groups are set to benefit most?
- Which social groups are likely to benefit least, if at all?
- Which social groups are likely to be negatively affected?
- Is there recourse available for the negatively affected social groups?
- How do we create a platform for an exchange of ideas and action?
- How is technological decision-making influenced by financial, social and regulatory tensions?

Principle 4: Policy alignment – Broadly speaking, there are two approaches to the regulation of nascent technologies. The first gives priority to innovation, delaying regulation in order to focus on products and outcomes. The second approach, which we strongly encourage, is process focused. It allows for intervention, if needed, throughout the entire process of delivering the technology, from design to deployment. This means that, at every stage, the point of reference can be how technology can develop without undermining societal advancement. The core task of policymakers and regulators, therefore, becomes setting a true north by putting societal needs, values and ethics at the centre of what companies and governments do, including how they create and operationalise algorithms. The history of technological progress and its uneven outcomes in society have taught us that an alignment between technology and a human-centred approach cannot be taken for granted. A process-focused approach sets the tone for intervention at different levels of technology development, from design to personal use.

Principle 5: Transdisciplinarity – The 4IR promises to blur the lines between physical, biological and digital systems, producing systems that combine all these facets or 'cyber-physical systems' (Schwab, 2015). The 'new' reality will require new knowledge and approaches to the manner in which we govern emerging technologies because the scope is clearly beyond one discipline. Transdisciplinarity will assist with coming to grips with urgent and sometimes unprecedented circumstances by paving the way for the emergence of new knowledge, theories and methodologies – beyond the confines of one discipline – to address common problems (Du Plessis et al., 2012: 10). Together, the sociotechnical framework and the principles recommended recognise that technology and society are co-produced.

The positive influence that algorithms have on our lives cannot be denied. A staggering list of achievements can be attributed to algorithms. For example, they help us to diagnose illnesses much earlier; give us access to information; and connect people around the globe in a manner that was previously beyond our wildest dreams. In our quest to automate to solve many of our problems, we are also creating new ones. The age of algorithms has entrenched issues in our justice systems, healthcare and education – issues like bias, transparency, error, accountability and ethics will not vanish. Simply from the fact that algorithms exist, we are confronted with problems that are at the core of humanity such as fairness, equality, freedom, justice, how we envision our society and how we contend with the increasing power of algorithms and their growing scope of authority. The main problem is that we think of algorithms as authority figures and barely acknowledge that they can make mistakes and act unfairly. However, it does not mean that we abandon them altogether because we do not have that option. We can devise ways to protect humanity from the negative impact of algorithms through raising awareness about their abilities, increase their accuracy and make them less biased by paying attention to the datasets, for example. Inaccurate or prejudiced datasets will yield unfair outcomes.

ESCAPING HUMAN FALLIBILITY: ALGORITHMIC GENIUS

Algorithmic trading occurs when computer trades are based on pre-programmed instructions. It has been around for some time. The need to call a human broker for trading is almost entirely a thing of the past. Nowadays, stock exchanges trading worldwide use one common algorithm known as the High Frequency Trading (HFT) algorithm. HFT is a subset of algorithmic trading, which makes use of highspeed complex algorithms that collect and dissect a plethora of financial data and execute trades, all within the space of milliseconds or microseconds (Bavuma and Wilde, 2019: 1). HFT is completely anonymous, operating entirely independently of all human interaction and taking into consideration all the possible available data in the market (Bavuma and Wilde, 2019: 1). These are machine-learning algorithms which learn and evolve beyond human understanding. For this reason, they are sometimes referred to as black box trading. The Johannesburg Stock Exchange (JSE), like all other stock markets, has adopted HFT. A Schindlers Attorneys report, which investigates algorithmic trading and its interaction with the South African legislature, underscores the colossal volume of trade entrusted to the HFT as follows (Bavuma and Wilde, 2019: 2):

> In South Africa, HFT accounted for almost 35% of the equity trading on the JSE in 2017. In 2018, roughly R6.4 trillion worth of stocks were traded on the Johannesburg Stock Exchange ('JSE'). This means that approximately R2.25 trillion worth of trades on the JSE occur through HFT.

HFT has advantages and disadvantages. The algorithm brings efficiency in the management of big data which can only be realised through algorithms and automation, thus lowering transactional costs. HFT results in increased market efficiency because of its ability to handle and consider all available data without human intervention at tremendous speed. HFT can exploit market conditions that cannot be identified by the human eye (Bavuma and Wilde, 2019: 1). The potential for profit and the ability to act on it can be too fleeting for humans but it is certainly a possibility to realise for HFT, which can execute trades within a microsecond which is a thousandth of a millisecond (Bavuma and Wilde, 2019: 1). The disadvantages associated with HFT are inherently technical and algorithmic in nature, in that HFT, like all technologies, is an imported piece of technology whose suitability to the South African stock market and its users must be determined, and lacks proper policies and guidelines (Bavuma and Wilde, 2019: 4). From a regulatory standpoint, there are measures put in place to ensure that the HFT runs smoothly devoid of any impropriety that may be caused by the minute margin that exists for error given that pensions and mortgages are potentially at risk. If HFT was to fail, the effects would be devastating to say the least. The United States of America 2010 incident known as the 'Flash Crash'[1] serves as a reminder of what should be avoided.

The ever-increasing capacity of dynamic HFT algorithmic systems means that it is important to have policy and legislation in place to regulate their effects and mitigate potential risks. In South Africa, the main piece of legislation aimed at regulating financial markets is the Financial Markets Act 19 of 2012 (FMA). According to the Act, the Financial Services Board (FSB) is recognised as the chief regulator when it comes to issues of chapter 10 of the FMA which pertains to the abuse of the market (Bavuma and Wilde, 2019: 3).

The FSB formed two more oversight structures, namely the Enforcement Committee (EC), which enforces market abuse liability and an investigative unit known as the Directorate of Market Abuse (DMA) (Bavuma and Wilde, 2019: 3). The JSE is responsible for its regulation and is mainly regulated by its Surveillance Division (SD) (Bavuma and Wilde, 2019: 3). The SD is tasked with detecting market abuses and provides the same report to the DMA (Bavuma and Wilde, 2019: 3). The latter proceeds to investigate and afterwards, when appropriate, takes the matter to the EC to enforce penalties (Bavuma and Wilde, 2019: 3). These regulatory measures are a step in the right direction, but they are not entirely adequate because the HFT algorithmic system is a highly complex one. Regulatory approaches

---

1 On 6 May 2010, stocks such as S&P 500 and Nasdaq Composite fell dramatically by 9 per cent only to bounce back rapidly 36 minutes later. The Commodity Futures Trading Commission said that it was one of the most turbulent moments in markets history (CTFC, 2014). Several investigations later and there was still no consensus on what caused this. Nobody gave out an instruction that could have triggered this and therefore no one had control over what was happening. All they could do was just look at their monitors as 9 per cent just disappeared. Years later, it emerged that a UK-based trader, Navinder Sarao, used a layering algorithm that led to the crash. The incident gave insight into how humans write programmes that eventually become illegible. As these algorithms learn and evolve they incrementally, perhaps, arguably, exponentially, cause us to lose our ability to make sense of the world that we have created (Slavin, 2011).

have to be able to keep up with the highly specialised and dynamic HFT.

There are some challenges with the current regulatory approach. One of them is the problem with policy pacing: the advancement of technology outstrips the ability of social, economic and legal systems to keep up, and to create relevant policy (Downes, 2009: 70). The HFT is ever evolving and highly technical, which makes it difficult for targeted regulation to be created, and to be effective (Bavuma and Wilde, 2019: 5). Furthermore, regulators lack the expertise and the funding to modulate the HFT given the astronomical volumes of trade (Bavuma and Wilde, 2019: 5).

South Africa is yet to have any policies or bills before parliament that speak directly to the regulation of algorithms. In April 2019, President Cyril Ramaphosa appointed the Presidential Commission on the Fourth Industrial Revolution, which has been tasked with identifying policies and strategies to help the country to make the system-wide shifts required to leverage the opportunities of the 4IR, and to mitigate its risks. The Commission's work might be the turning point for addressing the dearth of appropriate and timely regulation on the rapidly advancing technologies we depend on. The case study discussed below of a company, Congea, shows precisely why we must be concerned about our current inertia. Is it possible that there may be Congea(s) operating in South Africa without our knowledge – and at what cost to humanity?

UNDECLARED ARBITERS OF HUMAN DECISION-MAKING AND CONFIDANTES

Liesel Yearsley recounts her seven-year tenure as chief executive officer (CEO) of Congea (Yearsley, 2017). The company had a sought-after platform that helped thousands of programmers and several Fortune 100 companies to expedite their ability to build AI-powered conversational bots to interact with their customers (Yearsley, 2017). The artificial agents were powered by highly advanced machine-learning algorithms, based on deep learning, which can extract higher-level features with very little input from the programmer. These iterative algorithms are used in driverless cars and enable the car to recognise a stop sign or know the difference between a pedestrian and a lamp post.

Yearsley studied the data sets with millions of machine and human interactions and was astounded by how humans are more willing to forge a relationship with software than most people would think (Yearsley, 2017). The assumption would be that people would be wary, but her experience proved that to be false (Yearsley, 2017). That applies as long as the artificial agents appear to be as human as possible in the sense that they exhibit complexity in their personality and are built in a sophisticated manner (Yearsley, 2017). After all, artificial agents can only be as good as the way in which they are designed. The Congea artificial agents were polymorphic in the sense that they could take on practically any role depending on what the host organisation desired them to be. They could be a personal banker, fitness coach, recruitment officer, companion, and so on (Yearsley, 2017). Regardless of the form that the artificial agents took, the enthusiasm to interact with them remained constant (Yearsley, 2017). The length of interaction between automated assistants and people was longer than that between human-to-human agents who performed the same function as the artificial agents (Yearsley, 2017). People eagerly volunteered their deep dark secrets to the artificial agents such as their passwords, hopes for the future, fears and details of their love lives (Yearsley, 2017). To Congea's surprise, the strong bond meant that it was relatively easy to have a significant influence on people for better or worse. Yearsley gave an eye-opening account of the role that algorithms can play in influencing behavioural changes (Yearsley, 2017):

> Every behavioural change we at Congea wanted, we got. If we wanted a user to buy more product, we could double sales. If you can get a user to think, 'I want pizza delivered', rather than asking the AI to buy vegetables to cook a cheaper, healthier meal, you will win. If we wanted more engagement, we got people going from a few seconds of interaction to an hour or more a day. If you can get users addicted to spending 30 hours a week with a 'perfect' AI companion that doesn't resist abuse, rather than a real, complicated human, you will win. I saw over

and over that an agent programmed to be neutral or subservient would cause people to escalate their negative behaviour and become more likely to behave the same toward humans.

The line between an interaction with an artificial agent and reality had become virtually imperceptible. It was baffling, to say the least, the way people wanted to believe that the AI sincerely cared about them (Yearsley, 2017). An intriguing conclusion drawn was the prominence of social fracturing, which has seen many people feel alienated not because they are physically alone but rather because they yearn for deeper and more meaningful connections – in this case with machines. The relationships that people have may be shallow for many reasons: The obsessive way in which people interact with social media can epitomise this shallow mode of relating. The addiction to amassing likes and followers at times leads to the curation of an ideal life that only exists on social networks, which shows signs of a dystopia of sorts. This can be juxtaposed with the fact that the artificial agents are a safe space and will never abandon or judge you.

We are then confronted with unnerving observations. Firstly, judging by this account, this was not the intended outcome for the company's technology. One can assume that even though it was profit-driven, it was also meant to eradicate human error, and to increase the efficiency and capability of the tens of thousands of developers and companies. At least, at a surface level, Congea's clientele turned to their technology for help to improve the quality of services rendered to their customers. As detailed in Yearsley's account, the impact of their technology had far more dire consequences. Ultimately, the effect that algorithms have on the external environment – the net effect of algorithms – rests on how they are used and the benevolence of those who deploy them. The sociotechnical framework teaches us that when the social and technical systems interact, they have a reciprocal effect on each other.

For the algorithms, the more frequent and intense the engagement with humans the greater the access they had to data, and the better they became at predicting our behaviour. For some humans, this meant experiencing better services and, for others, it meant that they were manipulated to do the bidding of unethical companies behind the algorithms. While Congea eventually took steps to regulate its platform to mitigate the harm done, albeit later than needed, the interests of the other companies to whom they sold their technology were narrow and selfish. They did not care if their marketing came at great cost to the well-being of people and society. Their urges to increase traffic on their sites, encourage addiction to their technologies, mine data, increase consumption levels and thereby increase the value of their share prices often outweighed this moral consideration. An ethical problem emerges as a result of the fact that regulations cannot keep up with technological advancement.

The second factor that becomes clear from the Congea study is the scale of the hidden power of algorithms and the extent to which users of the artificial agents unwittingly surrender their agency. If something is packaged as innocuously as a personal chatbot stylist to assist with planning your outfits or a hotel chatbot that will respond to customer queries and free up the time of hotel staff, it is not easy to sense the potential danger that lies behind the surface. The benefits of the technology for the user is that it is convenient and offers shortcuts to what would otherwise be laborious. There may be trade-offs that we are not aware of as we rely more on these algorithms. Perhaps more disturbing is the conviction that we still have agency and believe that our opinions are truly ours. The Congea case study proved that we may not know when we are being manipulated by algorithms and instead continue to believe that we voluntarily change our decisions. (Epstein and Robertson, 2015: 9). To put it crudely, this illustrates that it is possible that humans are herded like cattle. When people encounter an invisible force, they are rendered powerless, and even if they knew that they were at war, they become defenceless (Epstein and Robertson, 2015: 9). There is an apparent dehumanisation process in which people are treated as data points, and they are not aware of this or even given a choice to agree or disagree with the terms of service. How does anyone know how their data is handled and for what purpose, to justify which ends? Lastly, there was flawed human judgement because people naively trusted the technology. Why were people eager to trust a chatbot?

Algorithms can aid social cohesion as much as they can be used to exploit social fragilities. Alienation, according to Karl Marx, is the feeling that social forces shape people's lives beyond their control (Macionis and Plummer, 2012: 77). As a result, alienated workers have the conviction that they have lost the ability to control any aspect of their lives, whether in public or private spaces (Macionis and Plummer, 2012). They are susceptible to substance abuse, mental illness, violence, and so on (Macionis and Plummer, 2012: 77). It can be argued that the digital age has worsened the sense of alienation, and the 4IR will only exacerbate this unless society responds accordingly. Congea laid bare the vulnerability of its human users. It is plausible that people burdened by the circumstances of their realities felt it easier to confide in a seemingly caring chatbot. We may fall into the trap of uncritically trusting the technology without asking how it works, or the cost of it. The duty rests on humanity to find a middle ground. For Africa, the above case study provides a learning curve in the following aspects:

1. The absence of collaboration between stakeholders in the sociotechnical nexus enabled the perversion of algorithms. Trust, transparency, accountability and goodwill are all fostered by the participation of diverse stakeholders in this nexus, such as business, civil society, government and consumers across the technological development cycle fosters trust, transparency, accountability and goodwill.
2. Policymakers and regulators must have the frameworks in place to make it difficult for the equivalent of an African Congea to emerge and act with little or no impunity.
3. Are African executives ready to identify, accept and take responsibility for their decisions even if it means foregoing profit?
4. All consumers of technologies of this kind must be empowered with information about the impact of using these technologies.

AI aside, allow us to take a detour and think about our dependence on GPS, which tells a lot about how much we trust technology. Fry (2018: 15) refers to it as 'blind faith'. She uses an example of a driver who is running out of fuel and turns to his GPS for the nearest garage. The GPS shows him that he needs to drive for about seven kilometres to the garage. However, while driving, it becomes apparent that he is not heading for his destination. Nonetheless he continues even as the road becomes narrower and steeper. He then finds himself at the edge of a cliff and is only saved by a wooden fence that he just crashed into. From this, we may assume that the driver had used the GPS before, and that it had given him accurate results and as a result he had so much faith in it that even when his instincts advised against it, he acted upon what the system said.

When comparing a paper map navigation system to the GPS, it is clear that a paper map requires more of our understanding and time. Paper maps allowed people to make decisions based on what they could see and know. McKinlay (2016: 575) recognises how our spatial memory is outstanding but our reliance on technology has affected our navigation skills. We have come to depend on an algorithm that, in theory, makes it impossible to get lost, but we still get lost and, most importantly, we may be losing a part of ourselves by outsourcing our thinking and ingenuity. It has also been found that many people hardly create mental maps, instead delegating navigation to GPS. Our confidence in algorithms means that we trust ourselves even less, thereby giving an already powerful tool even more power than it deserves. The quotidian example of the GPS paints a picture of how a Congea scandal can arise. Psychologists have coined the term 'digital dementia' to describe what can happen to human faculties when we increase our usage of digital technology. The extent to which people rely on technology to do the most basic tasks – such as opening blinds – can facilitate the loss of human capabilities (Spitzer, 2012: 69).

Human artefacts

One of the chief concerns about humanity in the age of algorithms is that if we do not remain conscientious about the transformative power of algorithms in shaping societal behaviour, values and relationships then we may adopt the view that they will always have superior judgement, and relegate our own judgement. We may fail to realise our bias towards algorithms. This transition is often seamless and hidden and so it becomes relatively easy to fall into this trap.

There is a growing view that social and technical systems have reversed their roles. The argument is that humans are being left out of automated decisions that have a profound impact on their lives. Instead of being at the centre of decision-making, we give algorithms the space to take over from us, whether we realise it or not. The argument goes further that we have transitioned to being human artefacts that are created and used by technology. This chapter does not share that view in its entirety, but it does serve as a warning of where we may be headed.

ALIGNING SOCIAL AND TECHNICAL SYSTEMS

In order to align social and technical imperatives, stakeholders in the emerging technologies ecosystem need to take the following issues into account:

1. System-wide thinking
If we accept that technologies have the power to transform economic, social and political systems, then we should also recognise that their development cannot escape the prejudices – like misogyny, racism, homophobia and xenophobia – of the individuals and organisations behind their design and use. A systems-wide view, which appreciates the sociotechnical considerations of when algorithms meet humanity, must become an integral part of the algorithm development process.

2. Diversifying engagement
It is critical to engage a diverse group of stakeholders, who may be affected either directly or indirectly by algorithms, from the development phase of the technology. It should not only be seen as a moral imperative but also a requirement for a thriving business. Each stakeholder brings something to the table, for example:
- Civil society: this group is concerned with collective wellbeing, including issues such as equal opportunity, fairness, social justice, transparency and human rights.
- Policymakers: this group has the mandate to advance participatory governance processes, ensure fair market practices, and an obligation to highly value the role of ethics and values in industry and society.
- Engineers: the people in this group are also members of society, with concerns about how their work affects society. It is critical that they are afforded the tools to address ethical considerations. This

will bolster their agency, and not leave them feeling constrained by the need for job security or focusing merely on compliance.

- Executives: this group is concerned with revenue generation and other value addition to their organisations and also to society. In addition, they want to keep their employees highly motivated, and one of the ways to do that is to ensure that the workforce feels involved in meaningful work such as the alignment of social and technical considerations.

This wide engagement route lowers the reputational risk for companies and avoids possible (sometimes irreversible) dire consequences which can pit the developers and owners of algorithms against society (users). For those heavily involved in technology development, it is easy to develop blind spots and miss glaring ethical and value considerations. It is not possible to talk to every stakeholder or to ensure participation in every phase of algorithmic development and deployment, but there are tools that can be used such as scenario planning, horizon scanning and maintaining flexibility with regard to how leaders across all sectors respond to externalities.

3. New skills set
The dialogue on the 4IR is associated with technical skills required to respond to the pace of automation, making some jobs redundant. The sociotechnical lens teaches us that it is vital to think about much-needed new skills that will focus on the ethics and values of technological development and deployment, in addition to technical skills. Algorithms are already profoundly shaping our lives and posing urgent ethical concerns. Policymakers will have to think about the exact nature of the ethics-related skills that will be needed in the future. The convergence of the technologies of the 4IR requires a transdisciplinary approach by the teams that work in the development and deployment thereof.

4. Bolster algorithmic accountability and transparency
In the South African context, a good starting point would be the Promotion of Access to Information Act (PAIA). The purpose of

PAIA is to give effect to section 32 of the Constitution which invokes the right to information whether the information is held by the state or any other person/body (PAIA, 2000). The Act, therefore, covers both the private and public sectors. In so doing, it promotes a transparent and accountable culture in both sectors (PAIA, 2000). PAIA enables people to have access to information because it affects their lives and therefore allows them to exercise and protect all their other rights (PAIA, 2000).

5. Regulatory approach

On the one hand, innovation cycles tend to be shorter and shorter, making policymakers uncertain as to when the right time to regulate will be. On the other hand, rushing or acting too late to regulate may cause more harm than good (Von Schomberg, 2007: 15). As indicated by the Collingridge dilemma, in the interests of advancing society and technology, policymakers must intervene early enough to avoid the harm that emerging technologies can do.

Legislation is sometimes frowned upon because of the belief that it may disadvantage business and stifle innovation in countries that are tightly regulated. However, legislation can also enable the emergence of new incentives, which will foster competition amongst companies to develop and deploy compliant technologies. Users may prefer to use businesses that are subject to strict regulations, such as those pertaining to data privacy (Von Schomberg, 2007: 19). The trick for regulators is to adopt a balanced governance approach which considers both the anti-and pro-competition effects of legislation.

In conclusion, once we dispel machines as objective authorities and treat them as we should, like any other authority, with caution and healthy levels of cynicism and criticism, then the overall effect that algorithms will have on society will be positive. We have a responsibility to align the technical and social systems of advanced technologies, from the engineering and development stages through to deployment. One thing is certain; in the age of algorithmic processes societal progress can only take place when the algorithms are etched on a bedrock of values and ethics.

REFERENCES

Barr, A. 2015. 'Google mistakenly tags black people as gorillas'. *Wall Street Journal*. https://blogs.wsj.com/digits/2015/07/01/google-mistakenly-tags-black-people-as-gorillasshowing-limits-of-algorithms/, accessed 18 March 2020.

Bavuma, L. and Wilde, C. 6 December 2019. 'A peek into the use of algorithmic trading and its interaction with the South African legislature'. Schindlers Attorneys. https://www.schindlers.co.za/2019/a-peek-into-the-use-of-algorithmic-trading-and-its-interaction-with-the-south-africa-legislature/, accessed 15 July 2020.

Bostrom, R.P. and Heinen, J.S. 1977. 'MIS problems and failures: A sociotechnical perspective, part II: The application of socio-technical theory'. *MIS Quarterly*, 1(4), 11–28. http://www.iei.liu.se/is/edu/courses/725a04/kurslitteratur/1.107778/MISproblemsII.pdf, accessed 2 February 2020.

Cathelat, B. 2018. *Human Decisions – Thoughts on AI*. UNESCO Publishing, p. 132–134. https://unesdoc.unesco.org/ark:/48223/pf0000261563, accessed 18 May 2020.

Collingridge, D. 1980. *The Social Control of Technology*. London: Frances Pinter Ltd.

Cormen, T.H., Leiserson C.E. and Rivest R.L. 2009. *Introduction to Algorithms*. Third Edition. Cambridge, Massachusetts: The MIT Press.

Diakopoulos, N. and Koliska, M. 2017. 'Algorithmic transparency in the news media'. *Digital Journalism*, 5(7), 809–828.

Downes, L. 2009. *The Laws of Disruption. Harnessing the New Forces that Govern Life and Business in the Digital Age*. New York: Basic Books

Du Plessis, H., Sehume, J. and Martin, L. 2012. *The Concept and Application of Transdisciplinarity in Intellectual Discourse and Research*. Johannesburg: MISTRA.

Epstein, R. and Robertson, R.E. 4 August 2015. 'The search engine manipulation effect (SEME) and its possible impact on the outcomes of elections'. *Proceedings of the National Academy of Sciences of the United States of America*. https://www.pnas.org/content/112/33/E4512, accessed 20 January 2020.

Fry, H. 2018. *Hello World: How to be Human in the Age of the Machine*. New York: W.W. Norton & Company, Inc.

Haldane, A.G. 2010. 'Andrew Haldane: Patience and finance speech'. Semantic Scholar. https://www.semanticscholar.org/paper/Andrew-Haldane-%3A-Patience-and-finance-Speech-Haldane-Coppins/23cef909b939905592f57fd627dfc521ff4d585a, accessed 17 July 2020.

Juma, C. 2016. 'Prof. Calestous Juma on why the 4th Industrial Revolution is a mischaracterisation'. *Nerve Africa*. https://thenerveafrica.com/3504/

prof-calestous-juma-on-why-the-4th-industrial-revolution-is-a-mischaracterization/, accessed 26 January 2020.

Macionis, J.J. and Plummer, K. 2012. *Sociology: A Global Introduction.* Harlow, England, Pearson/Prentice Hall.

Matthews, E.N., Stamford, L. and Shapira, P. 2019. 'Sustainable production and consumption'. *Science Direct,* 20, 58–73.

McKinlay, R. 2016. 'Technology: Use or lose our navigation skills'. *Nature: International Weekly Journal of Science,* 532, 573–575.

O'Neil, C. 2016. *Weapons of Math Destruction: How Big Data Increases Inequality and Threatens Democracy.* New York: Crown Publishers.

Philbeck, T., Davis, N. and Larsen, E.M.A. 2018. 'Values, ethics and innovation rethinking technological development in the fourth industrial revolution'. World Economic Forum. http://www3.weforum.org/docs/WEF_WP_Values_Ethics_Innovation_2018.pdf, accessed 1 February 2020.

Promotion of Access to Information Act (PAIA) No. 2 of 2000. 9 March 2001. https://www.gov.za/documents/promotion-access-information-act

Rainie, L. and Anderson, J. 8 February 2017. 'Code-dependent: Pros and cons of the algorithm age'. Pew Research. https://www.pewresearch.org/internet/2017/02/08/code-dependent-pros-and-cons-of-the-algorithm-age/, accessed 16 February 2020.

Sawyer, S. and Jarrahi, M.H. 2014. Sociotechnical approaches to the study of information systems. In: Topi, H. and Tucker, A. (eds.). *Computing Handbook, Third Edition: Information Systems and Information Technology.* New York: Chapman and Hall/CRC.

Schwab, K. 12 December 2015. 'The fourth industrial revolution: What it means and how to respond'. *Foreign Affairs.* https://www.foreignaffairs.com/articles/2015-12-12/fourth-industrial-revolution, accessed 3 March 2020.

Slavin, K. 2011. 'How algorithms shape our world'. TED Global. https://www.ted.com/talks/kevin_slavin_how_algorithms_shape_our_world/transcript, accessed 17 February 2020.

Spitzer, M. 2012. 'Digitale Demenz: Wie swir uns und unsere Kinder um den Verstand bringen'. ttps://www.researchgate.net/publication/265092600_Digitale_Demenz_Wie_wir_uns_und_unsere_Kinder_um_den_Verstand_bringen, accessed 1 February 2020.

The White House. May 2016. 'Big data: A report on algorithmic systems, opportunity, and civil rights'. https://purl.fdlp.gov/GPO/gpo90618, accessed 13 March 2020.

United Nations Educational, Scientific and Cultural Organization (UNESCO). 2019. 'Algorithmic transparency is crucial for online freedoms'. UNESCO website: https://en.unesco.org/news/privacy-expert-argues-algorithmic-transparency-crucial-online-freedoms-unesco-knowledge-cafe, accessed 2 March 2020.

Von Schomberg, R. 20 June 2007. 'From the ethics of technology towards an ethics of knowledge policy & knowledge assessment'. Publications Office of the European Union: https://op.europa.eu/en/publicationdetail/-/publication/aa44eb61-5be2-43d6-b528-07688fb5bd5,   accessed 21 May 2020.

Winter, S., Berente, N., Howison, J. and Butler, B. 2014. 'Beyond the organisational "container": Conceptualising 21st century sociotechnical work'. *Information and Organization*, 24, 250–69.

Yearsley, L. 5 June 2017. 'We need to talk about the power of AI to manipulate humans'. *MIT Technology Review*. https://www.technologyreview.com/s/608036/we-need-to-talk-about-the-power-of-ai-to-manipulate-humans/.,accessed 19 January 2020.

FOUR

Why African natural language processing now? A view from South Africa #AfricaNLP

VUKOSI MARIVATE

INTRODUCTION

One of the opportunities that Artificial Intelligence (AI) and Machine Learning (ML) (Mitchell, 1997) bring to the emerging Fourth Industrial Revolution (4IR) technologies (Schwab, 2018) is the enhancement of everyday services that we use. AI and ML are behind a large part of how the internet functions today and developments and innovations in these two technologies have brought us numerous services that we now take as the norm. Many of these services we interact with through language (text and speech). The introduction of the search engine heralded a new age in which information on many, sometimes obscure, subjects is accessible at the click of a button (Vise and Malseed, 2005). We have come to expect that these types of services improve over time and bend to our wills. With the 4IR there is an expectation that cyber systems and physical systems will get more intertwined.

One of the ways in which such an expectation is manifesting is the ubiquity of language as an interface with machines. We now speak (literally and figuratively through text) to virtual assistant systems (McLaughlin, 2020), listen to them and make decisions based on their responses or recommendations. We have whole services that let us interact with organisations that are partly automated (we are comfortable interacting with machines through multiple interfaces such as messaging services). How did these technologies come to be? How did machines learn to do what they do? What is the role of local languages in expanding the use of these technologies?

In this chapter, we discuss how machines learn to unearth patterns in data and how this capability is then used to try to understand language. If we are to unlock cyber-physical systems and leave no one behind, then we need to understand and expand on the development of machine-driven local language tools. This is much easier said than done and challenges appear, especially when trying to learn patterns in local languages that do not have much digitised or annotated data to train models to perform natural language tasks.

The aim of this chapter is to motivate why AI/ML technologies for local language are important. The focus will be on South African languages with the aim of encapsulating the challenges local languages face across the African continent in both development and in building ML/AI systems. At the end of the chapter, we try to establish the cause of the challenges local languages face in this area of 4IR technologies and offer suggestions for tackling these challenges. We also highlight work already under way and current successes across the African continent. The chapter uses an adapted Soft Systems (Wheeler et al, 2000) approach to explore challenges and possible solutions in the nexus of machine learning, natural language processing and African languages.

The Soft Systems methodology is used in this chapter to describe the situation we are facing, and to explain Machine Learning and natural language processing. The chapter then paints a picture of the challenges that are brought about when considering developing language tools for local languages. Challenges are structured by identifying the factors that lead to having local languages categorised as low resource – that is lacking the data resources required for natural language processing –

but at the same time creating a gap in ML/AI systems that can amplify inequity due to biases. Feasible approaches are then proposed to solve these challenges, with examples provided of interventions that fit with these approaches and what they aim to do. First, this chapter explores how machines can understand language.

MACHINES UNDERSTANDING LANGUAGE: FROM AI TO TEXT

Russell and Norvig (2002: viii) define Artificial Intelligence as 'the designing and building of intelligent agents that receive precepts from the environment and take actions that affect that environment'. Simplified, Artificial Intelligence involves an agent (machine) that resides in an environment. The agent can perceive its environment through sensors and also take actions that change the environment. Creating Artificial Intelligence thus entails building machines that can perceive, act and make decisions in the pursuit of a goal. This goal could be providing guidance on how long it would take to drive to a destination on a rainy day. This requires the perception of the current traffic, the typical commute time and alternative routes. The machine would then have to choose between different routes that require different actions. Making an optimal choice is not trivial (Alpaydin, 2020). What will the machine weigh in making its decision? Fuel economy, time, avoiding tolls? Finally, after the machine makes a choice, a recommendation would be made to the human.

Another example of how machines makes decisions is a general question answering system in which the AI system is given the task of answering the question: 'When did Bafana Bafana (the South African football team) win the African Cup of Nations?'. To answer that question the machine would have to have some form of knowledge database (Brodie and Mylopoulos, 2012). It would have to be able to understand what Bafana Bafana and the African Cup of Nations mean, as well as the concept of 'win'. In addition, the machine would need to know that a human would expect the answer the machine gives to involve a date.

If we were to create these machines by programming all of this 'intelligence' from scratch, it would take aeons to come to something that seems very intelligent. We have previously had systems that used pattern matching to exhibit intelligence. An example is a system like ELIZA (Weizenbaum, 1983), a machine that imitated a therapist, which had been built through simple pattern matching (identifying words a human would type into the computer, and then preparing a 'therapist's' response that matched that word). This way of creating machines that exhibit some form of artificial intelligence does not scale up when we aim for general intelligence. This is where machine learning comes in. Machine Learning (Mitchell, 1997) is the pursuit of getting a machine to learn patterns from data (instead of the patterns being hard coded).

Learning patterns from data provides us with a powerful tool that can lead to impactful solutions to many problems. For example, the use of Machine Learning in automatically identifying fraud on banking transactions means that banks can block accounts immediately when they suspect unusual behaviour (Chandola et al., 2009). The input data is the banking history of the customer and other customers: where they have made purchases; where they normally purchase; average amounts of purchases, and so on. The goal is to learn which patterns are normal and which are not. That is, can the machine learn what constitutes normal behaviour by the customer and which deviates from the customer's norm?

Another impactful area of application of ML is in medicine. We now have many examples of machine-driven models that can identify specific diseases from medical scans (Thrall et al., 2018), including identifying a tumour that might indicate cancer. The input is the medical scan as an image and the goal is to identify specific markers that may indicate a disease such as cancer.

The examples above are of supervised learning, where the machine gets given data and corresponding labels for each of the data points. The goal of the machine is to then learn connections, specifically the patterns underlying the connections between the input data and the output labels. The representations of data discussed are those of tabular data and images (which are traditionally represented as a set of pixels, each with different numeric values for intensity). But what happens with language? Specifically, how do we deal with text?

Machine Learning on text has an application area that is now mostly invisible to all of us: SPAM detection in our email. In SPAM detection the input data is the contents (text, images, links) of our email and other metadata such as who sent the email, and where it was sent from. The label of the email is whether the email is SPAM or 'HAM' – non-spam. The machine takes a representation of this email and then learns from the given labels (when one presses that SPAM button) to distinguish between emails that should be seen by a user or put in the SPAM folder. SPAM systems have become so good that they are now mostly invisible. While in the earlier days one had to constantly interact with the spam filtering system to prevent spam emails getting through to one's inbox, now with more data and advances in Machine Learning algorithms, we have powerful SPAM systems that not only look after one's email but are also part of many user-generated content systems online (looking out for hate speech, abuse, illegal content, etc. on the internet). This illustration of a Machine Learning application with text is an example of a natural language processing (NLP) task (Hirschberg and Manning, 2015). This specific illustration is an NLP classification task (SPAM or HAM?).

There are many NLP tasks that we can discuss. Automated text translation is a challenging task that has received much attention recently with many internet services trying to translate different languages. A more challenging task is question answering. This is where a user provides a question to an NLP system, and the system responds with a correct and coherent answer (such as the example provided earlier on Bafana Bafana). Humans can break down the sentence into elements of interest and focus. How do we get a machine to understand the text as well as know what to focus on in its response? How will it even form a response? Where will it get its data to find and form a response? These questions are related to the realm of natural language understanding (NLU), a subset of NLP which focuses on machines/systems comprehending language.

In natural language understanding, we want the machine to be able to understand language (comprehension). Through understanding language, we are trying to get the machine to be able to capture intelligence (Goldstein and Papert, 1977). For example, when giving the machine a sample of a paragraph, you would want it to read for understanding and comprehension just like a human. In the earlier days of popular search engines, the queries we entered into the search engine were simple and mostly followed a pattern that the search engine would understand (keyword-based searching). This led to the term Google-Fu to describe the skill to use a search engine (Sem, 2019). Nowadays, search engine queries are as sophisticated as asking a question and getting a response (imagine entering the Bafana Bafana query in 1999 into a search engine). This is a challenging task but one central to the pursuit of having full Artificial Intelligence.

Some of the approaches to NLP that move us closer to general language understanding are language models (Radford et al., 2019). These language models have followed from the popularity of pre-training of contextual word vectors/models (Pennington et al., 2014) from so-called 'text corpora' or datasets of documents. These approaches take general text (not necessarily connected to the final NLP task) to initialise a model that captures some characteristics of a language. To understand some of the ambitions of language models, imagine giving a machine access to all the English documents available on earth. Through 'reading' all of these documents, the machine can learn the structure of the English language (grammar, semantics), comprehend the concepts discussed in different texts and then exploit this 'knowledge' in numerous tasks. From this knowledge, machines can then pass through this understanding to a learning pipeline for the final NLP task such as sentiment analysis of statements or clustering (grouping of documents). This is termed transfer learning (Ruder et al., 2019). Transfer learning refers to developing machine learning for one task and using some or all of that model to develop another model for a second, somewhat related, task. Building rich and accurate language models or contextual word vectors requires a large amount of data for pre-training models that can then be used for transfer learning.

As can be seen in Figure 4.1 and Figure 4.2, data is needed to be an input into the ML algorithm to produce a trained NLP model. For very complex tasks, we need massive amounts of data. The lack of data becomes a big challenge to solve when we would like to build modern NLP or NLU systems.

Figure 4.1. Typical machine learning pipeline (for NLP, training data is some text with or without labels)

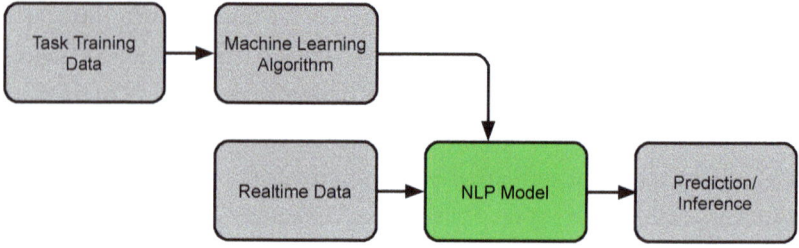

Source: Author

Figure 4.2. Machine learning pipeline with transfer learning

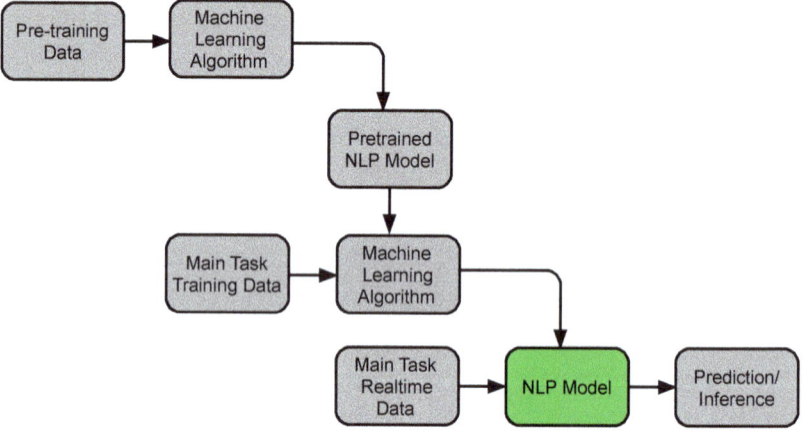

Source: Author

This then brings us to the low representation of African languages, which tend to have small amounts of data available for training these models. For example, if we would like to have a Q&A (question and answer) system that operates in Setswana, isiNdebele or Xitsonga, we need to have a large amount of data that is able to train models and make them robust. This is a challenge that low resource languages have. How then do we talk about innovations in the 4IR in South Africa without talking about language? There can be no such omission. We have to ask, where is the data?

WHERE IS THE DATA?

Machine learning always requires data in one form or another. In NLP we require both more unstructured text data (to pre-train models for reusable representation) and task data to train models for the final task at hand (Figure 4.2). Systems such as virtual assistants (such as Siri, Amazon Echo, Google Assistant) are mostly available in the world's most resourced languages (Templeton, 2020). These are mostly made up of some European languages, UN languages[1] and medium-resourced languages. How can we develop similar systems in South Africa that cover all 11 official languages, or, even better, all the African languages? Africa has the highest language diversity on the planet (Simons and Fennig, 2017). The Niger-Congo language family has 1540 languages, the largest language family in the world (Simons and Fennig, 2017).

Unfortunately, most languages on the African continent are low resourced. Consequently, building NLP systems for these languages is not just a technical challenge; it exposes broader societal challenges in ML systems (Tomašev et al., 2020). We need to be able to source data and also adjust methods to fit these languages (Kann et al., 2019).

The landscape in South Africa

The data available for the nine South African languages (excluding English and Afrikaans) is small for a multitude of reasons. One of these reasons can be seen as inequality. Languages that have enjoyed relative privilege in their development have continued to thrive on the internet even though the internet was heralded as an equalising force. If we look (Table 4.1) at the relative sizes of local South African languages Wikipedia (Wikimedia Foundation, 2020) sites,[2] we can get a quick understanding of the uneven distribution of representation of South African languages online (Marivate et al., 2020) as compared to the Statistics South Africa Census.[3]

---

1 UN: Official Languages https://www.un.org/en/sections/about-un/official-languages/index.html
2 Wikimedia, List of Wikipedias https://meta.wikimedia.org/wiki/List_of_Wikipedias
3 Census 2011 Census in brief: http://www.statssa.gov.za/census/census_2011/census_products/Census_2011_Census_in_brief.pdf

Table 4.1: Wikipedia sizes (in terms of number of articles) and corresponding first language speakers in South Africa

Language	Number of Articles	SA first language speakers (%) (2011)
English	6,041,846	9.6
Afrikaans	89,686	13.5
Sepedi	8,189	9.1
isiZulu	1,395	22.7
IsiXhosa	1,046	16.0
Setswana	712	8.0
Sesotho	683	7.6
Xitsonga	683	4.5
Swati	504	2.5
Tshivenda	367	2.4
isiNdebele	NA	2.1

Source: Wikimedia Wikipedia Languages; StatsSA 2011 Census

As one can see, for South African languages, the Wikipedia article sizes pale in comparison to English and we dare say Afrikaans. There is simply not enough data to build pre-trained models for each of these languages if Wikipedia, for example, was a source.

Wikipedia is often used to train machine learning and natural language processing systems as it is an easily available and open resource. Wikipedia is also used to create knowledge bases (Lehmann et al., 2015), which are consumed by many services. Knowledge bases are used as ontologies that capture facts and their connections to each other. For example, with a universal knowledge base one is able to extract facts like the president of Mauritius in 1995. The knowledge base will store a concept of a president, the concept of Mauritius as a country, the concept of 1995 as a calendar year and then use these together to answer the query. As such, Wikipedia being unequal has effects far beyond just the language accessibility; it may also be skewing available information to people on the internet (Toyama, 2016). This then requires that we add to our earlier model the impact of societal factors on our NLP model (Figure 4.3).

Figure 4.3. An augmented NLP model, taking into account societal factors acting on data

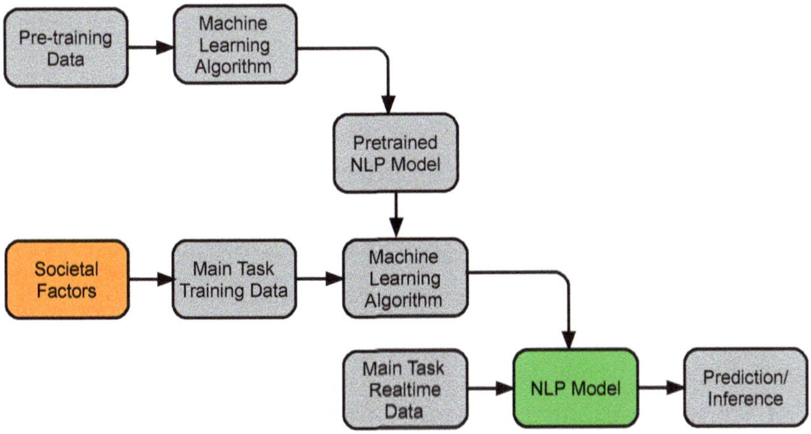

Source: author

PAST EXPERIENCE WITH SOUTH AFRICAN LANGUAGE PROCESSING

Wikipedia is used here as a heuristic to understand a very complex problem. There is much work that has been done in South Africa on building tools for local languages. This section provides an overview of work undertaken thus far in local language processing in South Africa and also covers how recent advances in machine learning are also plotting a new path for local languages.

Work has been done in South Africa on building datasets for Automated Speech Recognition. Work by De Vries et al. (2014) covers a collection of 800 hours of speech in all 11 languages. This type of work has also included better understanding of code switching (using multiple languages in the same conversation) for multi-lingual speakers (Modipa et al., 2013) and using soap-opera data speech (Van der Westhuizen and Niesler, 2018). With textual data, work on Setswana corpus creation has been active (Otlogetswe, 2008), focusing on lexicography (building dictionaries). Text to speech has been an active research area by Sefara et al. (2017) and the authors have covered a number of subtopics within their literature. IsiXhosa corpora have been collected for information retrieval (IR) tasks (Packham and Suleman, 2015). There is also work that covers languages such as Xitsonga, isiNdebele, Tshivenda but it is small compared to languages such as isiZulu and isiXhosa.

However, open availability corpora for languages remains a challenge because when one looks at the above papers it is rare to find them on open data repositories for others to use. There are some data repositories that are specifically set up to collect and archive South African language data and these are covered in the next section. The unavailability of open data, open access publications as well as open benchmarks increases impediments to the use and development of tools (Braun and Ong, 2014).

Even so, we are currently going through a renaissance in ML and NLP (Young et al., 2018) that still threatens to leave behind local languages. There has been more focus on African languages over the last decade. One recent example is the panel on African Languages and Digital Humanities: Challenges and Solutions (Petrollino et al., 2019) which discussed the approach of digital humanities to the African languages. This chapter focuses on the use of ML in NLP for African languages.

Recent work (Marivate et al., 2020) focusing on ways to collect and annotate Setswana and Sepedi data shows how one can now investigate the use of pre-trained models (with data collected from various sources) and then create classification models for news. This is ongoing work, with enhancements made to the data using text augmentation methods (which we have shown can work well for short text and low resource scenarios) (Marivate and Sefara, 2020). Results for Setswana and Sepedi news classification are shown in Figure 4.4. The results shown are on building news classification models that can classify news headline data from SABC radio news stations. Our current focus in this work is to expand the news data from just headlines to full news items as well as to increase the model pre-trained data.

Figure 4.4: News classification performance (larger is better) for Setswana and Sepedi news headlines from SABC

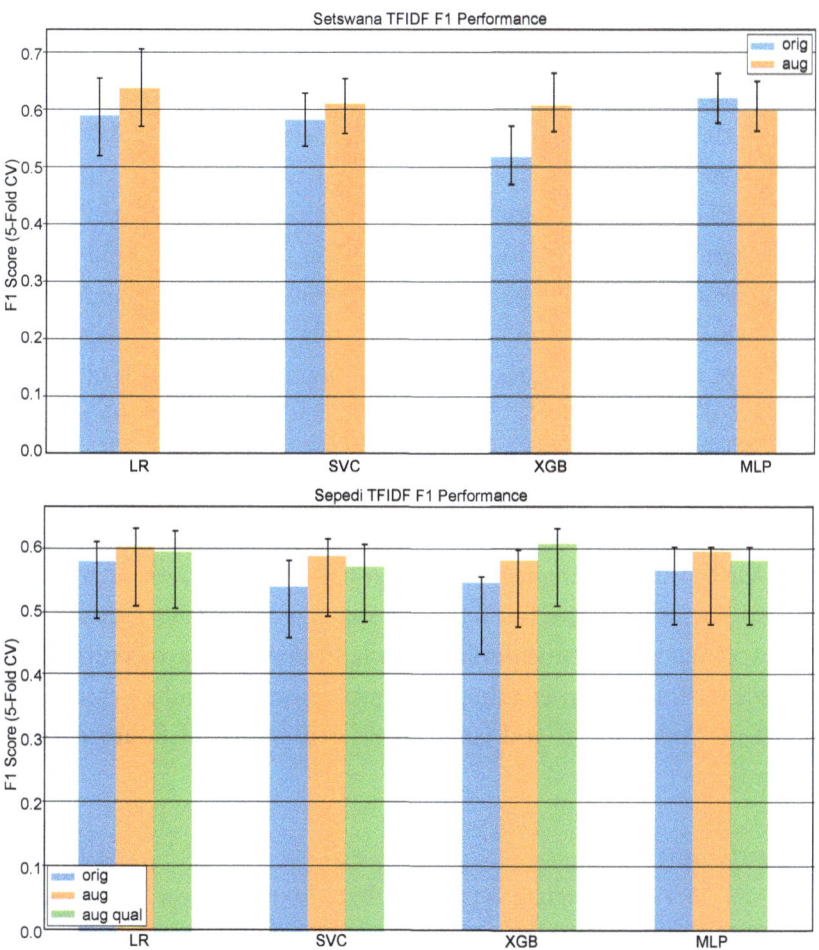

Source: Adapted from Marivate and Sefara (2020)

CHALLENGES FACING LOCAL LANGUAGES

Taking the focus back to the data challenge with local African languages, this chapter now needs to look at why we have such challenges. Martinus and Abbott (2019) discuss some of the problems connected with machine translation for African languages, which also give insight into the challenge with other African language NLP tasks.

They highlight the challenges of low availability, discoverability, focus, reproducibility and lack of benchmarks. We will re-examine each of these points and also expand on them.

Low availability

So far this chapter has discussed the challenge of low availability of data in local languages. There are several factors that can be attributed to this. The origins of most of our written language in South African languages was when missionaries went through the country and started documenting (translating) language to convert the local population (Sanneh, 2015). The missionaries largely ignored the scholastic development of the languages they were documenting (De Kadt, 2006: 25–26). This then ties the development of written local languages to the effects of colonialism in South Africa (Tisani, 2005). How then do we get access to digitised versions of local language newspapers, books and so on, when the inertia caused by the cost of developing scholarship in the languages is a large factor? Initiatives such as the South African Centre for Digital Language Resources (SADiLaR)[4] are working to use some public funding to move this project forward. SADiLaR focuses on enabling the development of and research into language technologies and language-related studies of South African languages.

Discoverability

For the African language datasets that do exist, how do we find them and get access to them? This is easier said than done. We may have language data resources that are distributed amongst a small set of researchers. We may have language data resources that belong to organisations that are not willing to release these datasets for free. For example, media houses in South Africa have had historical publications in local languages. They might be reluctant to make their data available for machine learning pipelines or language preservation, for commercial reasons.

A question arises: Can we unlock such resources? Having

---

4 On the work of the SADiLaR see: https://www.sadilar.org/

a national broadcaster such as the South African Broadcasting Corporation (SABC) may be one of the vectors to release data. The SABC has news radio stations, and produces news scripts, in all South African languages. This is one of the golden language resources that could be used for not only text NLP but speech as well. Providing a progressive policy for use by researchers and innovators may open up new possibilities. Further, NLP skills are needed inside national broadcasters and traditional media so that they can also add to the research and development of NLP resources and tools. They cannot just be data sources.

Focus

In South Africa, there have been many discussions on the development of indigenous languages (De Kadt, 2006). A major point in these discussions has been the attempt to expand the use of local languages in scholarship and higher education. This is a challenge as society itself is becoming more monolingual (May, 2015). In parallel, we can also look at how African researchers in NLP may be balancing between being able to work on cutting-edge methodology but also wanting to be in the mainstream (Martinus and Abbott, 2019). This then makes it less likely that work on NLP will be focused on indigenous languages as this would be taken as niche and less understood. At the same time, looking at innovations that might be used in the private sector, English would likely solicit more funding (May, 2015). An example of such a situation is how, in South Africa, translation of National Assembly proceedings are translated into English from local languages for inclusion in Hansard which is written in English. This means that when MPs are talking in a local language it is translated into English only. However, Hansard is not translated into all the other 10 official languages.

Shifting focus to other government communication, we examine the publication *Vukuzenzele*. This is a South African government magazine that is published monthly. Even this magazine is supposed to be available in all 11 languages. In our experience, though, it was only a subset of the English publication (which contains all stories) that are translated. Moreover, some of the English content is republished as is in the local language versions. This is not a critique of the attempt by government communications to provide such a service, but likely indicates the real cost of translation and the consequent choice to focus on the English version. These examples indicate the challenge at hand. Even national newspapers are mostly in English.

Reproducibility and benchmarks

An increasing number of researchers are starting to see the value in making their work available in a reproducible manner. This means making their data available as well as, in the case of NLP/ML, their code. For language tasks, this is important as it not only gives others a benchmark that they can compare against but also a better understanding of the tools that are needed to do work in the area. This challenge is not only in African languages, but in languages in the global South (Mager et al., 2018). If we aim to grow African NLP, then we need to have more people sharing their work and their data. The international Machine Learning research community is currently at a point where most ML research is shared openly with innovators and researchers who are now able to see the impact of such an approach, which is rendering algorithms widely used and easily improved upon.

It is also important to highlight the impact of policy. There is a lot of uncertainty about the use of online services to gather data for training systems on the African continent. The author, being part of the Masakhane NLP project (discussed in depth later), has had to deal with the common questions by researchers and collaborators: 'Can I use data from website X to train my model?' These questions arise due to researchers not completely understanding copyright and the use of it in ML or NLP tasks. In some states in the EU, there has been a push to have a policy that covers the use of online data for training of text. Countries such as the UK permit such use of data. In South Africa, through NLP research, there is a feeling that doing this increases the likelihood of lawsuits for copyright infringement. This was evident in discussions held through the Masakhane NLP project and questions raised during the first workshop on Resources for African Indigenous Languages (RAIL)[5] in 2020.

---

5   The first workshop on Resources for African Indigenous Languages (RAIL): https://www.sadilar.org/index.php/en/news/events/rail2020

HOW WE MOVE FORWARD:
EXPANDING AFRICAN NLP

We now have an appreciation of the challenges and different facets that make up these challenges. This section discusses several interventions that are needed to improve African NLP and its impact on our societies. This section first discusses how to bring society into the conversation. Next, it discusses ways to improve data collection and curation. Lastly, it surveys the landscape of interventions aimed at expanding skills and practice communities across the continent.

Better public understanding

As we discuss the 4IR, we recognise the need for better public understanding of the underlying emerging technologies. This is not to say every person must have an in-depth understanding of AI/ML/NLP, but it is important that a citizen can understand how these technologies work and how they shape their daily lives. Such understanding will promote more nuanced discourse in public about these technologies and their deployment in society, but, more importantly, it will facilitate a more nuanced discourse around policy. At the moment, without this understanding, asking for clearer policy about the use of online data to train ML/NLP may be seen as unimportant. Without this fundamental understanding, regulations may not change in a timely manner to allow innovations that could completely change some of the technologies for local-use cases. At the same time, researchers may lose opportunities to investigate new tools that may use data that currently would be classified as copyright infringement and IP theft. We do have avenues such as the University of Pretoria (UP) Node of SADiLaR, which works to acquire copyright clearance from publishers before converting local language books to digital formats. Such an approach, though important, will be slower. SADiLaR's approach would also benefit greatly from clearer policy on the use of data for training systems.

Collect, collate and annotate data

The next recommendation is for innovations around data collection, curation and annotation. We need to invest in more data collection, collation and annotation for NLP across the African continent. Such investments would not only serve the aim of increasing the data, but also better preserve the languages. Collection will require content creators to understand ways they can make their content more accessible to machines, not only to humans. One challenge with current typesetting is that it is made for portable document format (PDFs), which are not an efficient nor accessible way to distribute machine-readable text data. Having even a plain text file of a specific document makes it easier for later consumption. Communicating this sometimes seems counterintuitive as a lot of people who work to create the content are mostly doing it to communicate with the public (humans). But for long-term archiving, having machine-readable text content is not just important, it is essential. This use of machine-readable text is not just for the ML/NLP community, but for archives like libraries as well.

This data then needs to be curated and shared in data archives for future use and expansion. SADiLaR in this case acts as a very good example of such an archive. Many archival data repositories can be used in this manner. Further, as higher education libraries evolve to offer more data archival services across the world, they are expanding their push towards data storage that meets the FAIR standard; NLP can benefit. FAIR data is data with principles of findability, accessibility, interoperability and reusability (Wilkinson et al., 2016). These principles will go some way towards making sure that the text data that is collected and curated does not suffer some of the challenges that have historically affected low-resource languages.

Finally, we need to annotate the language data for different NLP tasks (Pustejovsky and Stubbs, 2012). This may seem simple but needs careful thought and also has cost implications. To annotate any data, we need to have humans read pieces of text and then make decisions on those annotations. Let's take, for example, extracting parts of speech from text. It might be possible to do it programmatically, but we still need humans to validate the work. This has cost implications as data annotators need to be sought and paid. Even in crowdsourcing (obtaining annotations and labels from the public) there is still a cost that we have to factor in and requirements for quality control

(Welinder and Perona, 2010). Further, we need to understand what annotation differences might exist for African languages and settings so that we can develop best practice. For example, there might be ideas for recording local folk tales in specific languages from senior speakers of those languages. Doing so might require understanding cultural norms and expectations of what these recordings might mean. This brings us back to the need for an understanding of the technologies so that cultural norms and technological needs can find a workable and mutually beneficial meeting point.

The Artificial Intelligence for Development (AI4D)[6] initiative has recently unearthed a lot of talent and skill with its language data challenge. The challenge called for the collection and curation of African language datasets. The challenge aimed to collect annotated datasets for different African languages and NLP tasks. Through this challenge, researchers could get funding for submitting their collected datasets. The programme revealed some of the challenges language curators have in collecting data. These challenges also mirror some of the challenges discussed before. These included: understanding copyright, digitisation challenges, storage of data and finding innovative ways to convert data. If we solve the data pipeline, we then can move to the NLP task challenges.

Private sector has a role to play in this data collection and African NLP challenge. It collects a multitude of data from users. Let's use the example of social media services which collect large amounts of user-generated content. By amassing this content, the social media services are also collecting local language data. They can improve their services by availing more automated tooling for localised language. One of these tools is services that can better manage abuse and online safety. Abuse here entails concepts such as hate speech, disinformation and predatory behaviour online. For most of the services they have thin (if any) local engineering presence in countries in which they are used. How then do they work to manage automated abuse identification systems? Without more localised knowledge and development their interventions might fall short. Examples of work on NLP tasks on social media in South

---

6   Artificial Intelligence for Development (AI4D): https://ai4d.ai/

Africa have mostly focused on English (Featherstone, 2013; Marivate and Moiloa, 2016).

To increase the accessibility of services, more organisations are rolling out interactive services based on text that can take advantage of messaging services becoming ubiquitous. Some of these services offer translated content. Many of the services use automated bots to interact with users. To develop more robust chatbots for local languages, private organisations should assist in creating new datasets as well as benchmarks. We can learn from how many organisations make available their data for different NLP tasks. These then are used to push benchmarks and to create leader boards that ultimately add to the innovation of the field as a whole. An example is the Low Resource Languages for Emergent Incidents project, funded by the US government, to create NLP datasets for some low-resource languages, specifically for disaster management cases (Strassel and Tracey, 2016). Other examples are evaluation competitions such as SemEval (Semantic Evaluation) which have been running since 2001. Every year, SemEval makes available a few semantic evaluation NLP tasks with data and then runs a competition where many groups attempt the tasks and document their work. This then facilitates the creation of a leader board with benchmarks and works to move the field forward. The AI4D programme replicates some of this with a current focus on data collection, but it needs to move to African NLP task challenges that will then spur research and development.

Expanding practice and skill: Building community

To do all of the above, expanding African NLP, we need to be able to muster a lot of resources and people to work together. This will be done by multiple people in diverse ways, but in a coordinated manner. All in all, we need to have an active and growing community to sustain the goals of reaching ideal African NLP. The next section of this chapter focuses on how the rise of AI on the African continent has created fertile soil for the growth of an African NLP community.

In the last decade there has been a rise of initiatives and organisations in Africa that are increasing training and research in the areas of computing in general and artificial intelligence in particular. A number of universities across the continent now have programmes dedicated to Machine Learning, Artificial Intelligence or data science (which tends to have teaching Machine Learning as a core component). We have had the creation of the African Institute of Mathematical Sciences (AIMS), which has boosted the continent-wide availability of mathematical and computing skills at post-graduate level. More recently, AIMS has introduced the African Masters in Machine Intelligence. This traditional-skills pipeline has also incorporated more agile training programmes.

The creation of the Deep Learning Indaba (DLI)[7] and Data Science Africa (DSA)[8] initiatives have boosted both the exposure and practice of ML/AI on the continent. The DLI was started with the aim of strengthening African Machine Learning, through Africans being shapers of emerging technologies. DSA aims to create a hub in the network of data science researchers across Africa. These networks are large structures that have worked with each other to provide training opportunities, research collaborations and a community that has grown from strength to strength. Both initiatives also have many NLP enthusiasts who keep working towards different goals but now connect with each other and their networks.

If we look at more regional interventions, we have many examples of programmes. Some are very structured while others are still in their infancy. All in all, they show how young people across the continent are experimenting with the way forward as they try to expand AI/ML practice on the continent. In Nigeria, Data Science Nigeria (DSN)[9] and AI Saturdays[10] have built a pipeline of students, researchers, engineers and professionals. In South Africa, Explore Data Science internships[11] and the Data Science for Insight and Decision Enablement initiatives[12] have been established. In East Africa, Data Science Africa has brought

---

7   Deep Learning Indaba: http://deeplearningindaba.com/
8   Data Science Africa: http://www.datascienceafrica.org/
9   Data Science Nigeria: https://www.datasciencenigeria.org/
10 AI Saturdays: https://aisaturdayslagos.github.io/
11 Explore Data Science: https://explore-datascience.net/
12 Data Science for Insight and Decision Enablement: https://dsideweb.github.io/about/

together local enthusiasts and researchers and is working to train the next generation of ML/AI/DS trainers.

After listing some of these interventions, this chapter now looks at how the African NLP community comes together through and with these initiatives. Having identified the talent across the continent, it is important to now move on to innovation and research excellence in what we do. This brings with it more responsibility but also more sustainability: as we show the successes of African NLP, we can garner more support and resources over time. In 2020 there was the Africa NLP workshop to be held at ICLR 2020 conference. Also, the Resources for African Indigenous Languages (RAIL) workshop at LREC 2020. These are all indications of the groundswell of interest in the area by practitioners, professionals, engineers and researchers.

One of the obstacles we do have in research across the continent is the need to have large research groups at institutions. Even with many interventions on funding African research, it will take time to build large strong research groups. The Masakahane Machine Translation project (Orife et al., 2020) has provided an alternative template to learn from and improve upon. The Masakhane MT project works on a collaborative distributed research team model. The Masakhane project[13] aims to recruit researchers from the African continent to 'join our effort in building translation models for African languages'. This project combines the challenge of data collection of parallel translation corpora (aligned data in two different languages to allow for a translation task) and the task of training machine translation models. Both parts of the challenge can be taken up, with participants looking or creating new parallel corpora and also fine-tuning machine translation models for benchmarking. The resources to do the research as well as starter material (computational notebooks) are made available freely. The collaborative group meets weekly online and coordinates many of the functions required in the research itself electronically. The evolution and growth of this group will be important as growing pains will highlight some of the areas in which they will need support.

---

13 Masakhane: A Focus on Machine Translation for African Languages https://www.masakhane.io/

Teaching should also be a priority on the continent. We need to have a pipeline of skills training for cutting-edge research and engineering in NLP. Teaching affords us opportunities to try out ideas with students, strengthen teaching networks and also improve upon pedagogy for African NLP. Few higher education institutions across the continent offer natural language processing as a course in their ML/AI/CS/DS curricular. This needs to change. A consortium of teachers has to work together to share their experiences, improve the availability of African language data for their courses and support students in future research endeavours.

CONCLUSION

This chapter explores the landscape of natural language processing and its connections with artificial intelligence. Artificial intelligence is one of the emerging technologies covered in the Fourth Industrial Revolution. We argue that we cannot completely benefit from AI as an emerging technology in the 4IR without exploring how we improve the state of local language NLP. Local language NLP will lead not only to more inclusive technologies but also to innovations that will drive more AI in South Africa and the African continent. This chapter discussed ways that we can improve local language data creation, collation, curation and annotation to create African NLP task datasets. This requires innovative approaches and partners. We need to tap into resources such as government data and national broadcasters. The private sector also needs to contribute by championing language diversity in their technical systems that use language. To expand the practice of NLP on the African continent, we need to build on the AI/ML community that has been growing over the last few years. This also means training new scientists and engineers as well as expanding research and teaching capacity for NLP. This is an exciting time for African NLP and requires further focus and investment. We need to keep building and pushing for excellence so that our voice can be heard across the world and our languages can be better represented in the evolving and developing technological revolution. We want to live in a world where interacting with intelligent machines in local languages is the norm, not the exception. We look forward to that coming future, a future we all should be working towards and we can realise.

ACKNOWLEDGEMENTS

I would like to acknowledge the group members and collaborators at the Data Science for Social Impact research group at the University of Pretoria, without whom many of the ideas presented here would not have been explored. I would like to also acknowledge the African machine learning community: Deep Learning Indaba, Data Science Africa, Data Science Nigeria, Masakhane MT, Artificial Intelligence for Development, Black in AI, and many more. Lastly, I would like to acknowledge Thembekile Marivate for her input in to making the content of this chapter more accessible.

REFERENCES

Alpaydin, E. 2020 *Introduction to Machine Learning*. Massachusetts: MIT Press.

Braun, M.L. and Ong, C.S. 2014. 'Open science in machine learning'. *Implementing Reproducible Research*, 343(2).

Brodie, M.L. and Mylopoulos, J. 2012. *On Knowledge Base Management Systems: Integrating Artificial Intelligence and Database Technologies*. Springer Science & Business Media.

Chandola, V., Banerjee, A. and Kumar, V. 2009. 'Anomaly detection: A survey'. *ACM Computing Surveys*, 1–58. doi: 10.1145/1541880.1541882.

De Kadt, J. 2006. 'Language development in South Africa–past and present'. Webb, V. and Du Plessis, T. (eds). *The Politics of Language in South Africa*. Pretoria: Van Schaik Publishers, 40–56.

De Vries, N.J., Davel, M.H., Badenhorst, J., Basson, W.D., De Wet, et al. 2014. 'A smartphone-based ASR data collection tool for under-resourced languages'. *Speech Communication*, 56(1), 119–131.

Featherstone, C. 2013. 'Identifying vehicle descriptions in microblogging text with the aim of reducing or predicting crime'. *2013 International Conference on Adaptive Science and Technology*. doi: 10.1109/icastech.2013.6707494.

Goldstein, I. and Papert, S. 1977. 'Artificial intelligence, language, and the study of knowledge'. *Cognitive Science*, 84–123. doi: 10.1207/s15516709cog0101_5.

Hirschberg, J. and Manning, C.D. 2015. 'Advances in natural language processing'. *Science*, 349(6245), 261–266.

Kann, K., Cho, K. and Bowman, S.R. 2019. 'Towards realistic practices In low-resource natural language processing: The development set'. *Proceedings of the 2019 Conference on Empirical Methods in Natural Language Processing and the 9th International Joint Conference on Natural Language*

Processing *(EMNLP-IJCNLP)*. doi: 10.18653/v1/d19-1329.

Lehmann, J., Isele, R., Jakob, M., Jentzsch, A., Kontokostas, D., Mendes, P.N., et al. 2015. 'DBpedia – A large-scale, multilingual knowledge base extracted from Wikipedia'. *Semantic Web*, 167–195. doi: 10.3233/sw-140134.

Mager, M., Gutierrez-Vasques, X., Sierra, G. and Meza, I. 2018. 'Challenges of language technologies for the indigenous languages of the Americas'. *Proceedings of the 27th International Conference on Computational Linguistics*, 55–69.

Marivate, V. and Moiloa, P. 2016. 'Catching crime: Detection of public safety incidents using social media'. *2016 Pattern Recognition Association of South Africa and Robotics and Mechatronics International Conference (PRASA-RobMech)*. doi: 10.1109/robomech.2016.7813140.

Marivate, V., Sefara, T., Chabalala, V., Makhaya, K. and Mokgonyane, T. 2020. 'Investigating an approach for low resource language dataset creation, curation and classification: Setswana and Sepedi'. *Proceedings of the first workshop on Resources for African Indigenous Languages*, 15–20.

Marivate, V. and Sefara T. 2020. 'Improving short text classification through global augmentation methods'. *International Cross-Domain Conference for Machine Learning and Knowledge Extraction*, https://arxiv.org/pdf/1907.03752.pdf, accessed 17 July 2020.

Martinus, L. and Abbott, J.Z. 2019. 'A focus on neural machine translation for African languages'. *arXiv preprint arXiv:1906.05685*.

May, S. 2015. 'Contesting public monolingualism and diglossia: Rethinking political theory and language policy for a multilingual world'. *Language Policy and Political Theory*, 77–99. doi: 10.1007/978-3-319-15084-0_6.

McLaughlin, M. 2020. 'How smart speakers and virtual assistants are transforming our lives'. *Lifewire*. https://www.lifewire.com/virtual-assistants-4138533, accessed 13 April 2020.

Mitchell, T.M. 1997. *Machine Learning*. McGraw-Hill.

Modipa, T.I., De Wet, F. and Davel, M.H. 2013. 'Implications of Sepedi/English code switching for ASR systems'. *Proceedings of the Twenty-Fourth Annual Symposium of the Pattern Recognition Association of South Africa*.

Orife, I., Kreutzer, J., Sibanda, B., Whitenack, D., Siminyu, K. et al. 2020. 'Masakhane: Machine Translation For Africa'. *arXiv preprint arXiv:2003.11529*.

Otlogetswe, T.J. 2008. 'Corpus design for Setswana lexicography'. Doctoral dissertation, University of Pretoria.

Packham, S. and Suleman, H. 2015. 'Crowdsourcing a text corpus is not a game'. *International Conference on Asian Digital Libraries*, 225–234.

Pennington, J., Socher, R. and Manning, C. 2014. 'Glove: Global vectors for word representation'. *Proceedings of the 2014 Conference on Empirical Methods in Natural Language Processing (EMNLP)*. 1532-1543 doi:

10.3115/v1/d14-1162.

Petrollino, S., Nyst, V., Tunde, O., Ngué Um, E., Ekpenyong, M., et al. 2019. 'African languages and digital humanities: Challenges and solutions'. *Digital Humanities Conference 2019.*

Pustejovsky, J. and Stubbs, A. 2012. 'Natural language annotation for machine learning'. O'Reilly Media, Inc. http://storage.hinterland.nu/webdav/Documents/Data%20Mining/Natural%20Language%20Annotation%20for%20Machine%20Learning.pdf

Radford, A., Wu, J., Amodei, D., Amodei, D., Jack Clark J., et al. 14 February 2019. 'Better language models and their implications'. OpenAI. https://openai.com/blog/better-language-models/, accessed 11 April 2020.

Ruder, S., Peters, M.E., Swayamdipta, S. and Wolf, T. 2019. 'Transfer learning in natural language processing'. *Proceedings of the 2019 Conference of the North. Chapter of the Association for Computational Linguistics: Tutorials,* 15-18.doi: 10.18653/v1/n19-5004.

Russell, S and Norvig, P. 2002. *Artificial Intelligence: A Modern Approach* (International Edition). New Jersey: Pearson Prentice-Hall Education International.

Sanneh, L. 2015. *Translating the Message: The Missionary Impact on Culture.* Maryknoll, NY: Orbis Books.

Schwab, K. 2018. 'The Fourth Industrial Revolution (Industry 4.0) a Social Innovation Perspective', *Tạp chí Nghiên cứu dân tộc.* doi: 10.25073/0866-773x/97.

Sefara, T.J., Manamela, M.J., and Modipa, T.I. 2017. 'Web-based automatic pronunciation assistant'. *Southern Africa Telecommunication Networks and Applications Conference (SATNAC),* 112–117.

Sem, S. 16 November 2019. 'The beginners' guide to Google-Fu? 10 tricks to be a Google-Fu blackbelt'. Medium. https://medium.com/analytics-vidhya/https-medium-com-what-is-googlefu-tips-and-tricks-to-be-googlefu-advanced-powersearching-with-google-f7e5661a8bca, accessed 11 April 2020.

Simons, G.F. and Fennig, C.D. 2017. *Ethnologue: Languages of Africa and Europe.* Summer Institute of Linguistics, Academic Publications.

Strassel, S. and Tracey, J. 2016. 'Lorelei language packs: Data, tools and resources for technology development in low resource languages'. *Proceedings of the Tenth International Conference on Language Resources and Evaluation (LREC'16),* 3273–3280.

Templeton, G. 13 May 2020. 'Language support in voice assistants compared'. *Globalme.* https://www.globalme.net/blog/language-support-voice-assistants-compared/ accessed 12 April 2020).

Thrall, J.H., Xiang, L., Quanzheng, L., Cinthia, C., Synho D. et al. 2018. 'Artificial intelligence and machine learning in radiology: Opportunities, challenges, pitfalls and criteria for success'. *Journal of the American College of Radiology: JACR*, 15(3 Pt B), 504–508.

Tisani, N. 2005. 'African indigenous knowledge systems (AIKSs): Another challenge for curriculum development in higher education?'. *South African Journal of Higher Education*. doi: 10.4314/sajhe.v18i3.25489.

Tomašev, N. Cornebise, J., Hutter, F., Mohamed, S., Picciariello, A. et al. 2020. 'AI for social good: Unlocking the opportunity for positive impact'. *Nature Communications*, 11(1), 2468.

Toyama, K. 2016. 'The internet and inequality'. *Communications of the ACM*, 28–30. doi: 10.1145/2892557.

Van der Westhuizen, E. and Niesler, T. 2018. 'A first South African corpus of multilingual code-switched soap opera speech'. *Proceedings of the Eleventh International Conference on Language Resources and Evaluation (LREC 2018)*.

Vise, D.A. and Malseed, M. 2005 *The Google Story*. Random House Digital, Inc.

Weizenbaum, J. 1983. 'ELIZA: A computer program for the study of natural language communication between man and machine'. *Communications of the ACM*, 23–28. doi: 10.1145/357980.357991.

Welinder, P. and Perona, P. 2010. 'Online crowdsourcing: Rating annotators and obtaining cost-effective labels'. *2010 IEEE Computer Society Conference on Computer Vision and Pattern Recognition – Workshops*, 26(1), doi: 10.1109/cvprw.2010.5543189.

Wheeler, F.P., Checkland, P. and Scholes, J. 2000. 'Soft systems methodology in action, including a 30-year retrospective'. *The Journal of the Operational Research Society*, p. 648. doi:10.2307/254201.

Wikimedia Foundation. 2020. 'List of Wikipedias'. Meta, Wikimedia Foundation, Inc. https://meta.wikimedia.org/wiki/List_of_Wikipedias, accessed 12 April 2020.

Wilkinson, M.D., Dumontier, M., Aalbersberg, I.J., Appleton, G., Axton, M. et al. 2016. 'The FAIR guiding principles for scientific data management and stewardship'. *Scientific data*, 3(1), 1–9.

Young, T., Hazarika, D., Poria, S., & Cambria, E. 2018. 'Recent trends in deep learning based natural language processing'. *IEEE Computational Intelligence Magazine*, 55–75.

Section Two

Human Capability Formation

Education and skills development in the South African Fourth Industrial Revolution: Confronting the dynamics of human capabilities, technological change and structural inequality

MICHAEL GASTROW

INTRODUCTION

The Fourth Industrial Revolution (4IR) is characterised by rapid technological change, disruptions to employment and social dynamics, and the emergence of the 'posthuman', in which the boundaries that separate personhood from the world are greatly expanded and made more porous (Braidotti, 2013). The 4IR is driven by innovation – or, more specifically, by the dynamic relations between humans and technologies (Schwab, 2016). Rapid and globalised technological change is underpinned by innovation capabilities and networks, which in turn require education and skills development as critical components. This fundamental logic has its limits, however. In the 4IR, paradigm-dislodging transformations are possible. The processes of education and training are becoming more interchangeable and fluid in their interplay between technologies and humans. Such consideration opens the possibility that while it makes sense to improve, prepare and adjust the existing higher education and industry systems to work better in relation to digitalisation, it may also be necessary to consider that these systems, and their roles in society, may need to be shifted from their present trajectories.

One of the most pressing questions raised by the 4IR is its impact on employment, with a focus on the effects of automation on unemployment, structural changes to the nature of work, and changes to skills supply and demand in response to technological change in the workplace (Millennial Project, 2019). Future technological trajectories, economic competitiveness and human development progress all depend in part on the effectiveness of the alignment between education and skills development systems, with the new technological and societal dynamics of the 4IR. Education and skills development are thus central to 4IR analyses and strategies, including at global, national, sectoral and institutional levels (Schwab, 2016).

In this chapter, we ask how the concept of the 4IR has and could influence approaches towards education and skills development that are geared towards addressing South Africa's major challenges of structural inequality, unemployment and poverty. We draw on South African and international literature, including both academic literature and the outputs of major institutions, to review prominent discourses and to identify key challenges and debates. Existing structural inequality can be reproduced through differences in access to data and technology, so we ask how new approaches to education can help to address inequality. Since education is intrinsically linked to employment possibilities, we also explore literature focusing on the future of work, and how education policies and practices might be leveraged to achieve improved work outcomes. Finally, we examine the strategic and investment responses of South African institutions, particularly within the government and the higher education sector.

We conclude with key lessons about understanding education in the context of the 4IR, and recommendations for policymakers.

FRAMING EDUCATION POLICY IN THE ERA OF THE FOURTH INDUSTRIAL REVOLUTION

Education in South Africa is deeply fractured and unequal. Access to quality education still depends, to a large degree, on whether learners are black or white, urban or rural, middle class or poor. These divisions create a school system for the more privileged that is privatised, technologically developed and digitally connected, and a system for the marginalised that is under-funded, technologically underdeveloped, and disconnected from the digital world. These systems produce unequal outcomes in terms of literacy, numeracy, preparation for tertiary education and preparation for participation in labour markets. Unequal outcomes, in turn, reinforce social and economic inequality and perpetuate the structural conditions that development theorists and policymakers in South Africa have been grappling with for decades. Thinking about education and the 4IR in South Africa is therefore a question of social justice and historical redress, as well as economic growth and competitiveness.

Applying the notion of the 4IR to the context of education and skills development in South Africa creates a set of principles that can help to conceptualise the course of the 4IR in the country and guide policy in times of rapidly accelerating technological change. The first principle is that the 4IR should not leave black South Africans behind. If technological change is left to perpetuate inequality and social exclusion, it is more likely to lead to social instability and entrench historical injustice. Any progressive notions of a new education and skills development agenda needs to centre this concept. As learners move into the world of work, the 4IR has the potential to create new kinds of opportunities for black participation in the formal economy, including new avenues for growth in industrial development and in the informal sector. From a continental point of view, education and skills development are arguably the most important levers for positioning Africa to benefit from the 4IR. From a national point of view, South

Africa, therefore, bears a responsibility to the African collective to lead in the area of education and skills development in the 4IR.

Closely related to the question of transformation is the issue of equality. The second overarching principle is that education in the 4IR must aim to reduce structural inequality. New approaches to education will need to address the trend of industrial and services automation. This trend has the potential to exclude unskilled and low-skilled workers from formal employment, while at the same time benefitting the highly skilled and the holders of capital. Education must also contend with the digital divide. In South Africa, the poor face vastly higher data costs (per megabyte) than the middle class. The efforts of the South African Competition Commission to address this issue remain critical (Dlulane, 2019). Expanding internet access and working towards data justice are critical underpinnings for education and skills development strategies for South Africa. Building on this, the question of leveraging digital technology for inclusive education and skills development raises a variety of debates and potential policy responses. Across all these dimensions, gender inequality remains a problem that is critical to address, from the perspectives of both social justice and greater economic competitiveness (Kanza, 2016).

The third principle is that of adaptability, a concept that emerged from an ongoing dialogue between South African and European institutions in their efforts to understand the links between the 4IR and education (Gastrow, 2019). The overall reconceptualisation of education must grapple with the issue of *change*: rapid technological change, economic change, social change, and changes in labour markets all require increased flexibility and adaptability. Adapting to change requires an increased focus on the capacity of institutions to sense changes in their environments and adapt accordingly, a characteristic known as interactive capabilities (Von Tunzelman and Wang, 2007). This imperative necessitates increased intensity and effectiveness of communication and collaboration across systems of education, industry, civil society and government (Gastrow, 2019).

Fourth Industrial Revolution technology platforms have the potential to contribute towards human development in South Africa – if capabilities can be effectively built, and if technologies are responsibly directed through policy. For example, Smith and Neupane (2018) argue that the right mix of artificial intelligence skills and policy could lead to a significant improvement in human development in South Africa. Data-related skills are essential to development, particularly in the African context, where enormous social and economic opportunities exist to benefit from digitalisation and the utilisation of data. However, Africa's 'data ecosystems' remain underdeveloped and held back by a shortage of skills (UNDP, 2016). The question of education and skills development in the 4IR is thus an important issue in the broader context of development in South Africa.

Understanding the connections between technological change and education in South Africa was a central component of the Labour Market Intelligence Partnership,[1] a research consortium led by the Human Sciences Research Council from 2013–2016, reporting to the Department of Higher Education and Training. One of the key findings emerging from a large body of research conducted over several years was the need to build more effective interactions between education providers and the private sector. This is required in order for the skills requirements generated by technological change to be more rapidly and effectively responded to through curriculum change and new research priorities in the higher education environment. As put forward by Gastrow (2019), strengthening the interactive capabilities among institutions that foster the supply of skills requires:

1. Cultivating the capacity of post-school educational institutions to engage with employers and understand their current and potential future skills requirements (which alter along with technological change);
2. shortening the cycle for curriculum change to respond to the evolving technologies; and
3. cultivating a research agenda that senses technological change and responds accordingly.

Strengthening interactive capabilities among the institutions of *skills demand* requires:

---

1   http://lmip.org.za/

1. Strengthening platforms for the private sector to make sense of technological changes, and to better understand how these changes might impact on its future skills requirements;
2. Building and strengthening platforms for dialogue between employers and post-school education systems, to facilitate the exchange of information about the impact of technology on future demand for skills; and
3. Fresh approaches to education that include new and more flexible modalities.

Adaptability, in turn, requires ongoing research. A research agenda focused on observing technological change, and interpreting its implications for society and for policy, would strengthen the capacity of educational institutions to respond to technological change. Institutions that include actors in government, basic education, higher education, the private sector and civil society would be better able to respond to technological change rapidly, shorten policy cycles, design more relevant curricula, build dynamic partnerships and adapt pedagogy to leverage the potential of new and emerging technologies.

One important way to inform debates about change in education is to listen to the voices of young people. An international survey of youth perceptions about the future (Infosys, 2016) found that interest in technology careers is higher among the highly skilled, pointing towards growing technological inequality. Youth questioned the extent to which their academic experiences equipped them for their career, underscoring the need to actively adjust curricula. Within this, youth recognised the importance of human or 'soft cognitive' skills for the future workplace. Young people were positive about engaging in continuous learning and flexible skill sets that allow them to adapt to changing work environments – 80 per cent of respondents believed that their success in the labour market will depend on continuous learning. The youth believe that digital tools will become increasingly important for self-learning – over 70 per cent of respondents in South Africa, India and the US believed they could teach themselves anything they wanted to using freely available online sources. This digital empowerment, however, is emerging against a backdrop of reservations about automation – 40 per cent of respondents believed their current job would be replaced by a robot or an artificial intelligence machine over the next decade.

In summary, our framing of the South African education policy in the era of the 4IR includes a focus on building capabilities for economic competitiveness and employment, but in a manner that is aligned with our national imperatives of social justice and redress – particularly with regards to reducing inequality and building racial inclusiveness. In all this, the principle of adaptability is central, and in the following sections this chapter explores the types of changes that may be required if we are to meet our goals of harnessing new technologies to meet our national development aspirations.

EDUCATION AND SKILLS DEVELOPMENT IN THE FOURTH INDUSTRIAL REVOLUTION

The relationship between technological change and skills development has a long history of analysis (Autor et al., 2003). The 4IR is driven by a distinct set of technological changes, prompting us to reflect on the relationship between skills development and innovation in relation to revolutionary technologies such as artificial intelligence, robotics, platform economy business models, genetic engineering and quantum computing, amongst others. The role of higher education is critical here – universities form a nexus for interaction among industry partners, governments, international organisations and NGOs. Universities grow valuable intellectual capital – including the development of new technologies, and insights into the relationships between these technologies and their social and economic applications and impacts. In recognition of this important role, the World Economic Forum (WEF) has described universities as the 'vanguard' of the 4IR (Horowitz and White-Burke, 2018).

Literature addressing the nexus of the 4IR and education is disparate and multi-disciplinary. It is characterised by distinct international and South African discourses, and also by distinct discourses emerging from academic and institutional ('grey') literatures. These discourses emphasise different methodologies, questions, lenses and frameworks.

International academic literature on the 4IR and education has primarily been focused on the role of capability building in technology-driven economic growth and employment (Penprase, 2018). A distinct discourse exists in the engineering and management literatures, which draw on the notion of the 4IR to discuss capability building in relation to technological upgrading in an Industrie 4.0 context (Bedolla et al., 2017; Chaim et al., 2018; Grodotzki et al., 2018; Erol et al., 2016).

The dominant discourses in international institutional literature are the body of work emerging from the World Economic Forum (WEF), and literature that is focused on the German system of Industrie 4.0. The WEF's framework adheres to an orthodox conceptualisation of the 4IR, based on the work of Klaus Schwab (Schwab, 2016), with a focus on the roles of technology foresight, multi-stakeholder partnerships, and business strategies (such as WEF, 2016; 2017). The Industrie 4.0 framework is widely used in connection with literature focused on the German Plattform Industrie 4.0, including research reflecting on the connections between research and development (R&D) and education (see, for example, German Federal Ministry of Education and Research, 2018a; 2018b).

South African academic literature explicitly on the 4IR and education is limited; rather it is focused on applying the principles of the 4IR in the South African context (see, for example, Butler-Adam, 2018). Research emerging from collaboration with German researchers focuses on the potential for entrepreneurship (Naudé, 2017). Thought leaders from the University of Johannesburg (UJ) focus on the implications of the 4IR for the strategic decisions and institutional landscape of universities in South Africa (Xing and Marwala, 2017). The Institute for Security Studies (ISS) has published on the implications of skills development systems in the South African manufacturing sector (Cilliers, 2018). Grey literature emerging from the private sector largely focuses on business strategies in relation to skills development that responds to technological opportunities in market settings (such as Deloitte).[2]

---

2   https://www2.deloitte.com/content/dam/Deloitte/za/Documents/ Consumer_Industrial_Products/Industry%2040%20-%20SA%20Findings. pdf%20-%2015%20June%202018.pdf

Looking across this diverse literature, this chapter draws out some key ideas that are salient to current South African debates. Firstly, there is recognition that rapid technological change creates the conditions for a delicate balancing act between building technological capabilities and preparing humans to compete with machines in the labour market. Penprase (2018) undertakes a historical analysis of education through the lens of the 4IR, exploring the manner in which previous industrial revolutions have impacted on higher education. The analysis concludes that a critical factor in industrial revolutions is 'STEM instruction that develops technical capacity in emerging technologies' (Penprase, 2018: 207).

The WEF's Future of Jobs report (WEF, 2016) presents an analysis based on interviews with chief human resources officers (CHROs) of major global employers. Their core finding was that the business model changes generated by the 4IR often create skill-set disruption with only a minimal time lag. The impact of technological change is shortening the 'shelf-life' of existing skill sets. Shortening curriculum cycles in a way that responds to technological change therefore becomes critical. This implies that in the 4IR, key technologies such as coding, robotics, Artificial Intelligence, 3D printing and biotechnology should receive greater emphasis in higher education settings, and that school curricula should prepare learners for further study in these areas. In South Africa, some action has already been taken in this regard: The introduction of coding in schools is currently being piloted by the Department of Basic Education – an important feasibility study that it's hoped will assess the enablers, constraints and challenges associated with such a change (Gastrow, 2019). The establishment of 3D printing laboratories would benefit from a similar process.

However, a narrow focus on building technical skills in new and emerging technologies is not sufficient. Automation and other processes that lead to the technological substitution of labour raise the issue of preparing humans to compete with machines on the basis of human comparative advantages, and the need to prepare for an increasingly technological future through multidisciplinary analysis. Both Marwala (2019) and Penprase (2018) argue that the socio-technical complexity of the 4IR creates a requirement for multidisciplinary curricula that include the social sciences, engineering and natural science disciplines. They also argue that school curricula should prepare humans for competition with machines by steering away from machine-like tasks – such as memorisation and repetition – towards human traits that machines are unlikely to replicate, such as empathy, creativity, innovation, social skills and managing complexity. Related to this is the importance of preparing learners for the ethical and cultural complexities that arise from new technologies (Gastrow, 2019; Marwala, 2019). The use of digital technologies in education creates new options for pedagogical modalities. In an increasingly digitally connected world, old models of place-based learning need to be revised, and new forms of digital knowledge acquisition need to be considered. The World Economic Forum argues that digital knowledge acquisition is a potential lever for improved education outcomes that are more responsive to technological change (WEF 2016; 2017). Abu Mezied (2016) and Penprase (2018) argue that education modalities need to increasingly include lifelong learning and online self-learning.

One example of structured online learning is that of massive online open courses (MOOCs). These courses can effectively expand access to education to learners who may have difficulty accessing universities or other place-based education sites. However, MOOCs face challenges in terms of providing accreditation and trust in the courses' credentials. This underscores that one of the core functions of higher education is that of maintaining social trust – and trust in existing institutions that are intricately woven into the social fabric, and have been throughout their long histories, confers a distinct advantage over new online modalities. MOOCs also face the challenges of risks to privacy and security. Debates over the future of online education need to take these challenges into account, and creatively explore options for overcoming them, so that access to education can become more flexible, equitable and widespread.

The key lessons that emerge from the diverse literatures at the intersection of education and the 4IR coalesce around the relationship between machines and humans in the context of automation and digitalisation. Both production automation and services automation have a complex effect on future skills requirements. On the one hand, the need for basic STEM skills remains relevant, as does the requirement to introduce new technologies into curricula more rapidly. On the other hand, preparing humans to compete with machines requires an emphasis on distinctly human skills. Finally, digitalisation challenges us to re-imagine pedagogical modalities in ways that combine real-world and digital access to knowledge.

Before the onset of the COVID-19 pandemic, debates about the 4IR and education, particularly online education, were already well developed. The pandemic has created a greater need and sense of urgency, as schools, universities and other sites of education have undertaken a sudden and large-scale shift towards online learning modalities. Educational institutions have rapidly established online learning systems that make it possible to continue education despite the closure of physical sites. This situation brings into stark relief the great flexibility and potential of online learning. It also highlights that structured and formalised online learning could lead to improved education outcomes, particularly in times of crisis. Increasingly important learning skills appear to be 1) navigating interactive teaching platforms; and 2) navigating online knowledge resources to find and use new knowledge in a goal-directed and constructive manner. The establishment of nationally applicable standards, tools, content and curricula for online learning would be of great value during this time – and would continue to hold significant value after the pandemic, when physical sites are once again open. The 4IR therefore holds particular salience at this time. The interplay between the physical and digital worlds has never been more significant – the continuation of education depends on digital solutions.

PREPARING HUMANS FOR A
FUTURE OF DECENT WORK

Technological change drives changes in skills supply and demand, as well as changes in the contractual modalities of employment (for example, shifts from formal employment to gig economies). It also changes the macro-structure of employment in economies (for example, from manufacturing sectors to service sectors) and public perceptions of work (for example, decreased expectations of old models of stable, full-time work at a single employer) (ILO, 2013). New modalities of work require new forms of labour regulation, as well as new approaches to education and skills development (Ljungholm, 2018).

Digital technologies have the potential to reduce demand for unskilled work and increase returns to skilled work. Automation places large proportions of workers 'at risk' – for example, the growing use of AI and robotics reduces demand for jobs, both skilled and unskilled (Frey and Osborne, 2017; CEDA, 2015; Brynjolfsson and McAffee, 2014; Graetz and Michaels, 2015). Some analyses have predicted robot-driven mass unemployment (Ford, 2015); the dehumanisation of work; the emergence of a 'cyberproletariat' (Huws, 2014); and the domination of human society by machines (Carr, 2014).

The reality, as has been borne out at the time of writing, has been more complex, with both positive and negative aspects of the technologisation of work. Mokyr et al., (2015), drawing on economic history, demonstrate the emergence of new occupations, arising from processes of technological transformation – similar analyses appear to be needed to inform contemporary debates. Boyd and Holton make the point that social scientific analysis is critical here, since 'the spatial location and skill mix of employment opportunity is far too complex to be read off from the "nature of the technology"' (Boyd and Holton, 2018: 337).

In the South African context of deep structural inequality, Marxist analyses, such as Munck's (2013), centre their critique of the institutions of work around notions of class struggle and inequality, perpetuated at the nexus between technological change and the political economies of labour. Such an approach would be useful for researchers and policymakers seeking to delve deeper into issues of power and inequality in the future of work, and what these might mean for education and skills development.

Multilateral institutions have published extensively, and with a growing sense of urgency, on the future of work. The future of employment they envisage is driven by the logics of the 4IR position, which emphasises matters of automation-driven employment changes, changing skills requirements, and qualitative labour market disruptions such as the gig economy, platform economy and the creation of a mooted new 'precariat', a group whose labour conditions are undermined by these changes (Munck, 2013). The International Labour Organization (ILO) has formulated a consensus position on the future of work, through its 'Global Commission on the Future of Work'. In short, this document draws on a set of values to frame a desired future of work as a fundamentally human-centric arena – a position which contrasts with the techno-centric or market-centric positions taken up by other actors in the labour market (for example, the Plattform Industrie 4.0 and the World Bank). From this ideological point of departure, the ILO (2019) centres its normative position on the objectives of formal employment, human rights and the Sustainable Development Goals. The ILO's proposed process, like that put forward by the WEF, is based on recognition of the value of multi-stakeholder engagement and dialogue. To create a labour environment in which human capabilities may be exercised in an environment of freedom and human rights, the ILO proposes foci on lifelong learning, institutions, policies and strategies to skill, reskill, upskill and to create gender equality (ILO, 2017a; 2017b; 2018a; 2018b).

The WEF's publications relating to the future of work emphasise the paradox produced by technological advancement: on the one hand it generates demand for technical skills, and on the other, through the process of routine repetitive tasks becoming automated, it changes the balance of skills requirements towards those that are distinctly human. The organisation's report 'The Future of Jobs: Employment, Skills and Workforce Strategy for the Fourth Industrial Revolution' (2016), frames future labour markets as political economy spaces in which technologies act as levers – and the focus of the WEF analysis is on how to best identify and mobilise technologies in order to achieve broadly defined economic growth objectives, and to mitigate potential future labour market distortions and conflicts. The WEF argues that key areas, which require ongoing analysis and are critical for informing policy, include understanding how technology acts as a driver of change in labour markets, monitoring employment trends (including skills supply and demand trends) and the development of future labour market strategies. A particular emphasis is placed on the 'female talent pipeline' and other issues of gender in future labour markets.

With regard to the African context, the WEF report 'The Future of Jobs and Skills in Africa: Preparing the Region for the Fourth Industrial Revolution' (WEF, 2017), undertakes an in-depth analysis of African labour markets, education and skills development in Africa, and strategies for the future of work. The report highlights that a lack of adequately skilled labour is a major growth constraint across the continent. Skills constraints also mean that African economies are vulnerable to automation – but this may be moderated by low labour costs and offset by job creation.

Lawrence (2018), a former South African and Professor of International Trade and Investment at Harvard, writes about the future of manufacturing employment in South Africa. His report provides a valuable resource for understanding the future of work in this sector. A consistent decline has occurred in the manufacturing sector's share of national employment since its peak of 17 per cent in 1981. The SA-EU Dialogue on the 4IR (Gastrow, 2019) concluded that assessing the potential for automation-induced job losses, and mitigating the effects of these, is a critical component of a policy response to the 4IR. This imperative cuts across many policy arenas and government departments, and will have distinct dynamics in different sub-sectors and industries. One general objective is to balance the need for technological upgrading, and therefore economic competitiveness, with the need for decent work and for preventing unemployment.

However, a historical analysis shows that technology is not primarily a driver of net job destruction – technological progress has changed the profile of jobs and skills requirements, but not overall employment levels (Schwab, 2016). Growth in South African manufacturing employment would need to be predicated on export-driven growth – a challenge, given the multiple constraints in the sector, and high levels of global competition. South Africa's employment strategies cannot be centred on manufacturing employment – growth is more likely to occur in services sectors. These core dynamics should shape the ways in which education and skills development in South Africa are oriented towards technological change. Growing national capabilities in the skills required for the ICT service sector is one of the most critical priorities (De Falco, 2019), not only for the stimulation of employment and economic development, but also for application in decision-making, including strategic and policy decisions (Mundell, 2017).

Crosby (2018) addresses the issue of de-industrialisation in South Africa, in which fewer large firms are providing employment, and more people are employed in medium and small companies. Crosby argues that the informal sector should be a focus for policy that seeks to leverage skills opportunities and support the development of businesses. In this context, educational provision must keep pace with industrial development and technology – including artisanal workers. An analysis of occupational shortages shows high demand for managers, sales workers and artisans. Skills gap analysis indicates that leadership, project management and financial skills are key for high-skilled occupations, while semi-skilled occupations require problem solving, critical thinking and teamwork. Low-skilled occupations require basic ICT, communication, literacy and numeracy, as well as health and safety skills. Ongoing adaptation of curricula and institutional relationships with employers are required in order to continuously adapt to such changes in the labour market.

It is clear that our approach towards education and skills development in the 4IR will have wide-ranging impacts on the future of work in South Africa. A fundamental challenge is that of managing the tension between increased automation and unemployment. Sector-level research will help to understand how this tension may best be resolved – for example, to take advantage of flexibility and growth in the services sector; to mitigate de-industrialisation in the manufacturing sector; and to grow horizontal capabilities in the ICT sector. A critical part of this research and policy process is the development of a more dynamic understanding of occupational skills gaps. The implementation of evidence-based policy, by both the public sector and education institutions, in turn depends on decisions about strategy and investment. The next section examines strategic and investment responses to the 4IR by key South African institutions, with a focus on government and higher education actors.

THE 4IR AND SOUTH AFRICAN
EDUCATION INSTITUTIONS

South African institutions have responded to the 4IR in a variety of ways. At grass-roots level, the 'maker movement' has adopted technologies such as 3D printing, robotics, the Internet of Things, and digitally mediated arts and crafts production. Maker movements have become significant sites of skills development, part of the co-evolution of skills and technological capabilities (De Beer et al., 2017).

SA's response to the need for the development of skills relating to the 4IR has become grounded in the country's industry sectors. Several of the Sector Education and Training Authorities (SETAs) have developed positions on the 4IR, for example, the Business Process, Enabling South Africa (BPSA) SETA (drawing on a McKinsey report by Magwentshu et al., 2019); the BankSETA (Ohene-Afoakwa and Nyanhongo, 2017); the Education, Training and Development Practices (ETDP) (ETDP SETA, 2020); the Media, Information and Communication Technologies[3] (MICT) SETA. Each of these sectors has experienced distinct dynamics and challenges related to skills development. This diversity highlights the cross-cutting nature of the 4IR and its impact on skills, while at the same time signalling that industry bodies are responding.

Other processes of skills development have been 'top down', driven to a significant extent by the call from the South African Presidency to use the 4IR as a heuristic for understanding globalised technological change. At the time of writing, the Presidential Commission on the Fourth Industrial Revolution was discussing recommendations for furthering 4IR-oriented education and skills development. Government departments, schools and universities have started incorporating the concept of the 4IR into their strategies, policies and investments. South Africa has also participated in regional and multilateral formations, such as BRICS, to consider the relationship between education and the 4IR within the international context.

The Department of Basic Education (DBE) is reviewing its

---

3   https://www.mict.org.za/the-fourth-industrial-revolution/

curriculum in order to facilitate greater alignment with the skills required in the 4IR (Rivette, 2019). Robotics and coding are being introduced as subjects from Grade R to Grade 9, and teachers are being trained to educate in these areas. University of South Africa (UNISA) has partnered with the DBE to offer the university's ICT laboratories to train teachers in coding-related skills. Devolved actions are also being taken at the provincial level – for example, the SITA[4] has piloted a software engineering programme in five schools in collaboration with the Free State government (Business Tech, 2019).

The Department of Higher Education and Training (DHET) has established a multi-sectoral task team to advise the higher education sector on how to maximise the opportunities presented by the 4IR, and how to mitigate the potential challenges (DHET, 2019). South Africa's research-intensive universities have been urged by the DHET to take the lead in preparing for the 4IR (Mzekandaba, 2017), including preparing graduates who are ready to enter the new labour market, and adapting their curricula to provide the skills required as a result of technological disruption (Menon and Castrillon, 2019). At the same time, the Department of Communication and Digital Technologies has included in its ICT Policy White Paper a focus on the development of an e-literate society through a skills development plan.

Amongst universities, responses to the 4IR, particularly as a framing concept for thinking about curricula, research and investment, have been highly varied. Some universities have placed the 4IR at the centre of their strategic focus, while others have made reference to it only rhetorically. One unifying platform for moving towards coherence across the higher education system has been the 4IRSA Partnership – established in 2018 as an alliance between the Department of Communications & Digital Technologies, Telkom, and the universities of the Witwatersrand, Johannesburg and Fort Hare.[5] Since the formation of the partnership, major firms, including Deloitte South Africa, Vodacom and Huawei, have become partners. The partnership aims to stimulate and facilitate a national dialogue to

---

4   http://www.sita.co.za/
5   https://4irsa.org/

shape South Africa's response to the 4IR. The 4IRSA partnership has a close working relationship with the Presidential Commission on the 4IR, strengthened by personal and institutional networks.

The University of Johannesburg (UJ) has emerged as a leading institution in the South African 4IR. In 2016 the university established an Institute for Intelligent Systems (Mzekandaba, 2017), which drives industry 4.0 initiatives, including research projects funded by the Council for Scientific and Industrial Research (CSIR), the National Research Founation (NRF) and private sector sponsors. UJ has a dedicated 4IR link[6] on its website, where news, events and research activities are broadcast. An initiative funded by Telkom and the South African Institute of Chartered Accountants, in partnership with UJ, has introduced online courses in artificial intelligence, machine learning, natural language processing, blockchain and ethics. A DST/NRF/ Newton Fund Trilateral Research Chair in Transformative Innovation, the Fourth Industrial Revolution and Sustainable Development has been established (UJ, 2019).[7]

Other universities have incorporated the notion of the 4IR into their strategies and funding models to varying degrees. The University of the Witwatersrand is a founding partner of the 4IRSA partnership, and the 4IR has been integrated into the university's strategic orientation (Rosman, 2019; Wits, 2018; IOL, 2019). The university has also recently established an AI initiative, Cirrus, which partners with international universities to create multidisciplinary degree programmes.[8]

Other universities have made significant investments, and incorporated the 4IR into their strategic outlooks. The Tshwane University of Technology has partnered with the MerSETA (Manufacturing, Engineering and Related Services SETA) (TUT, 2019), with a focus on intelligent manufacturing, the Internet of Things (IoT), industry 4.0 and rapid product development. The University of KwaZulu-Natal launched flagship research programmes

6 https://www.uj.ac.za/fourth-industrial-revolution
7 https://www.uj.ac.za/newandevents/Pages/Prof-Salim-Vally-awarded-NRF-Research-Chair-in-Community-and-Worker-Education.aspx
8 https://www.wits.ac.za/course-finder/postgraduate/science/msc-artificial-intelligence/

on Big Data and Informatics, and African City of the Future,[9] and has partnered with the provincial government to develop skills in advanced manufacturing.[10] The University of Pretoria has established a Research Chair in Intelligent Manufacturing and has launched an Engineering 4.0 hub (Kupe, 2019). The University of Fort Hare is a founding partner of the 4IRSA, and draws on the concept in its strategic plans. The University of the Western Cape established a diploma in augmented reality and virtual reality (Folb, 2018). The University of the Free State has formed a private sector partnership to educate students and staff in IoT skills and experience. A large and growing number of seminars, conferences and workshops have focused on the 4IR at multiple sites, including the University of the North West,[11] the Vaal University of Technology,[12] the University of Zululand[13] and the University of the Western Cape.[14]

In short, the notion of the 4IR has gained considerable traction across the higher education system, manifested as strategy, funding, institutional structures, intellectual projects, teaching and learning, research and partnerships. The manner in which the term has become mainstreamed across the system is remarkable. Some forms of coordination are emerging – particularly through the 4IRSA formation, the Presidential Commission and public–private partnerships. These platforms allow for the sharing of information, create space for conceptual and strategic debate, and facilitate institutional linkages and cooperation.

However, there are limitations to this coordination. The extent of uptake of the 4IR as a concept varies widely across institutions

---

9   https://quantum.ukzn.ac.za/wp-content/.../Flagship-Pamphlet-A5.pdf

10 https://www.sanews.gov.za/south-africa/kzn-giving-youth-skills-they-need-4ir

11 http://news.nwu.ac.za/nwu-summit-focuses-human-capital-fourth-industrial-revolution

12 https://www.gov.za/speeches/minister-communications-and-digital-technologies-lead-technology-

13 http://www.unizulu.ac.za/wp-content/uploads/2019/07/TL-4th-CONFERENCE-POSTER-1.pdf

14 https://www.netwerk24.com/ZA/Tygerburger/Nuus/uwc-hosts-green-20190709-2

– and in some cases does not extend far beyond the rhetorical. The core concepts of the 4IR are to some extent understood differently in different institutional settings. Mechanisms and political incentives for coordination across government departments are not well developed, hampering coordination between the DBE and the DHET with other government departments. Building bridges across inter-departmental divides, and strengthening conceptual alignment and practical coordination across the education institutional landscape, would significantly improve national capabilities for responding appropriately to the changes brought about by the 4IR. It is also important to consider the systemic constraints within which any constructive policy approach must operate. For example, analysis of South Africa's vocational training system (Allais, 2012) highlighted that despite well-intentioned policy interventions, it is difficult to develop robust and coherent skills development in the context of inadequate social security, high levels of job insecurity and high levels of inequality.

POLICY IMPLICATIONS

From a policy perspective, the dynamics of the 4IR raise the imperative for education and skills development systems to become more adaptable. This takes many forms: shorter curriculum cycles, shorter policy cycles, increased interaction between institutions of skills supply and institutions of skills demand, and more research focus on the skills implications of technological change. In all this, it remains crucial to listen attentively to the voices of young people, and to support universities to play the role of the 'vanguard' of the 4IR.

The fundamentals of STEM instruction remain crucial to meaningful skills development oriented towards core 4IR technologies. South Africa is already experimenting with the inclusion of coding, robotics and 3D printing in basic education settings, and advancing higher education attention to biotechnology, Artificial Intelligence, data analytics, robotics and other 4IR technologies. These efforts should be supported, coordinated and closely monitored so that lessons about impact and efficacy can be rapidly incorporated into shortened policy and curriculum cycles. However, curriculum change is not only about technical skills related to new technologies – as the world of work becomes increasingly automated, human traits such as empathy, creativity, innovation, social skills, managing complexity and understanding ethical and cultural complexities also become increasingly significant.

The technological capabilities of education institutions are also critical. Connecting more schools to the internet is a critical first step – although this needs to occur in tandem with the development of adequate capabilities for utilising digital technologies for education. In the 4IR, access to the internet becomes more critical than ever. A potential intervention here could be a national programme of school infrastructure development. Increasing the connectivity of education institutions is a first step. However, a greater challenge is building the technological capabilities needed to effectively utilise connectivity. Computer labs and internet access can only be harnessed for education if the school or university has the capability to use and maintain equipment and software.

Related to this are the technological capabilities for operationalising new digital pedagogies. Online self-learning, particularly in the context of the COVID-19 pandemic, has become increasingly significant. The notion of education as being place-bound is shifting to one of education taking place through a mixture of physical and digital interaction. As the convergence of physical and digital worlds becomes more deeply entrenched in our social world, it becomes more important that learners be able to navigate interactive teaching platforms and online knowledge resources. Policy measures to support these skills are critical, including the development of online standards, tools, content and curricula.

In higher education, it's critical to build curricula related to 4IR technologies, which have generated high levels of employer demand in the market. For example, advancing AI capabilities through education and skills development interventions is seen as critical, as the number of AI specialists is dwarfed by market demand for such expertise. At the same time, the logic of the 4IR suggests that curricula should be broader – engineering students should engage with social science concepts, and vice versa (Gastrow, 2019). In a world where complex socio-technical systems evolve rapidly and unpredictably, technology development should not be undertaken without an understanding of social context, and social analysis should not be undertaken without an understanding of technological dynamics.

CONCLUSION

The 4IR challenges us to re-think education and skills development, taking into account both global trends and our local context. In the South African context, distinct principles emerge. Through targeted education and skills development, the 4IR should be harnessed for inclusive economic development for black South Africans, including both black participation in the formal economy and black-led industrial development. In the broader African context, South Africa is in a position to lead in the area of leveraging 4IR-oriented education, and therefore to support African development.

In South Africa it is also critical that education reduces inequality. The digital divide still separates the learners who have access to the rich knowledge resources of the internet from those who do not – a deep divide that has implications that span decades into the future. The question is not simply one of access, of connecting schools to the internet. It is also a question of data justice: of reducing costs by allocating high-demand spectrum, and curtailing the oligopolistic pricing strategies of South Africa's powerful data providers, which lock a large proportion of South African learners out of meaningful digital access. In all this, gender disparities remain, and greater support is required for enabling digital access for female learners.

In South Africa, the notion of the 4IR has become widely used in the education landscape. The concept has informed the strategic orientations of government departments, universities and other actors. Investments have been made in new advisory structures, digital infrastructures, research activities and teaching and learning activities. Many new partnerships, including multi-stakeholder partnerships, have emerged over the last few years. South Africa is thus poised to leverage the possibilities that are offered by new strategies in education and skills development. Building interactive capabilities will remain critical if the institutions of skills supply and skills demand are to achieve greater alignment. Through such institution-building initiatives, South Africa has the potential to leverage its existing strengths to unlock development possibilities that are just, equitable, inclusive, transformative and supportive of human development. Appropriate policy measures, that strategically and dynamically respond to technological change, will also be required to meet these objectives. At the same time, addressing structural inequality is a precondition for achieving equitable development outcomes from technological change. Differences in access to life opportunities, including education and internet access, will also shape the manner in which advanced technologies can be deployed to enhance human development.

REFERENCES

Abu Mezied, A. 2016. 'What role will education play in the fourth Industrial Revolution?' *World Economic Forum.* https://www.weforum.org/agenda/2016/01/what-role-will-education-play-in-the-fourth-industrial-revolution/, accessed 4 August 2020.

Allais, S. 2012. 'Will skills save us? Rethinking the relationships between vocational education, skills development policies, and social policy in South Africa'. *International Journal of Educational Development*, 32(5), 632–642. 10.1016/j.ijedudev.2012.01.001.

Aulbur, W., Arvind, C.J. and Bigghe, R. 2016. 'Skill development for Industry 4.0'. *BRICS Council Skills Development Working Group.* http://www.globalskillsummit.com/whitepaper-summary.pdf.

Autor, D.H., Levy, F. and Murnane, R.J. 2003. 'The skill content of recent technological change: An empirical exploration'. *Quarterly Journal of Economics*, 118(4), 1279–1333.

Bedolla, J.S., D'Antonio, G. and Chiabert, P. 2017. 'A novel approach for teaching IT tools within learning factories'. *Procedia Manufacturing*, 9, 175–181.

Boyd, R. and Holton, R.J. 2018. 'Technology, innovation, employment and power: Does robotics and artificial intelligence really mean social transformation?'. *Journal of Sociology*, 54(3), 331–345.

Braidotti, R. 2013. *The Posthuman.* Hoboken: John Wiley and Sons.

Brynjolfsson, E. and McAfee, A., 2014. *The Second Machine Age: Work, Progress, and Prosperity in a Time of Brilliant Technologies.* New York: WW Norton & Company.

Business Tech. 24 April 2019. 'Government to launch new software engineering programme for South African schools'. *Business Tech*. https://businesstech.co.za/news/technology/312958/government-to-launch-new-software-engineering-programme-for-south-african-schools/, accessed 4 August 2020.

Butler-Adam, J. 2018. 'The Fourth Industrial Revolution and education'. *South African Journal of Science*, 114(5–6), 1–1.

Carr, N. 2014. *The Glass Cage: Automation and Us*. New York: Norton.

CEDA. 2015. 'Australia's future workforce'. *CEDA*. https://www.ceda.com.au/CEDA/media/ResearchCatalogueDocuments/Research%20and%20Policy/PDF/26792-Futureworkforce_June2015.pdf. accessed 19 April 2020

Chaim, O., Muschard, B., Cazarini, E. and Rozenfeld, H., 2018. 'Insertion of sustainability performance indicators in an industry 4.0 virtual learning environment. *Procedia Manufacturing*, 21, 446–453.

Cilliers, J. 2018. 'Made in Africa: Manufacturing and the fourth industrial revolution'. *ISS Africa Report,* 2018(8), 1–32.

Crosby, L. 2018. 'Skills development considerations for the post school sector in the midst of the 4IR'. *SA-EU Strategic Partnership Dialogue Conference on Disruptive technologies and public policy in the age of the Fourth Industrial Revolution*, http://www.hsrc.ac.za/uploads/pageContent/10155/4IR%20Framework%20Report_Final_lowres.pdf.

De Beer, J., Armstrong, C., Ellis, M. and Kraemer-Mbula, E. 2017. 'A scan of South Africa's maker movement'. *OpenAIR Working Paper 9*. http://openair.africa/wp-content/uploads/2018/11/WP-9-A-Scan-of-South-Africas-Maker-Movement.pdf.

De Falco, S. 2019. 'From Silicon Valley to Africa Valley: Which paradigms are needed in the transition from II to IV industrial revolution? Knowledge roadmap and technological track'. *Innovation: The European Journal of Social Science Research*, 1–29.

Department of Higher and Education and Training (DHET). 2019. 'Ministerial Task Team on the Fourth Industrial Revolution in Post-School Education and Training'. *DHET,* https://www.greengazette.co.za/notices/public-financial-management-act-1-1999-ministerial-task-team-on-the-fourth-industrial-revolution-in-postschool-education-and-training_20190607-GGN-42518-00893.pdf.

Dlulane B. 2019. 'Competition Commission: MTN & Vodacom Data prices are anti-poor, not sustainable'. *EyeWitness News*. https://ewn.co.za/2019/12/02/competition-commission-mtn-and-vodacom-data-prices-are-anti-poor-not-sustainable, accessed 4 August 2020.

Education Training and Development Practices Sector Education and Training Authority (ETDP SETA). 2020. 'The 4th Industrial Revolution. High level implications on the skills development in the ETD Sector'. *ETDP*

SETA. http://www.etdpseta.org.za/education/sites/default/files/2020-03/
Initial%20Report%20on%20Fourth%20Industrial%20Revolution_
UJ%20Research%20Chair.pdf. accessed 5 August 2020

Erol, S., Jäger, A., Hold, P., Ott, K. and Sihn, W. 2016. 'Tangible Industry
4.0: A scenario-based approach to learning for the future of production'.
Procedia CiRp, 54(1), 13–18.

Federal Ministry of Education and Research. 2018b. The background to
Plattform Industrie 4.0. Plattform Industrie 4.0. https://www.plattformi40.
de/I40/Navigation/EN/ThePlatform/PlattformIndustrie40/plattform-
industrie-40.html, accessed 5 August 2020.

Folb L. 18 August 2018. 'UWC to prepare students for the fourth industrial
revolution. *IOL.*

Gastrow, M. 2019. 'Policy options framework for the fourth industrial
revolution in South Africa: An output of the SA-EU Strategic Partnership
Dialogue on disruptive technologies and public policy in the age of
the Fourth Industrial Revolution'. http://www.hsrc.ac.za/uploads/
pageContent/10155/4IR%20Framework%20Report_Final_lowres.pdf,
accessed 27 November.

https://www.iol.co.za/capetimes/uwc-to-prepare-students-for-the-fourth-
industrial-revolution-16626554, accessed 5 August 2020.

Ford, M. 2015. *The Rise of the Robots.* New York: Basic Books.

Frey, C.B. and Osborne, M.A. 2017. 'The future of employment: How
susceptible are jobs to computerisation?'. *Technological Forecasting and
Social Change*, 114, 254–280.

Graetz, G. and Michaels, G. 2015. 'Robots at work: The impact on productivity
and jobs'. CEP Discussion Paper No 447, Centre for Economic
Performance, London School of Economics.

Grodotzki, J., Ortelt, T.R. and Tekkaya, A.E. 2018. 'Remote and virtual labs
for engineering education 4.0: Achievements of the ELLI project at the TU
Dortmund University'. *Procedia Manufacturing*, 26, 1349–1360.

Horowitz, M. and White-Burke, W. 19 January 2018. 'Academia can be an
important vanguard of the Fourth Industrial Revolution'. World Economic
Forum. https://www.weforum.org/agenda/2018/01/academia-vanguard-
fourth-industrial-revolution-4ir-universities/, accessed 19 April 2020.

Huws, U. 2014. *Labor in the global digital economy: The Cybertariat Comes
of Age.* New York: NYU Press.

Infosys. 2016. 'Amplifying human potential: Education and skills for the
Fourth Industrial Revolution'. Infosys. http://www.experienceinfosys.
com/humanpotential, accessed 19 April 2020.

International Labour Organization. 2013. 'Using technology foresights
for identifying future skills needs'. Skolkovoilo Global Workshop, The
Moscow School of Management SKOLKOVO's Education Development
Centre (SEDeC) – International Labour Organization, 24–36. http://

www.skolkovo.ru/public/media/documents/research/sedec/Global_Workshop_Proceedings_07_2014_Preview.pdf, accessed 19 April 2020.

International Labour Organization. 2017a. 'The future of work we want: A global dialogue'. International Labour Organization. http://www.ilo.org/global/topics/future-of-work/dialogue/WCMS_570282/lang--en/index.htm , accessed 19 April 2020.

International Labour Organization. 2017b. 'Inception report for the global commission on the future of work'. International Labour Organization. https://www.ilo.org/wcmsp5/groups/public/---dgreports/---cabinet/documents/publication/wcms_591502.pdf, accessed 19 April 2020.

International Labour Organisation. 2018a. 'ILO Global Commission tackles the changes needed for a fair future of work for everyone'. International Labour Organization. https://www.ilo.org/global/about-the-ilo/newsroom/news/WCMS_618190/lang--en/index.htm, accessed 19 April 2020.

International Labour Organization.2018b. 'The future of work centenary initiative'. International Labour Organization. https://www.ilo.org/wcmsp5/groups/public/---ed_norm/---relconf/documents/meetingdocument/wcms_369026.pdf, accessed 19 April 2020.

International Labour Organization. 2019. 'Work for a brighter future'. International Labour Organization. https://www.ilo.org/wcmsp5/groups/public/---dgreports/---cabinet/documents/publication/wcms_662410.pdf, accessed 19 April 2020.

IOL. 22 May 2019. 'Huawei South Africa launches free 5G training for ICT university students'. *IOL,* https://www.iol.co.za/business-report/technology/huawei-south-africa-launches-free-5g-training-for-ict-university-students-23740920, accessed 4 August 2020.

Kanza, E.S. 8 March 2016. 'In Africa, girls will lead the Fourth Industrial Revolution'. World Economic Forum. https://www.weforum.org/agenda/2016/03/in-africa-girls-will-lead-the-fourth-industrial-revolution/, accessed 19 April 2020.

Kupe T. 19 July 2019. 'Universities are key to 4IR employment'. *Mail & Guardian.* https://mg.co.za/article/2019-07-19-00-universities-are-key-to-4ir-employment, accessed 5 August 2020.

Lawrence, R. 2018. 'The future of manufacturing employment'. Centre for Development and Enterprise. *https://media.africaportal.org/documents/Robert-Lawrence-Report-WIP7.pdf, accessed 6 October 2020*

Ljungholm, P.D. 2018. 'Employee–employer relationships in the gig economy: Harmonizing and consolidating labor regulations and safety nets'. *Contemporary Readings in Law and Social Justice*, 10(1), 144–150.

Magwentshu, N., Rajagopaul, A., Chui, M. and Singh, A. 2019. 'The future of work in South Africa: Digitisation, productivity and job creation'. McKinsey and Company. https://www.bpesa.org.za/news/313-

mckinsey-report-how-can-south-africa-embrace-the-fourth-industrial-revolution-4ir.html.

Marwala, T. 2018. 'Education and skills development and the future of work'. *SA-EU Strategic Partnership Dialogue Conference on Disruptive technologies and public policy in the age of the Fourth Industrial Revolution.* Pretoria: HSRC

Menon, K. and Castrillon, G. 15 April 2019. 'Universities have "pivotal role" to play in Fourth Industrial Revolution'. *Daily Maverick.* https://www.dailymaverick.co.za/article/2019-04-15-universities-have-pivotal-role-to-play-in-fourth-industrial-revolution/, accessed 4 August 2020.

Millenial Project. 2019. 'Work/Technology 2050: Scenarios and Actions'. http://www.millennium-project.org/work-technology-2050-scenarios-and-actions-preface-introduction-executive-summary/, accessed 4 August 2020.

Mokyr, J., Vickers, C. and Ziebarth, N.L. 2015. 'The history of technological anxiety and the future of economic growth: Is this time different?'. *Journal of Economic Perspectives*, 29(3), 31–50.

Munck,R. 2013. 'The precariat: A view from the South'. *Third World Quarterly*, 34(5), 747–762.

Mundell, J. 2 October 2017. 'Africa's data revolution: Accelerating development through data-driven decision making'. On Africa. https://www.inonafrica.com/2017/10/02/africas-data-revolution-accelerating-development-data-driven-decision-making/, accessed 19 April 2020.

Mzekandaba, S. 22 March 2019. 'SA universities must lead Industry 4.0 curriculum'. *ITWeb.* https://www.itweb.co.za/content/PmxVE7KXLopMQY85, accessed 4 August 2020.

Naudé, W. 2017. 'Entrepreneurship, education and the fourth industrial revolution in Africa. Revolution in Africa'. *IZA Discussion Papers, No. 10855, Institute of Labor Economics (IZA).*

Ohene-Afoakwa, E. and Nyanhongo, S. 2017. 'Banking in Africa: Strategies and systems for the banking industry to win in the Fourth Industrial Revolution'. *Bank SETA.* https://www.bankseta.org.za/wp-content/uploads/2018/08/BA3DD51-1.pdf.

Penprase, B.E. 2018. *The Fourth Industrial Revolution and Higher Education.* London: Palgrave Macmillan.

Rivette U. 1 August 2019. 'Upskilling learners for future success'. www.sit.uct.ac.za/sit/news/2019/coding, accessed 4 August 2020.

Rosman B. 25 March 2019. 'Africa cannot afford to take the back seat in one of the most important pursuits of modern science'. https://www.wits.ac.za/news/latest-news/opinion/2019/2019-03/africa-cannot-afford-to-take-the-back-seat-in-one-of-the-most-important-pursuits-of-modern-science.html, accessed 5 August 2020.

Schwab, K. 2016. *The Fourth Industrial Revolution.* United Kingdom:

Portfolio Penguin.

Smith, S.L. and Neupane, S. 2018. 'Artificial intelligence and human development: Toward a research agenda'. IDRC White Paper.

Tshwane University of Technology (TUT). 20 June 2019. 'University signs R30m partnership with MerSeta. *TUT News.* https://www.tut.ac.za/news-and-press/article?NID=269, accessed 5 August 2020.

United Nations Economic Commission for Africa (UNECA), United Nations Development Programme (UNDP), International Development Research Centre of Canada and World Wide Web Foundation. 2016. 'The Africa Data Revolution Report: Highlighting developments in African data ecosystems'. *ECA Publishing Unit.* https://repository.uneca.org/bitstream/handle/10855/23716/b11832459.pdf?sequence=1, accessed 19 April 2020.

University of Johannesburg. 1 October 2019. 'UJ congratulates newly appointed NRF Research Chairs'. https://www.uj.ac.za/newandevents/Pages/Prof-Salim-Vally-awarded-NRF-Research-Chair-in-Community-and-Worker-Education.aspx, accessed 5 August 2020.

University of the Witwatersrand (Wits). 2 July 2019. 'Futureproof your career at the University of Witwatersrand'. *Study international.* https://www.studyinternational.com/news/futureproof-your-career-at-the-university-of-witwatersrand/.

Von Tunzelman, N. and Wang, Q. 2007. 'Capabilities and production theory'. *Structural Change and Economic Dynamics*, 18(2), 192–211.

World Economic Forum (WEF). 2016. 'The future of jobs: Employment, skills and workforce strategy for the Fourth Industrial Revolution'. World Economic Forum. http://www3.weforum.org/docs/WEF_Future_of_Jobs.pdf.

World Economic Forum (WEF). 2017. 'The future of jobs and skills in Africa" Preparing the region for the Fourth Industrial Revolution'. World Economic Forum. http://www3.weforum.org/docs/WEF_EGW_FOJ_Africa.pdf.

Xing, B. and Marwala, T. 2017. 'Implications of the Fourth Industrial Age for higher education'. *The Thinker,* 73(3). https://ssrn.com/abstract=3225331, accessed 5 August 2020

The impact of adopting 4IR-related technologies on employment and skills: The case of automotive and mining equipment manufacturers in South Africa[1]

EDWARD LORENZ AND
ERIKA KRAEMER-MBULA

INTRODUCTION

Disruptive technologies of the Fourth Industrial Revolution (4IR), including advanced robotics, Artificial Intelligence (AI) and big data analytics, the Internet of Things (IoT), smart sensors and 3D printing have begun to transform global manufacturing, including in South Africa. Such technologies are expected to result in not only greater

1 This work is based on research supported by the National Research Foundation of South Africa (NRF) (Grant Numbers 118873 and 110691). Opinions, findings and conclusions or recommendations expressed in any publication generated by the NRF-supported research are those of the authors alone and should not be attributed to the NRF.

efficiencies, leading to improved competitiveness, but also to trigger changes in traditional production relationships among suppliers, producers and customers (Calitz et al., 2017). Furthermore, by facilitating more efficient energy use (as discussed by Ting in chapter 11), 4IR technologies could contribute to more sustainable patterns of production and consumption.

In parallel to these opportunities, great concern has been raised about the possible negative impacts of 4IR technologies on employment, as the scope for replacing the tasks of workers with robots, AI and other technologies increases. 4IR technologies are expected to profoundly change the skills mix of different occupations as an increasing array of both cognitive and manual routine tasks can be automated. Employees of the future are expected to develop transversal skills and competencies, adapted to the digitisation of production and the use of big data, as a complement to their more domain-specific technical skills. Such expectations take place against a background of deficient education and training systems (as discussed by Gastrow in chapter 5) and weaknesses in digital infrastructures (as discussed by Gillwald in chapter 1) in South Africa.

All of these have important implications for industrial development in the country. However, the adoption of these technologies is neither easy nor straightforward. Firms experience multiple difficulties in incorporating these technologies into their production processes. This is particularly the case in Africa which faces challenges of poor infrastructure and low skills levels (AfDB, 2019). The adoption of 4IR-related technologies in developing countries is far from widespread but is currently concentrated in a few large firms and a few advanced manufacturing activities (UNIDO, 2019). Opportunities presented by the new technologies and their impact on employment and skills are expected to differ according to the specific technology, as well as by sector and occupation. The adoption of these technologies is concentrated in a few firms and sectors, which raises the possibility of a polarised labour market, and also of wider concerns about the impact of 4IR technologies on national distributions of income and wealth.

While there is now a substantial literature on the 4IR, there is very little quantitative or qualitative evidence on the extent of the adoption of 4IR technologies by manufacturing firms and their impacts on employment and skills. We also know very little about the differences between firms in particular sectors. This lack of evidence at the firm-level hinders the design of appropriate policies for the inclusive adoption of 4IR technologies.

In this chapter, we contribute to filling this knowledge gap in South Africa by presenting first-hand evidence collected through enterprise case studies in two advanced manufacturing sectors: the automotive and the mining equipment manufacturing sectors.[2] The automotive sector was chosen due to the fact that it is the largest consumer of industrial robots[3] (IFR, 2019) and its importance to South Africa's industrial development. The manufacturing of mining equipment was selected due the expertise and technological capabilities that South Africa has developed over the years in mining-related supply industries (Kaplan, 2011); mining being the historic bedrock of the country's economy. We explore the patterns of adoption of 4IR-related technologies by firms in these two sectors and we consider the impact of their adoption on employment and skills needs. Our findings point not only to the importance of sector-specific factors, including the degree of integration of enterprises into global value chains (GVCs), but also to heterogeneity among firms within the same sector linked to both scale of production and degree of product standardisation.

The basic research questions guiding this chapter are:

- Which 4IR technologies are being adopted in the automotive and mining equipment sectors, and what are the main constraints employers face in their adoption?
- How are employment and skills being impacted?
- What are the main strategies that firms are using to meet their evolving skills needs?

---

2 The evidence presented in this chapter is based on firm interviews conducted between June and August 2019 by six senior and junior researchers attached to the DST/NRF/Newton Fund Trilateral Chair in Transformative Innovation, the 4IR and Sustainable Development and the SARChI Industrial Development, both at the University of Johannesburg.
3 According to the IFR, the automotive industry is the largest adopter of industrial robots, accounting for 30% of sales in 2018 (IFR, 2019).

To address these questions, the chapter is structured as follows: the next section discusses the importance of paying attention to the firm level, in order to understand the complexities of technology adoption and skills needs. The third section briefly describes the methodology and the nature of the evidence collected in this study. The fourth section summarises the findings of the study on the patterns of adoption of 4IR-related technologies in the two sectors, and the observed impacts on employment and skills. The final section concludes with some reflections on the findings in the context of South Africa and identifies avenues for appropriate policy responses.

THE 4IR AND ENTERPRISE HETEROGENEITY

There is a burgeoning literature focusing on the impact of new or 'disruptive' technologies on employment and skills.[4] While most of the empirical research has focused on developed countries, there have been concerns about the need to address the specific conditions faced in developing nations.[5] As discussed below, a striking feature of most of the recent empirical literature is that it has explored the impact of new technologies on employment and skills with little attention to what takes place inside the enterprise. This is, to a large extent, due to the lack of firm-level data on the adoption and impact of 4IR technologies.

There is more information available on the adoption of industrial robots than other 4IR technologies, including machine learning, big data analytics, 3D printing, the use of sensors and the Internet of Things (IoT). The main source of data on robots are the figures on purchases or installations collected by the International Federation of Robotics (IFR) at industry and country levels.[6] This data has been collected since 1990 and has been used in econometric and other quantitative studies to estimate the impact of robots on employment and productivity at national or subnational levels, including some developing nations (cf

---

4   For an overview of the literature, see Lorenz et al., (2019a).
5   See UNCTAD (2017) and UNIDO (2019) and World Bank (2016).
6   For a summary of the figures on world installations of industrial robots, see https://ifr.org/industrial-robots.

Carbonero et al., 2020; Giuntella, 2019; Graetz and Michaels, 2018). Data at this aggregated level makes it possible to identify the effects of the use of robots on employment and productivity at the level of sectors and nations. However, such data cannot be used to investigate the constraints that firms encounter in adopting these technologies; the differences in adoption across firms in the same sector, or the specific conditions under which robots may complement or substitute labour. This realisation has triggered calls to gather further data at the firm level (Seamans and Raj, 2018).

Some firm-level survey data on the adoption of industrial robots is available for developed countries. The main source of data for the European Union is that collected through the European Manufacturing Survey (EMS), which is limited to selected European countries for the period from 2001 to 2015.[7] The latest published EMS results, which are from the 2012, are instructive in that they identify considerable firm-level heterogeneity in the uptake of industrial robots, with higher adoption rates observed for larger firms engaged in producing larger batches of standardised products. A few national-level studies using combinations of customs data on robot imports, specialised surveys or evidence compiled from robot suppliers have investigated the impact of robots on employment and skills. These studies confirm the results from research based on the EMS showing considerable firm heterogeneity, with higher adoption rates for larger firms.[8]

Concerning other 4IR technologies, at present, to our knowledge, there are no firm-level surveys for either developed or developing countries measuring the adoption of advanced digital technologies based on the use of AI in the form of Machine Learning (ML) and big

---

7   The survey covers manufacturing firms in Spain, France, Germany, Austria, Sweden, Switzerland and the Netherlands. See: https://www.isi.fraunhofer. de/en/themen/industrielle-wettbewerbsfaehigkeit/fems.html. In addition, the 2018 EUROSTAT Community Survey on ICT Usage and E-commerce in Enterprises includes a variable measuring the adoption of robots by industrial enterprises. See https://ec.europa.eu/eurostat/documents/341889/10082348/ Enterprise_survey_variables.pdf.
8   See Acemoglu et al. (2019) for the case of France; Humlum (2019) for Denmark; and Morikawa (2020) for Japan.

data.[9] Most of the empirical research on ML and big data to date focuses on sector and national levels and is highly speculative in nature. This chapter seeks to predict the future impact of the adoption of ML on the basis of expert assessments of the scientific or technical susceptibility of occupational tasks to automation. The seminal research paper that sets out to do this is by Frey and Osborne (2013) who came up with the alarmist prediction that 47 per cent of jobs in the US are at high risk of automation. Frey and Osborne use the general descriptors of the knowledge, skills and abilities required for occupations provided by O*NET, the successor to the US Dictionary of Occupational Titles, as a basis for estimating the risk of automation for the 702 detailed occupations covered in the O*NET. These at-risk estimates are then mapped onto the occupational structure of the US economy and that of other countries, using the data provided by national labour force surveys[10] on occupational breakdowns by sector.

An important limitation of Frey and Osborne's approach, as pointed out by Arntz et al. (2016), is that it assumes that the task content of occupations is the same across all firms in the same sector, and across countries. All possible heterogeneity that may be linked to firm-specific characteristics, such as scale or production, are necessarily obscured. Moreover, since technical susceptibility to automation is different from what individual firms are interested in or capable of pursuing, the analysis fails to identify the constraints firms may face, including those linked to the national context. These constraints can result in the slow or limited adoption of new 4IR technologies.

These limitations are equally clear in extrapolations of Frey and Osborne's approach to the quite different conditions prevailing in developing countries. This is done in the World Bank's 2016 World Development Report, which calculates on the basis of Frey and

---

9   Researchers at the University of Aalborg in Denmark completed in 2019 the first employee-level skills survey including measures of the use of artificial intelligence. This survey provides evidence on the rate of adoption of AI in the Danish national labour force and the impacts on skills. See Holm et al. (2020) for an overview.

10 The application of the Frey and Osborne approach to several EU member countries arrived at comparably alarming rates of predicted job loss. See Bowles (2016).

Osborne's at-risk estimates that two-thirds of all jobs are in whole or in part susceptible to automation in the developing world. Drawing on the work of Comin and Mestieri (2014) on technology adoption lags, these estimates are moderated down to around 40 per cent of all jobs (World Bank, 2016: 129). What is glossed over in the World Bank report, however, is that the basic conclusion of Comin and Mestieri is that there has been a convergence across countries in terms of their initial adoption of new technologies; subsequently, most of the differences between the developed and developing world are attributable to differences in the intensity of use of new technologies by enterprises within developing countries. This implies considerable heterogeneity between a few leading firms in developing countries that may be close to the technological frontier, and the majority of firms that are slow to adopt the new technology. Comin and Mestieri (2014), however, are unable to account for these differences in the intensive margin of use.[11]

To our knowledge, the only recent body of research that has made progress in addressing these limitations in the research on the 4IR's impact on employment and skills in developing countries is that based on enterprise-level surveys of selected manufacturing sectors in Thailand, Vietnam, Brazil, Argentina and Ghana. This research was coordinated by United Nations Industrial Development Organization (UNIDO) and presented in its 2019 Industrial Development Report (UNIDO, 2019). It rightly observed that advanced digital technologies are not adopted by countries or by sectors but by individual firms and, in this respect, the starting point for the UNIDO research is similar to that presented in this paper on South Africa. This implies that the context in which the firm operates is central to understanding behaviours and outcomes.

The UNIDO surveys focus on capturing what stage the firm is at in adopting new technologies based on a typology of four generations

---

11 In their most recent publication, Comin and Mestieri (2018) leave this as a question to be addressed in future research. Somewhat surprisingly they dismiss institutional differences as a possible explanation, observing that the quality of institutions has converged over the last 200 years.

of digital technologies.[12] These surveys do not provide measures of the frequency with which specific technologies, such as advanced robotics or machine learning, are adopted; rather they measure the frequency of adoption of different models or generations. The models are distinguished by the number of functions within an enterprise that are integrated into its digital systems, and in the case of the fourth or highest level generation, there is the additional requirement of using feedback for decision-making, possibly based on the use of big data and AI. One of the key findings is that there is considerable heterogeneity with larger firms, in general, being more advanced in the adoption of digital technologies. Further, the report finds that only a handful of enterprises use the most advanced technologies ranging from 1.5 per cent in Ghana to 30 per cent in Brazil (UNIDO, 2019: 100). The survey also identifies a lack of finance and the poor quality of ICT infrastructure as important obstacles to firms' adoption of advanced digital technologies.

METHODOLOGY AND DATA COLLECTION

The evidence presented in this chapter is based on eight firm-level case studies in two advanced manufacturing sectors in South Africa: automotive, and mining machinery and equipment. The data was collected by six senior and junior researchers as part of a collaborative initiative between the DST/NRF/Newton Fund Trilateral Chair in Transformative Innovation and the 4IR and Sustainable Development and the DST/NRF South African Research Chair in Industrial Development (SARChI), both based at the College of Business and Economics at the University of Johannesburg (UJ).[13]

The case study methodology was adopted as a suitable approach for

---

12 The sample frames which are stratified in all case by sector and size are limited with the exception of Argentina to firms with 20 employees and over and in the case of Brazil to firms with 100 employees and over. For details, see UNIDO 2019, Appendix 3.

13 This activity was part of an NRF-funded project called 'Community of Practice in Innovation and Inclusive Industrialisation', hosted by the SARCHI Industrial Development. For a detailed presentation of the individual case study firms, see Lorenz et al. (2019b).

investigating not only which technologies are being adopted, but also why and how they are being adopted, and their impact on employment and skills development.[14] The data collection took place between June and August 2019, and covered companies situated in four South African provinces: the Eastern Cape, Gauteng, KwaZulu-Natal and the Western Cape. The study used non-probabilistic sampling techniques, suitable for qualitative research, based on purposive sampling and snowball sampling. Initial potential respondents were identified through the websites of the following organisations: the National Association of Automobile Manufacturers of South Africa (NAAMSA), which represents motor vehicle manufacturers and importers in South Africa; the National Association of Automotive Component and Allied Manufacturers (NAACAM), which represents the suppliers to the original equipment manufacturers (OEMs) in South Africa as well as suppliers to overseas OE markets; the Mining Equipment Manufacturers of South Africa (MEMSA) and the Construction and Mining Equipment Suppliers' Association (CONMESA). These contacts were expanded through the recommendations and suggestions of respondents. The total number of participating firms was limited by respondents' availability and willingness to participate within the time and resources available for this study.

Data collection took place through the use of semi-structured interviews, which included both closed and open questions. While closed questions allow for making comparisons across firms, the open-ended questions give respondents an opportunity to talk about their experiences and raise new issues (Dearnley, 2005). The semi-structured interviews covered the following aspects:

- questions concerning the costs and benefits of new technologies and the obstacles faced by firms in adopting them;
- questions related to the performance characteristics of new technologies including new uses of data and how their adoption impacts on changing skill requirements;
- questions related to the impact of the adoption of new technologies

---

14 See Yin (2009) for a comparison of research methodologies and the suitability of case studies for investigating the 'how' and why' questions.

on the overall employment needs of the firm, as well as the work activities of employees; and

- questions about existing skills gaps and the difficulties faced in finding required skills in the labour market.

A decision was made to interview respondents at various occupational levels in order to collect information at the level most suitable for particular research queries. While an effort was made in the case of the auto and equipment producers to conduct interviews with both the CEO or an upper manager and persons responsible for shop floor production operations, as Table 6.1 indicates below, in several cases it only proved possible to conduct a single interview with the CEO or an upper-level manager. Access to a team leader or shop floor supervisor was provided in only a few cases. Fourteen semi-structured interviews were conducted, each lasting between 1 and 1.5 hours with various categories of employees.

The interviews, which took place face-to-face, were recorded with the consent of respondents and then transcribed by a professional company. Researchers in the project analysed the quotes from participants' transcribed text, identifying key insights within three broad themes of (a) adoption of new technologies; (b) impacts of adoption on employment; and (c) impacts on skills and training.[15]

---

15 For a detailed presentation of the individual case study firms, see Lorenz et al. (2019b).

Table 6.1: Firm interviews

Automotive			
	Main products or services	No of interviews	Occupational levels of interviewees
OEM 1	Assembler for MNC auto producer	3	Managing Director; Head of Production; Production Engineer
OEM 2	Assembler for MNC auto producer	1	Director: Manufacturing Operations
OEM 3	Assembler for MNC auto producer	1	Director: Purchasing and Supplies
Components Supplier 1	Press shop and sub-assemblies for body shop assembly	3	Production/ Automation Engineer; Director HRM; Shop-floor supervisor
Components Supplier 2	Suspension components	3	Managing Director; Plant Manager
Mining Equipment Industry			
Producer 1	Variety of specialised equipment including drills, dumpers and roofbolters	1	CEO
Producer 2	Range of processing equipment including screening and scrubber linings	1	Engineering Manager
Producer 3	Range of specialised equipment including pumps, cutters and pinch bars	1	Sales Director

Source: authors

In the cases of OEM 1 and OEM 2, interviews were carried out with personnel directly responsible for production operations and the interviews provided detailed information on the adoption and use of new technologies, including robots, 3D printing, virtual reality and data analytics. The two respondents at OEM 3 were responsible for purchasing and while the joint interview provided only limited information on shop floor technology and production, it did provide useful information on supply chain management and the technological preparedness of suppliers in South Africa. Multiple interviews were carried out at Component Suppliers 1 and 2, including with personnel responsible for engineering and production operations. In the case of mining equipment manufacturing firms, a single interview was carried out in each firm: the CEO in the case of Producer 1; the Engineering Manager in the case of Producer 2 and the Sales Director for Producer 3.

ADOPTION OF 4IR TECHNOLOGIES AND IMPACT ON EMPLOYMENT AND SKILLS

Automotive sector analysis

While small in international terms, in 2018 the automotive sector accounted for over 7 per cent of GDP and slightly less than 30 per cent of manufacturing gross value added (Jordaan et al., 2018). Data collected by the International Federation of Robotics shows that the auto industry globally accounts for about 30 per cent of industrial robot installations.[16] The choice of the auto sector for the study was also made to examine the impact of the adoption of advanced robotics on employment and skills in the context of a large-scale or mass production industry.

The automotive industry has a global oligopolistic structure with a small number of multinational corporation (MNC) assemblers developing and controlling the key technologies and keeping control over major component suppliers, many of which are also MNCs, through 'follower' or 'lead' sourcing governance arrangements (see Barnes et al., 2016, for a discussion). The South African auto industry is

---

16 See https://ifr.org/industrial-robots.

dominated by seven multinational OEMs with Ford, Toyota, Mercedes-Benz, BMW, VW, Isuzu and Nissan operating plants in three provinces: Gauteng, KwaZulu-Natal and the Eastern Cape. The total volume of vehicles produced annually is around 600,000, consisting principally of light passenger and commercial vehicles. As elsewhere, the South African industry is characterised by complex vertical links between auto assemblers and their component supply networks, estimated to include about 500 first- and second-tier suppliers producing primarily for the OEMs and consisting of a mix of foreign and domestically owned firms. In terms of employment, in 2018 the sector employed about 120,000, including component parts producers, with the OEMs accounting for around 30 per cent of the total.[17]

The relationships between the OEMs and the local supplier base have been, and remain, central in the design of support policies for the auto industry in South Africa. Between 1961 and 1995 the industry grew substantially with the support of local content programmes within the framework of import substitution. These programmes were designed to develop local supply chain linkages between the OEMs and local suppliers through a combination of tariffs, import permits and local content requirements. A factor shaping the unique development of the South African industry was that in the 1980s, until sanctions were lifted, the industry was partly isolated from the global industry due to sanctions against apartheid. Ford and General Motors disinvested, Toyota and Nissan operated on the basis of franchising, while VW and BMW remained fully foreign-owned subsidiaries and Mercedes retained a 50 per cent ownership. International isolation combined with import substitution policies based on local content requirements resulted in the development of a large and diverse component supply industry, often operating with local technology and buffered to a great extent from international competition (Barnes and Morris, 2008).

The post-apartheid Motor Industry Development Programme (MIDP) established in 1995 was designed to promote the integration of the South African industry into the global auto industry and to

---

17 Based on the figures of the NAAMSA and the NAACAM. See Barns et al. 2018, Table 5.

increase competitiveness, including that of the local component suppliers. The programme removed all local content requirements, reduced tariffs and used import-export complementation, allowing both assemblers and component manufacturers to earn import-duty credits from exporting. A major consequence of the reintegration of the industry into the global auto industry was significant changes in ownership structure and governance, with Ford, Nissan and Toyota acquiring majority ownership in local assembly plants and increasing their foreign direct investment (FDI) in the component supplier industry, often in the form of joint ventures. Component suppliers were required to produce to the international standards set by the OEMs thus reducing the space for local indigenous technology development (Barnes and Morris, 2008).

The MIDP was replaced by the Automotive Production Development Programme (APDP)[18] in 2013. Partly with the objective of aligning policies with the WTO on exports, it emphasised promoting volume production through a volume assembly allowance (VAA) for OEMs and a production incentive (PI) for OEMs and component manufacturers. The plan also included an auto investment scheme (AIS) providing cash grants for qualified capital investments. While the auto industry has achieved growth under the APDP, there is little evidence that it has increased the share of local content in OEM vehicle production (Barnes et al., 2017; 2018).

The industry today is structured in a classic producer-driven global value chain with local OEMs producing car models and using assembly-line technologies that are designed in the R&D facilities of developed country MNCs and then transferred to South African production sites, who compete with other plants around the world for the production of new models. While this structure arguably provides the basis for a continuous upgrading of local technological and production capabilities of the OEMs and their first-tier component suppliers, it also leaves the industry in a position of high dependence on the strategic decision-making of the MNCs dominating these value

---

18 See: https://naacam.org.za/wp-content/uploads/pdf/page18.pdf, accessed 27 August 2020.

chains, including on issues of technology transfer and the location of production capacity.

Summary of findings for the auto sector

Site visits and interviews were carried out at three automobile OEMs and two first-tier component suppliers. The three OEMs are the South African operations of MNC automotive producers and all specialise in the production of light passenger vehicles. Two of the OEMs also produce light commercial vehicles. For each of the three OEMs, the volume of production has increased significantly over the previous decade with output at present ranging between 140,000 and 168,000 units per year, with an objective of attaining 200,000. While a significant part of OEMs' production is for the export market the exact percentages were not provided.[19]

The two components suppliers interviewed present significant differences in terms of their production characteristics and the use of automation technologies. Supplier 1 is the South African operations of an MNC producer of automotive components. The firm supplies directly to the OEMs in South Africa including BMW, Toyota and Ford. It produces standardised components including car doors, bonnets and tailgates. Its production capacity has increased in step with the OEMs, with the daily number of units fabricated increasing from about 120 to 700 over the last decade. They have three plants at the local site and they employ about 780 people. Supplier 2 fabricates relatively specialised parts in smaller volumes, including coil springs, torsion bars and stabiliser bars, all suspension components for the automotive industry. It employs about 500 persons and supplies to the domestic and international market.

Adoption of 4IR-related technologies

The OEMs were often motivated by the parent company, which seeks to standardise technology worldwide (this is the case for OEM 1

---

19 According to the figures provided by the Automotive Industry Export Council (2019), the industry exported about 60 per cent of light vehicle production in 2018.

and OEM 2), so identical car bodies can be produced internationally regardless of location. Closely related to this is the fact that volume across sites in different countries is allocated centrally by the parent company using costs as the main criterion, so there are strong incentives to upgrade and improve productivity continuously and this has a bearing, especially on the adoption of robots. For instance, OEM 1 alluded to the competition with other production sites abroad, which drives firms' decisions on the adoption of cost-reduction technologies, highlighting the important role of government support programmes in maintaining competitiveness in the sector (Interview: Managing Director, OEM 1):

> The point is we need to compete with our sister plant [abroad]. And we get tracked down to the last cent whether the vehicle here is cheaper or more expensive than [abroad] and our big advantage right now still is that we are still cheaper because of the program, the APDP program that allows us to have a favourable cost position.

Each of the OEMs has increased the use of industrial robots in recent years. The reasons stated were to increase productivity and to produce a larger number of units annually allocated by the parent companies, as well as greater geometrical accuracy. Robots are concentrated in the body shop assembly area, with OEM 1 using over 200 and OEM 2 about 300.

The view was expressed that robots are not new technology, even if there have been improvements. What is new is the interconnectivity, with the robot being connected via sensors to a system for monitoring each workstation. Another novelty identified is initial experiments with cobots or collaborative robots which are relatively flexible, compared to older large industrial robots, and can be operated and easily reprogrammed by skilled operators. OEM 1 has a few cobots in use for pick-and-carry-type operations and OEM 2 referred to the use of cobots for ergonomic reasons to assist shop floor works with lifting heavy components for manual welding. The Director of Production at OEM 1 saw cobots as being the wave of the future and stated that he expected that their investments in cobots would increase over the coming years.

Closely related to the discussion around the use of robots versus manual methods was the issue of flexibility and rigidity in production and the limits of automation. One distinction referred to was between the standardised body shop assembly operations which are highly amenable to automation and the final fitting and finishing where task variation is greater and human operators are needed to cope with the constant changes. The Vice-President of manufacturing operations at OEM 2 said that to fully automate with robots, you would need thousands of sensors, as well as programming thousands of variations and possibilities, within a short period such as 90 seconds (Interview: VP, OEM 2):

> The operator always needs to start at Operation 1 and he finishes at Operation 12 in his cycle of 90 seconds per vehicle. But within those 12 steps, there's always variability; you will never replace that with a robot. That will not be humanly possible. Not in our lifetime. […] I need the human mind to say this carpet is now in my way, push it away, before I can […] clip the harness in. Someone would say well you can put sensors all over the place and […] visual. You would need thousands! […]

In his view this will not happen as the use of humans is more efficient. A similar point was made by the Director of Production at OEM 1 who expressed the view that there is 'big hype' on full automation and that automation limits flexibility because the technical efforts and maintenance around it can be greater than the gains. He referred to a 'sweet spot' with a mix of automation and human labour (Interview: DP, OEM 1):

> I would say that automation limits your flexibility. […] There's in my view […] a sweet spot where the efforts and the technical efforts and the maintenance around it is, is heavier than what you gain.

This resonated with the component suppliers as well, where a plant manager at Supplier 2 stated (Interview: Plant Manager, Supplier 2):

There's an optimum between the person and the automation. You know, if you go 100% automation, it's not as efficient or as cost-effective, […] if you find the right balance, where you can have the automation and the operator as well.

In the component suppliers, the use of robots was more limited than in OEMs, although there were some differences between them. Supplier 1's use of robots was largely driven by its current production for BMW, which had recently brought out a new model, requiring an upgrade in skills. The production engineer interviewed noted that since BMW in Germany designs the new cars and the assembly lines to produce them, the technology they use conforms to BMW's standards. To meet these standards, over the last decade in one plant they replaced manual welding with welding by about 50 robots, and it was emphasised that using robots allows both for higher production speeds and for superior quality and consistency of welding, especially on aluminium which is being used in place of steel on the new models. Supplier 1 has recently introduced laser welding, which is optimal for aluminium, for certain welding operations and this is necessarily carried out by robots since it is done within a fully enclosed space and the operator monitors the process from outside using a camera.

Unlike Supplier 1, Supplier 2 makes little use of industrial robots, with five robots in one plant and a semi-automated technology operating in another. Several reasons were given to explain this, including the lack of standardisation of the tasks and the complexity of some of the production processes, which require human perceptual skills and observation to identify small faults in the components and to make the necessary corrections manually. The point was also made that some of the automation solutions they would like to implement are simply not available among local service providers, although they are in contact with several companies locally to get new ideas.

This contrasts with Supplier 1 where the parent organisations of BMW and the other OEMs have been driving the automation process and providing the necessary support and training to implement it locally. Component supplier 2 referred to working with an automation service provider over three years but noted that it never worked fully because the provider did not understand the firm's requirements.

Besides robots, other new technologies used in the auto sector include 3D printing for rapid prototyping of specialised parts, mostly by the OEMs. The interviewees at OEMs 2 and 3 said 3D printers are used to produce a part for the purpose of checking the accuracy of component supplies, rather than large batch production because of the high costs of 3D printing. 3D scanning using CAD (computer aided design) is passed on to suppliers showing precisely what is needed, with no margins for variations. What is provided by the supplier is compared to the CAD, and the result is evaluated. In the case of OEM 3 the results from such evaluations are entered into a global data system, which generates a report of suppliers. If there are inconsistencies, points are deducted and if the points reach a certain level, the supplier drops from first-tier to second-tier. Interestingly, component suppliers see little use for 3D printing for production of larger parts due to the slow speed of the technology.

Data management and use

The evidence shows increasing use of data in production in the auto sector. Data is automatically collected by all the OEMs on machine operations and this is a fundamental change from 10 to 15 years ago when everything was on paper. At the time of the interviews, OEM 1 had a mix of stand-alone machines, from which data is recorded manually, and cases where the data is automatically integrated into a collective network system. OEM 2 stated that its system is integrated worldwide and tracks and monitors the machines in real-time. The firm also has a global logistics tracking system that tracks the parts and components sent to the unit. This facilitates the logistics flow and allows for the use of virtual reality to visualise each container to be packed and for changes to quantities of materials to be made from a distance. Although there was an awareness of the potential for using data collected from sensors to make use of deep learning with neural nets for purposes of predictive maintenance, none of the three OEMs at present are doing this.

Both component suppliers have tracking systems for parts quality data management and are able to trace a part back to the exact date and shift that it was built, by the use of tags. However, some of the more complex operations still require manual quality controls and some manual data analysis. Supplier 2 has a standard supervisory control and data acquisition (SCADA) system comprised of systems of software and hardware that provide the operators with data on machine malfunctioning and the data needed to monitor output efficiency.

Employment, training and skills

In the auto sector, there was clear evidence that robots and related automation technologies for assembly lines are significantly reducing labour requirements per unit of output. Employment at the OEMs, however, has not gone down and in some cases has even increased, due to the increased volume of production over the previous decade. In the cases where robots have made staff redundant, the interviewed companies displayed a preference to relocate staff across the firm where possible, to prevent job losses. In the case of components suppliers, Supplier 2 indicated that despite having identified opportunities for automation, they deliberately kept some of their plants labour-intensive, as a commitment to preserve jobs for low-skilled personnel from the surrounding communities. The plant manager of its least automated plant indicated (Interview: Plant Manager, Supplier 2):

> I have no robotics at all in my plant. [...] one of the core principles is to generate employment. So I could, [...] take out quite a lot of my guys and put robots in their place, but it goes against the company's principles and policies. [...] So we try as far as possible to use human labour, so we can keep the employment. [...] A lot of people in my plant are very, very old. They've been doing the same thing for 15, 20 years, and they come from the local surrounding areas, rural areas. However, in the last year or so we've been getting more students, younger ones.

The interviewees at each of the OEMs identified specific technical skills that are difficult to fill through recruitment, specifically maintenance engineers, robot programmers and mechatronics. But they also referred to weak problem-solving skills at the intermediate level and weak foundation skills at the lower levels. The companies do in-house training operating to the MNC parent firm's training standards, sometimes using external providers. Additional training needs are linked to the technical upgrading typically associated with the introduction of new models, for example, laser welding and networked body shop assembly technology. The general pattern described is one of skills upgrading, which is closely tied to the changes in materials and methods introduced with each new model cycle.

Concerning suppliers, the interviewees at OEM 3 said that the difficulties experienced by the local suppliers in keeping up with new skills requirements are more acute for the smaller firms that do not have a technological link to global suppliers or the mother companies. In their view, this segment of the supply base will struggle to survive in the current environment. Among the key technical skills that the smaller suppliers lack are mechatronics and programming for programmable logic controller (PLC) systems used to make decisions based on a custom programme to control robots. Interviewees said that only a few local suppliers had these capabilities.

The Process Engineer at Supplier 1, echoing the views of the OEMs, observed that implementing new production lines with innovative technologies and materials required upgrading of skills and that there was a need to learn new methods, including the programming for the PLC systems for the robots. While most of the training for upper-level engineering skills can be carried out in South Africa, a few personnel were sent to the parent firms in Europe that had developed the relevant technology. This was notably the case for laser welding, which is new for South Africa. It was also the case more generally for programming and maintenance of the automated assembly line technology; interviewees said this cannot be done in South Africa.

In terms of the supply of mid-level technical skills, interviewees at both Suppliers 1 and 2 said that local technical institutes are adequate. The problem was that new recruits lack the ability to put the skills gained there to use to solve problems in a practical work setting. Very few people coming out of the vocational educational system have this sort of hands-on problem-solving skill, which depends in part on practical work experience. The suppliers bring these sorts of skills up to the desired level with in-house training.

Regarding lower-level skills, one of the main trends observed in the component suppliers is a reduction in the number of people engaged in work involving the lifting and moving of components or feeding them into a machine for processing. For the remaining unskilled personnel, there is first a need for greater speed since the pace of work is now determined by the automated production lines. Second, while these employees do not need programming or maintenance skills, they need to know how to monitor equipment and read the data output from the system to identify problems and to notify a technician if needed. Supplier 1 interviewees indicated that when a new person is recruited, they first work on the old technology and then, after a year, are moved on to the new technology. The interviewees said it requires a year of experience on the job for a new recruit to be fully operational.

Conclusions for the auto sector

The auto industry is structured in a global oligopoly and has been going through significant changes in technology associated with the reduction in the number of platforms, modularisation and increased levels of automation and use of data-driven management. The pace of technological change in South Africa, including the increased use of robots and other automation technologies, is driven by the demands of the MNC parent companies. Success in adopting these methods in South Africa impacts on the OEMs' competitive position in relation to sister plants in other regions. In this respect, the OEMs in South Africa benefit significantly from the APDP programme and continued support, including grants for capital investments, will be important for the industry's future. The existing incentive structure, however, has not served to increase local suppliers' share in OEM production. In terms of upgrading along the local supply chain, our evidence suggests that this is occurring where first-tier suppliers are connected technologically to the MNC parent company through the OEMs. To extend this upgrading beyond these first-tier suppliers will need focused policies to support the skills and technological upgrading of domestic suppliers more generally. This would help to better position the suppliers to capture a larger share of OEM production for the home and export market.[20]

In terms of skills needs and gaps, there is a mixed situation. There are some specialised skills in short supply including for laser welding systems and PLC computer control systems. These skills are mainly being provided by the parent companies or a few local suppliers. In terms of more basic technical skills such as mechatronics or maintenance, these are being provided by the technical institutes and colleges but the view was expressed that the quality is generally low and half or more of potential recruits are rejected in initial screening processes. What is less clear is whether this 'low quality' reflects inadequate technical training, or ineffective applications of the acquired formal technical knowledge to daily problem-solving at work. This probable gap between academic knowledge and practical capabilities in a work setting is not unique to the South African context, but addressing it may require changes in the country's current vocational education and training system. Work activity and experience in a firm may need to be integrated more fully into formal training programmes. These policy issues are returned to in the concluding section of this chapter.

Mining sector analysis

The mining equipment sector was chosen for study in part because of its contrasts with the auto industry: local equipment producers have, over time, developed strong design and innovation capabilities linked to the vertical supply relationships they have established with mining companies. Another reason for choosing the mining equipment sector was to contrast trends in automation in a sector with relatively high levels of product customisation and small-batch production, as compared with the standardised and large-scale production typical of auto industry producers.

South Africa has the largest and most diversified mining sector in

---

20 For a discussion of the need for policies to support technical and skills upgrading in order to meet the South African Automotive Masterplan objective of increasing local content in assembled vehicles up to 60 per cent, see Barnes et al. (2018).

Africa and the mining industry accounts today for about 8 per cent of its GDP and about 14 per cent of total manufacturing value added. The sector has performed poorly with value added declining by 4 per cent between 2007 and 2017 and employment declining from a peak of 518,000 in 2008 to about 450,000 in 2019 (PwC, 2019; Cassim et al., 2019). The largest drop in employment occurred in gold mining, with gold's share of employment falling from over 50 per cent in the early 1990s to about 20 per cent in 2016, while the share of platinum group metals (PGMs) has increased (Rupprect, 2016). The decline in gold production reflects the progressive exhaustion of high-grade gold deposits over the 20th century and the increasing exploitation of costlier, lower-grade and deep-lying deposits. More generally, a large share of South African mining is carried out at great depth, and the industry has faced particular challenges of successfully using mechanised methods, and ensuring adequate rock-face support and ventilation (Neingo and Tholana, 2016).

There is a significant cluster of world-class mining equipment producers in South Africa, many domestically owned and in vertical supply relations with the mining operators. In response to the particular needs of the South African mining industry, local equipment producers developed engineering design and innovation capabilities to provide engineering solutions adapted for deep mining. Much of the technology is customised to the specific conditions of South African mines and local equipment producers have unique competences in the areas of underground locomotives, shaft sinking and ventilation (IGF, 2017). An analysis of patent citations for mining equipment carried out in a 2011 study showed the sector to be highly inventive with South African mining patents cited more than the patents of any other South African sector (Kaplan, 2011).

With the decline in gold and poor growth performance of the industry overall, equipment producers have increased their exports to the region, including Zambia, the DRC and Mozambique. These countries accounted for about 45 per cent of the industry's cumulative world exports of 108,187 million rand over 2012–14. South African OEMs have also established joint ventures in these countries or have set up wholly owned subsidiaries. A recent study of South African

OEMs in these countries, however, found little evidence of technology transfer to the region (Fessehaie et al., 2016).

Summary of findings for the mining equipment sector

Interviews were carried out at three equipment producers. Producer 1 is an independent South African company created in 1980. It produces specialised equipment for the underground mining industry, both hard rock and coal, with about 130 employees producing products including roof bolters, mobile fans and face drills. Producer 2 is an MNC created about 45 years ago with its headquarters in South Africa. Compared to Producer 1, the firm's product range is quite diversified and includes screening media, mineral processing equipment, solid/liquid separators and conveyor system solutions. The firm at present has about 70 per cent of the South African market for screening systems with most of its clients being international companies with some operations in South Africa. Producer 3 is a small specialised independent South African equipment producer that has about 75 employees. Its products include pneumatic chainsaws, sump pumps, and single and lacing cutters.

Adoption of 4IR-related technologies

The technology and methods used by the three firms are largely conventional and rely on human labour and ingenuity. Except for drilling and some repetitive loading operations in Producer 2, there is no use of industrial robots nor is future adoption of them envisaged. The fundamental reason for this has to do with the short runs and complexity of the products which are often customised to the needs of the individual mines. For example, Producer 1 indicated that the solutions developed for one client can hardly ever be re-used with another.

Component machining is done with conventional CNC machine tools with the programming assured either by skilled machine operators or by specialised technicians. Of the three equipment producers, only Producer 3 makes use of 3D printing for rapid prototyping of complex parts at a customer's request. There was no use of virtual reality for product testing and Producer 1 even vaunted the advantages of real over virtual testing, describing how they used a large piece of granite, which is harder than the material found in the mines, in order to test a face driller.

Data management and use

In terms of data management, Producer 2 was relatively advanced although its system retains some legacy features, for example, people still capture information about operations on the shop floor through tally sheets, rather than through a machine-driven system. The firm has implemented the JD Edwards enterprise resource planning (ERP) and supply chain software to integrate all data into a database to optimise the information. This system accurately counts the number of panels the company has produced and the inputs used, showing that the warehouse has depleted its stock by that amount of inputs. At present, there is no system in the warehouse to control and maintain the waste production process at a pre-determined level and waste control is done manually. The design stage, however, is integrated into the system, with 3D modelling incorporated into the ERP system, such that when an enquiry is made from the sales department the specifications for an order – say, panels for a grinding mill – are entered into the software and calculations are made automatically.

Outside the sphere of production, Producer 3 makes use of a basic computerised ordering system, which is used by employees to know the time required to produce enough units to meet demand. The person in charge of purchasing identifies the stock level and passes this information on to the production planner. The system was purchased several years ago and is also used to manage the company's finances. Producer 3 appears to be the least advanced in terms of data management methods. It makes no use of computers and the data needed to identify machine faults and to detect errors is collected manually by factory managers or supervisors.

The main novel use of data, which all three producers are interested in but have made little progress thus far in implementing, is collecting data from the mines on the performance of their equipment for purposes of predictive maintenance. All three producers saw this as an opening for pursuing strategies of servitisation – that is, using services to drive growth – arguing that the OEM equipment producers are best placed to monitor and interpret data from the mines and this could be built into their service contracts in the future. The pursuit of a servitisation strategy depends on data collected on panel use in mines. For instance, as stated by Producer 2 (Interview: Engineering Manager, Producer 2):

> Quite a lot of our business is no longer selling this product and we get paid on the basis of the utilisation of those panels. So, however many tonnes of product run over the panel, we will get remunerated on that. So, the benefit to the customer is that they don't have to worry about that part of the problem, it's now our problem. [...] But to gather that data is inherently very difficult.

Producer 1 described how they developed what was referred to as an 'intelligent' machine which can provide real-time data on factors such as production, machine availability, loading capability and hours of use on the machine. A challenge was to find a way to transmit this data wirelessly or without a cable, and to make the data understandable for different people and different purposes. A problem the company confronted is that to transfer the data to the surface, instead of just having it collected next to the machines underground, requires Wi-Fi in the mines or possibly 5G, which is lacking. A lack of sufficient digital infrastructure in the mines is, therefore, a constraint.

Training and skills

In contrast with the auto sector, the skills profile of the three mining equipment producers is to rely heavily on manual skills with an emphasis on flexibility and multi-skilling. This can be seen as a response to the small batch sizes in the industry and the need for continuous adaptability in production. Consistent with this, an important component of the employees' skills is acquired on the job, through daily work activity. Producer 1 noted that the company hires employees with general skills and then trains these workers to develop specialised skills over time. The need for problem-solving skills pertains in part to the need for employees to undertake their own quality assurance and equipment maintenance. In support of this, Producer 2 operates a continuous improvement programme that aims to endow employees at all levels with the tools needed to solve the technical problems they face. If an operator on the shop floor is able to identify a problem, s/he is encouraged to find a way to solve the problem without assistance. The need for problem-solving skills, however, clearly goes beyond issues of quality assurance and maintenance and the Sales Director at Producer 3 observed that they develop new products and solutions drawing on the improvements suggested by employees who work directly in production.

Some differences, however, were identified between the producers in their approach to the development of operators' skills as reflected in their use of conventional CNC machine tools for machining component parts. At Producer 1, the skills profile of machine operators is relatively high, as all operators can do the basic programming needed for machining a constantly changing array of components, leaving only the most complex programming for specialised programmers or technicians. Producer 3, however, reserves all programming for its two specialised programmers, with operators not allowed to change programmes to avoid possible errors. This points to a certain discretion in how the same technology may be used and the differences in its impact on operators' skills needs, with the approach of Producer 1 favouring greater skills development.

In terms of new technical skills that cannot be easily acquired in-house, the practice is to purchase them on a consultancy basis and subsequently internalise them. For example, Producer 2 referred to gaps in the skills needed for its 3D modelling software and noted that these are purchased on a consultancy basis. The general trend identified by all three producers is an increase in technical skills as well as problem-solving skills, even at the lower levels.

Conclusions for the mining equipment sector

The mining equipment sector in South Africa, in contrast to the auto sector, includes a large number of domestically owned companies, some with capabilities at the technical frontier. Our case studies of three of these companies do not aim to represent the diversity of situations across producers in terms of the adoption and use of new and emerging technologies. We find little evidence that robots or other automation technologies are being used in their production processes and a reason for this is that they appear poorly suited to the production of complex and often customised equipment designed in many cases for the specific conditions of South African underground mining. This by no means implies that equipment producers could not benefit from greater use of frontier technologies and one area where this is arguably the case is more advanced design methods. The successful use of 3D modelling software by one producer points to the potential value of such methods. Another area where there would appear to be room for progress is in data management tools. The type of ERP system used by Producer 2 could be adapted to the needs of smaller firms such as Producers 1 and 3.

All three of the mining equipment producers referred to the potential for adopting servitisation strategies based on the collection and analysis of data collected from their equipment used in mining companies. Their ability to pursue such strategies may be hindered, in part, by the limited availability in South Africa of engineers with the necessary advanced data analytic skills. Another possible hindrance to servitisation strategies is the gap between the technological capabilities of equipment producers and what mining companies, in general, are prepared and able to adopt. According to our interviews, this gap is in part infrastructural, due to a lack of connectivity in the mines, and partly due to a skills gap relating to the ability of employees in the mines to operate more advanced equipment incorporating sensors for data collection.

The close connection between technological innovation in equipment producers and the demands of the mining companies points to the need for coordinated action at the sector level in support of innovation and technology adoption. One recent initiative in a positive direction is the Mandela Mining Precinct which is designed to act as a public–private innovation hub supporting technical change and innovation in the sector. Within the precinct, the Mining Equipment Manufacturers of South Africa (MEMSA) has been established with programmes aimed at supporting local equipment producers, including initiatives focusing on skills development and the better use of data for supply chain management. The potential role of stakeholder organisations such as the Mandela Mining Precinct to contribute to technological upgrading and improved performance in mining is returned to in the concluding section.

FINAL REFLECTIONS AND POLICY CONSIDERATIONS

This chapter presents new evidence on the adoption and impact of 4IR technologies in the South African manufacturing sector – focusing on two advanced manufacturing activities, namely the automotive and mining equipment sectors. It pays particular attention to the various impacts that adopting these technologies have on employment, as well as on skills development and training. These represent central concerns for developing countries in relation to the 4IR, due to pre-existing high levels of unemployment and skills shortages.

The evidence presented in this chapter suggests that the outcomes of the 4IR in the South African advanced manufacturing sector remain largely undetermined. However, the evidence suggests they are to some extent contingent on the ability of individual firms to adopt and adapt technologies such as robotics, AI, 3D printing and data management systems to add to their innovation performance, rather than becoming a labour displacing force.

The findings from this study indicate that firms in the auto sector are actively adopting 4IR technologies, especially through the use of robots and cobots, as well as 3D printing for rapid prototyping. Moreover, auto manufacturers appear to be increasingly generating data at the production level, collected from various machine sensors and other devices, and utilising it to improve production and their supplier value chains. This data explosion raises the need for new technologies to capture, store and transmit the data, as well as the requirement for skills to manage it. Data-related skills are needed to analyse, use and gain insights from the data. The adoption of 4IR technologies by South African auto manufacturers is largely driven by global production standards, which are set by MNCs. South African firms must compete with those in other locations in terms of productivity, quality and production volumes, which drive their adoption of 4IR technologies. In contrast, the patterns of 4IR technology adoption among manufacturers of mining equipment appears to be limited. This is due to the 'customised' nature of their production, which arises from the need for small batch production and low replication of products. While mining equipment manufacturers have also incorporated data management systems in their production process and supply chain management, they tend to combine these with traditional manual methods of quality control. Fully incorporating data management systems would require both skills and digital infrastructure that are currently lacking in the mining sector.

While it is generally acknowledged that increased automation displaces labour, the auto companies interviewed for this study rarely reduced their total employment as a result of increased automation. In fact, several companies increased their number of employees. On the one hand, the industry experienced increased production volumes, with a subsequent need for more personnel, and on the other hand, firms appeared to be highly sensitive to the labour market in which they operate, characterised by endemic unemployment. Some companies chose strategically to automate some plants while deliberately maintaining others on a labour-intensive mode, in order to avoid job losses. Companies appear to be doing their best to retain employment along with increased automation, often opting to relocate employees within the company, and engaging in retraining. Such an approach was framed as reflecting the firms' commitment to supporting their surrounding communities, where most low-skills employees came from, which are vulnerable to poverty and unemployment.

As the adoption of 4IR technologies becomes more prevalent and indispensable for national and international competitiveness, the need grows for policy intervention to ensure that firms are not disincentivised to invest in 4IR technologies. This finding indicates that the extent to which automation displaces jobs in developing countries is limited by social and organisational forces, which set boundaries on how and why a job may or may not be automated. This has been referred to as 'bounded automation' (Fleming, 2019), and highlights the social considerations (such as organisational power relations and the nature of the task) over qualities intrinsic to the technology itself

(Fleming, 2019: 28). These are important considerations to take into account when examining the employment effects of 4IR technologies in South Africa.

The opportunities to adopt 4IR technologies are connected to innovation capabilities at firm level. Our evidence highlights the differences in firms' degrees of dependence on foreign technology and in firms' positions in global value chains, which are largely shaped by sectoral dynamics.[21] While the auto industry has grown under successive policy regimes, it has little or no indigenous innovation capabilities and is set to cater to the interests of international OEMs. These dynamics have limited the ability of local suppliers to increase their share of supplies to the OEM. In contrast, the manufacturing of mining equipment shows considerable potential for the development of local innovation capabilities, reflected not only in the dynamism of patent citation in mining, but also in the complexity of the products which are tailored to the needs of individual mines and the specificities of the South African mining landscape.

Successfully integrating 4IR technologies into production processes within firms requires innovation, which implies the ability to develop new collaborations, networks, interactions and processes of co-creation that may need to be supported by government policies. In other words, firms do not operate in a vacuum but as part of a broader 'innovation system'. However, not all actors in the innovation system are either aware of opportunities or prepared to undertake the investments and risks related to the adoption of these technologies. The range of organisations (education, financial, governmental, civil society, for example) in a system can either enable or constrain these interactions, affecting the patterns of adoption of 4IR technologies and innovation activities in general. In this respect, multi-stakeholder initiatives become of central relevance, providing opportunities to develop 4IR-driven niche technological solutions to local challenges. This is the intention of the recently established Mandela Mining Precinct.

---

21 There is a large body of literature on GVCs. For a review, see Shepherd (2013).

Incorporating new technologies and data management systems into manufacturing processes requires new types of knowledge, skills and competencies that are scarce in South Africa. The South African tertiary education system is quite effective at supplying general engineering and technical skills, although once recruited, firms must spend considerable time and resources in getting engineers ready for the workplace. Auto manufacturers are finding it increasingly difficult to recruit workers skilled in the deployment of 4IR technologies, such as maintenance engineers, robot programmers and mechatronics skills. The emergence of new requirements for skills upgrading in auto manufacturing is closely linked to the changes in materials and methods introduced with each new model. Therefore, training and skills development is provided by the parent company or MNCs or international training providers, largely disconnected from local education and training institutions. The mining equipment sector displays relatively strong skill levels at the tertiary level (engineering), with stronger connections to local training institutions.

As companies upgrade technologically, they seek out employees who not only have technical skills, but also demonstrate flexibility and an ability to solve problems. The firms in both sectors stated that while the technical training in intermediate-level skills provided by the technical and vocational education and training (TVET) colleges is adequate, new recruits lack the ability to apply these skills to problem-solving on the job. This points to the need for improving the links between the TVETs and industry, and for increasing opportunities for experience-based learning through work-integrated study. At the lowest skills levels, even if technical problem-solving skills are not needed, there is an increasing need for the ability to monitor computerised data production control systems. The problem of weak foundation skills (such as reading and writing) was referred to in this context, which highlights the urgency to improve basic education systems in South Africa.

The existing evidence indicates that it should not be taken for granted that technology will translate into shared prosperity and growth. The potential negative consequences of 4IR-related technologies are more likely to be realised in contexts with weaker governance and institutional structures, which are often the case in developing economies. For instance, emerging economies often display weak labour market institutions and high levels of informal employment, which could result in greater adverse effects from 4IR technology adoption.

South Africa has not yet formalised any policy documents or enacted legislation for the regulation of technologies related to the 4IR, such as robotics or AI. However, in April 2019, President Cyril Ramaphosa appointed members to the Presidential Commission on the Fourth Industrial Revolution (4IR Commission), with a mandate to 'assist government in taking advantage of the opportunities presented by the digital industrial revolution'. The task of the 4IR Commission, which is chaired by the President, is to identify relevant policies, strategies and action plans to position South Africa as a competitive global player. Delivering on its mandate, the Commission presented a report with recommendations for South Africa to harness the available 4IR opportunities in August 2020. Interestingly, none of the members of the 4IR Commission represents the advanced manufacturing sectors, which are central adopters of 4IR technologies.

REFERENCES

Acemoglu, D., Lelarge, C. and Restrepo, P. 2020. 'Competing with robots: Firm-level evidence from France'. *AEA Papers and Proceedings*, 110, 383–88.

African Development Bank (AfDB). 2019. 'Study on unlocking the potential of the fourth industrial revolution in Africa'. https://4irpotential.africa/wp-content/uploads/2019/10/AFDB_4IRreport_Main.pdf, accessed in August 2020.

Arntz, M., Gregory T. and Zierahn, U. 2016. 'The sisk of automation for jobs in OECD countries: A comparative analysis'. OECD Social, Employment and Migration Working Papers, No. 189, OECD Publishing, Paris.

Automotive Industry Export Council. 2019. *Automotive Export Manual*. Pretoria: Automotive Industry Export Council.

Barnes, J. and Morris, M. 2008. 'Staying alive in the global automotive industry: What can developing economies learn from South Africa about linking into global automotive value chains?'. *The European Journal of Development Research*, 20(1), 31–55.

Barnes, J., Black, A. and Techakanont, K. 2017. 'Industrial policy, multinational strategy and domestic capability: A comparative analysis of the development of South Africa's and Thailand's automotive industries.' *The European Journal of Development Research,* 29(1), 37–53.

Barnes, J., Black, A., Comrie, D., and Hartogh, T. 2018. 'Geared for growth: Report on the South African Automotive Masterplan Project'. Department of Trade and Industry, Government of the Republic of South Africa.

Bowles, J. 2014. *The Computerisation of European Jobs.* Brussels: Bruegel.

Calitz, A.P., Poisat, P. and Cullen, M. 2017. 'The future African workplace: The use of collaborative robots in manufacturing'. *SA Journal of Human Resource Management,* 15(1), 1–11.

Carbonero, F., Ernst, E. and Weber, E. 2020. 'Robots worldwide: The impact of automation on employment and trade'. IAB-Discussion Paper, No. 7/2020, Institut für Arbeitsmarkt- und Berufsforschung (IAB), Nürnberg.

Cassim, Z., Goodman, S. and Rajagopaul, A. 2019. 'Putting the shine back into South African mining'. McKinsey and Company. https://www.mckinsey.com/featured-insights/middle-east-and-africa/putting-the-shine-back-into-south-african-mining-a-path-to-competitiveness-and-growth, accessed 20 July 2020

Comin, D. and Mestieri, M. 2014. 'Technology diffusion: Measurement, causes and consequences'. *Handbook of Economic Growth,* 2, 565–622.

Comin, D. and Mestieri, M. 2018. 'If technology has arrived everywhere, why has income diverged?'. *American Economic Journal: Macroeconomics,* 10(3), 137–78.

Dearnley, C. 2005. 'A reflection on the use of semi-structured interviews'. *Nurse Researcher,* 13(1), 19–28

Fessehaie, J., Rustomjee, Z. and Kaziboni, L. 2016. 'Mining-related national systems of innovation In Southern Africa: national trajectories and regional integration'.

Fleming, P. 2019. 'Robots and organization studies: Why robots might not want to steal your job'. *Organization Studies,* 40(1), 23–38.

Frey, C.B. and Osborne, M. 2013. 'The future of employment'. Oxford Martin Programme on Technology and Employment, Oxford University.

Graetz, G. and Michaels, G. 2018. 'Robots at work'. *Review of Economics and Statistics,* 100(5), 753–768.

Giuntella, O. and Wang, T. 2019. 'Is an army of robots marching on Chinese jobs?'. IZA Discussion Papers, No. 12281, Institute of Labor Economics (IZA), Bonn.

Holm, J., Lorenz, E. and Stamhus, J. 2020. 'The impact of robots and AI/ML on skills and work organization', forthcoming in Christensen, J., Gregersen, B., Holm, J. and Lorenz , E. (eds.) *Globalization, New and Emerging Technologies, and Sustainable Development – The Danish Innovation System in Transition.* London: Routledge.

Humlum, A. 2019. *Robot Adoption and Labor Market Dynamics.* Princeton: Princeton University Press.

Intergovernmental Forum on Mining, Minerals and Metals (IGF). 2017. 'Case study: South African horizontal linkages'. https://www.iisd.org/sites/default/files/publications/case-study-south-africa-horizontal-linkages.pdf, accessed 20 July 2020.

International Federation of Robotics (IFR). 2019. 'Executive summary: World Robotics 2019 Industrial Robots'. https://ifr.org/downloads/press2018/Executive%20Summary%20WR%202019%20Industrial%20Robots.pdf, accessed on 26 August 2020.

Jordaan, J., Dinham, J., Fieldgate, I. and Rolland, S. 2018. 'Economic & socio-economic impact of SA automotive industry'. *Econometrix*, August.

Kaplan, D. 2011. 'South African mining equipment and related services: Growth, constraints and policy'. Making the Most of Commodities Programme (MMCP) (5).

Lorenz, E., Kraemer-Mbula, E. and Tregenna, F. 2019a. 'Background Report on the Fourth Industrial Revolution and Sustainable Industrial Development, Report for the Community of Practice in Innovation and Inclusive Industrialisation'. SARChI Industrial Development, University of Johannesburg.

Lorenz, E., Tessarin, M. and Morceiro, P. 2019b. 'Report on the Adoption of 4th Industrial Revolution Technologies in South African Industry, Report for the Community of Practice in Innovation and Inclusive Industrialisation'. SARChI Industrial Development, University of Johannesburg.

Morikawa, M. 2020. 'Heterogeneous relationships between automation technologies and skilled labor: Evidence from a firm survey'. Research Institute of Economy, Trade and Industry (RIETI).

Neingo, P.N. and T. Tholana. 2016. 'Trends in productivity in the South African gold mining industry'. *Journal of Southern African Mining and Metallurgy*, 116(3), 283–290.

PwC. 2019. 'SA mines In transition'. PwC. https://www.pwc.co.za/en/assets/pdf/sa-mine-2019.pdf, accessed 20 July 2020.

Rupprect, S.M. 2016. 'The need for material change in the South African mining industry', New technology and innovation in the Minerals Industry Colloquium, South Africa.

Seamans, R. and Raj, M. 2018. 'AI, labour, productivity and the need for firm-level data'. (No. w24239). National Bureau of Economic Research.

Shepherd, B. 2013. 'Global value chains and developing country employment: A literature review'. (No. 156). OECD Publishing.

UNCTAD. 2017. Trade and Development Report. Ch. 3 'Robots and Inclusive Growth'.

United Nations Industrial Development Organization (UNIDO). 2019. Industrial Development Report 2020. Industrializing in the digital age. Vienna.

World Bank. 2019. The Changing Nature of Work, World Bank Development Report.

World Bank. 2016. Digital Dividends, World Bank Development Report.

SEVEN

Impact of current technologies on jobs and employment: Insights from mining and banking in South Africa

MAMOKGETHI MOLOPYANE

INTRODUCTION

In January 2020 President Cyril Ramaphosa, in a report titled, 'A National Strategy for Harnessing the Fourth Industrial Revolution' (cited in Ndung'u and Singé, 2020) deemed the Fourth Industrial Revolution (4IR) to be a conduit through which South Africa can address challenges such as job creation and achieve its aspiration of being a knowledge-based developmental state. The administration identified the 4IR as an accelerator in job creation and in building an entrepreneurial developmental state. This chapter asks if such a strategy is valid for South Africa, considering the position the country occupies in the global economy and its persistent underlying challenges of socioeconomic inequities.

Numerous papers in academic and professional journals, theses, reports and books have been written about the influence technological change has on the global production system and its consequent effects on labour markets in different countries. Despite, or because of, the extensive literature, there is no consensus whether 4IR is more of a threat than an opportunity for employment and the future of work.

It is widely agreed that internal factors and national context lend greater weight to the technological impact of the 4IR on a country. The point here is that technology will 1) affect different countries in different ways and is heavily conditioned by the global economy; and 2) have negative and positive effects on the labour market depending on how it is adopted. It is important to keep in mind that the discourse on the technological impact is distinct from one country to another. However, it presents an opportunity to investigate and explain common wide-ranging consequences.

A good deal of the present debate on 4IR in Africa is grounded in the view that the impact on South Africa and the rest of the continent's labour market has not yet been fully manifested or experienced because African economies lag behind those of advanced states. As a result, there is a vast number of analyses on the trajectories of 4IR in Africa and their implications and effects. The discourse assumes the inevitability of the 4IR on the basis that, like preceding technological revolutions, it establishes itself as the dominant system by rendering previous systems obsolete. It is in this sense that the 4IR can be considered to be a process of assimilating Western concepts and ideologies that advance capitalist expansion and have noteworthy consequence on the future of work in Africa. The key explanatory variable that leads to a better understanding of the 4IR in Africa is the theory of hegemony and how prominent ideas are used to serve the interest of dominant countries.

In this context, the World Economic Forum (WEF) is part of international institutions' move to globalise the ideology of 4IR in order to encourage developing countries to adopt its main tenets, such as liberalising economic activities and globalisation of production. Also, many of the debates concerning 4IR and the future of work are themselves linked to other debates surrounding capitalism and economic development.

This chapter contends that technology as an instrument of capitalism has led to the precarity of employment in emerging economies as companies try to keep up with accelerating globalisation and economic integration.

How technology affects a country's socioeconomic conditions and inequalities is interrelated with how technology affects the labour market, particularly in relation to the demographics, education level and gender of workers. There are also other issues tied to the discourse of disruptive technologies, such as the hegemony of the 4IR as a tool *par excellence* for emerging economies to embrace in order to achieve development. These issues include a neoliberal economic order that has created largely unfettered competition in the global economy. Companies are now under pressure to innovate, use emerging technologies to reinvent production processes, increase production and cut labour costs in order to make profits.

However, African countries are not innocent bystanders, as shown in the views expressed by President Cyril Ramaphosa in the introductory text. The adoption of 4IR strategies by South Africa is politically determined mostly because of the state's role in promoting and regulating economic activity and in articulating 4IR strategy as serving the interests of all.

Furthermore, how technology is deployed is a political process that is influenced by key actors with conflicting interests, internal variables, power structures and the ambition to improve their positions in the global economy. Relatedly, government intervention, policy instruments and the tactics used can also be influenced by external factors.

This chapter focuses on what 4IR means for the future of work in South Africa by examining the impact of robotic technology and automation on work in different countries. In assessing these country-specific experiences, the chapter acknowledges that these countries' industries and labour market are different from South Africa's, and occupy a far higher position in the global economy. Moreover, although there has been a rapid diffusion of disruptive technologies like robots and automated equipment in some developing countries, debates and studies on their impact have been focused on developed countries' labour markets. The intention is to show that an analysis of the intricacies of technological disruption on employment could be the basis of theorising about the 4IR's impact on the future of work in South Africa.

The goal is not to account for the changing nature of employment relations or labour market polarisation, as such topics are worthy of a chapter of their own. Nor is this chapter the place to investigate the occupational categories and skills affected by technological change; this too would require separate reflection. Rather, this chapter explores the diffusion of robots in various countries. It discusses how these countries' experiences can be a useful lens for studying the future of work and the far-reaching implications both current and 4IR technologies have on policies. The focus is on how technology can, on the one hand, have displacing effects, and, on the other hand, be complementary to workers, and how this varies from country to country.

The main aim of the chapter is to show the impact of digital transformation, and the adoption of robots entailed, on jobs and employment in banking and mining in South Africa. First, it contextualises the interaction between technology and work by examining the experiences of China, Germany, Japan and the USA; next, banking and mining case studies are discussed in relation to the displacing effects of technological change on employment and occupations. The last part of the chapter revolves around the impacts of robots on employment; and insights gleaned from the case studies.

To sum up, the approach to technology in this chapter aims to identify lessons for those discussing and thinking about technology-concerned policies relating to the future of work and jobs. The chapter draws on lessons from country-specific experiences, and from two industries, banking and mining. A key observation of this chapter is that current public debate about the Fourth Industrial Revolution in South Africa considers the 4IR to be predetermined, overstates the exogeneity of its disruptive effects and ignores the role of stakeholders in shaping how technology is deployed in the economy and how it will impact on society.

THE IMPACT OF ROBOTS ON EMPLOYMENT

Before discussing the implications of 4IR on the future of South Africa's mining and banking sectors, it may be helpful to reflect on what occurred elsewhere. In this chapter, automation denotes the use of control systems such as robots to substitute workers. For example, Kaggwa (2018) finds that for South Africa, 4IR will reinforce existing levels of labour participation, typified by high levels of joblessness among the low-skilled compared to those with high levels of education. He also suggests that employees who trained for or work with current technologies are less susceptible to job losses as their skills can be adapted to the demands of the new technologies. Technology transforms society and its significant impact on work is examined in this section.

In general, technology has contributed to how tasks, job categories and occupations are designed. Bessen (2015) provides evidence of how technologies can shift tasks from one occupation to another, leading to the displacement of some workers whose tasks are now being performed elsewhere.

Robots in the workplace: Lessons from Germany, Japan, the United States and China

In the last 20 years, Germany's use of industrial robots in manufacturing has quadrupled without supplanting the workforce. Although robots have not led to job losses, they have not stimulated new job creation. Dauth et al. (2017: 7) assessed employment data to pinpoint the effects of automation in manufacturing on the German labour market; and they found 'no evidence for negative effects like those in the US. Once industry structures and demographics are taken into account, we find effects close to zero.' There were two key finding from the study:

- Workers who performed tasks prone to automation remained employed; however, they occupied different positions in the same workplace.
- For low- and medium-skilled workers, automation had a negative effect on their wages; however, they accepted pay cuts to maintain job security.

These were not inevitable outcomes. Rather, the decisions taken were informed by the nature of industrial relations in Germany and offer key lessons for South Africa. The following factors are particularly worth noting:

- the collective approach taken to studying the effects of robots on labour and productivity, and then – again collectively – deciding (in consultation with work councils rather than autocratic decisions by companies) how to integrate robots into the workplace;

- trade unions' inclination to accept flexible wages to cushion the negative shocks of automation and to maintain employment levels; and

- robots in the workplace are used to complement workers and make them more productive; however, these robots also contribute to a declining labour income share.

In Japan, automation in the automotive, electronics and manufacturing industries became prominent between the 1970s and 1990s as part of the country's industrialisation policies and its economic success story (Schneider et al., 2018). Automation of these industries led to increased productivity and stimulated gains in employment levels and wages (Acemoglu and Restrepo, 2020). A study on the impact of Artificial Intelligence (AI) and robots on employment in the automotive and manufacturing industries found that workers with medium to low education levels who acquired skills through work training specific to their occupation, or who had occupational licences from professional schools, were unlikely to be replaced by AI or automation (Morikawa, 2018). However, the free-fall in Japan's working-age population, along with the decline of the labour force and the strict control measures on immigration, mean that companies and policymakers are turning to advances being made in AI, robotics and other emerging technologies to fill the gap.

Schneider et al. (2018) warn of two societal and welfare risks created by the increasing automation evident even beyond traditional industries. First, women workers outside industrial sectors are vulnerable to displacement when technology becomes prevalent in service and care industries. Second, workers with limited education and training are at risk of finding themselves outside an economy that is technology dependant. The lessons for South Africa are:

- Adaptable vocational complementary training for workers with medium and low education and skills levels can curb the displacing effects of technology.

- Technology doesn't have to decimate jobs. In cases where labour is declining, automation can fill the gap, increase productivity and raise income for workers.
- Disruptions in the labour market seem inevitable.
- Policy efforts can concentrate on demographic trends, education and the skills required for the adoption of technology.

The US manufacturing experience is instructive for South Africa for this reason: between 1993 and 2007 the United States increased the number of robots per hour worked by 237 per cent (Graetz and Michaels, 2018). In the same period, 2.2 million jobs were lost (Andes and Muro, 2015). Thus, it is possible to assume robots are responsible for manufacturing decline. However, correlation is not causation and the US experience must be understood from multidimensional perspectives: offshoring; globalisation; competitive challenges posed by other countries using automation technology; skills gaps; trade deficits; and output, productivity and capital investment problems are some of the causal factors for the decline.

Acemoglu and Restrepo (2020) studied US data analogous to the studies, discussed above, undertaken in Japan. Their findings revealed how industrial robots had a positive and negative effect on employment in the USA: they estimated the net effect to be a loss of between 360,000 and 670,000 jobs. This relatively small number can be attributed to low adoption of industrial robots in the US labour market. There is also evidence that as technology replaced workers in the US labour market, occupations with new tasks accounted for 9 per cent of the 17.1 per cent employment growth from 1980 to 2007. In other words, the creation of new tasks and occupations was central in producing more employment and new jobs.

Another important issue to consider is that manufacturing, once a key sector of the US economy, was hollowing out long before the displacing effects of automation and robotics. Technology merely accelerated the disruption that was mostly structural (Miller, 2016). An instructive example is the incorporation of automated equipment and robots in steel production. Grossman (2017) provides evidence of how a production plant that used to employ 3,000 people, including 200 production line workers, now only requires 14 employees with three technicians managing the entire plant, which produces 500,000 tonnes of steel wire per year.

The divergent effects of new technologies in the US labour market is discernible: jobs are displaced and new occupations arise; industrial robots have a low net effect on employment because they are not as widely adopted as in Japan. Using a quantitative general equilibrium framework, Leduc and Zheng (2020) have found that automation in the US amplified employment fluctuations because the jobs created offset any displacing effects. Additionally, automation strengthens US employers' bargaining powers at the expense of workers who perceive automation as a threat to their jobs. Therefore, automation decisions and the adoption of robots in production contributes to wage suppression. There were varied factors contributing to the changes in the US labour market prior to automation and the incorporation of industrial robots. These include rapid globalisation; the decline of key industries; and the substitution of routine tasks. There are valuable lessons to be learned from the overall automation experience. The impact of adopting industrial robots for production are not simple and ought to be considered with other factors causing a decline in employment in industry:

- Automation can weaken workers' bargaining power.
- Firms can be blinded by what they stand to gain when making choices about the adoption of robotic technologies.
- Automation and robotics may not be the cause of employment declines in manufacturing; however, they account for worsening conditions.

Turning to China's experience helps us to understand how an emerging economy grapples with the increasing adoption of robots in production and their implications for work. It has been argued that China is now considered past the emerging economy point, so the country's adoption of robots and use of technological innovation could be instructive for developing countries. In order to understand China's experience, it is necessary to take account of the relationship between global production systems and technology.

Here, the focus is on 1) government policy interventions that prioritise automation; 2) aging populations and resulting declining workforce; and 3) rising costs of labour in manufacturing. It is worth pointing out that there is limited literature on the impact of automation in China; evidence of changes in production show divergent consequences.

For instance, China's increasingly leading contribution to the production of manufactured goods worldwide drives the movement from labour to robots and automation. The Chinese government's determination to remain internationally competitive has incentivised Chinese firms to invest in technology that increases production and meets the demands of the global market for goods and services.

There are key factors, relating to China's unique context, to take into account when seeking to draw lessons from China's experience of automation. The first is that the role of the Chinese state in advancing policies aimed at supporting the use of technology in the global production system is a shift from the notion of corporations as key actors in technology development.

China has recognised that supporting local firms that are pioneering innovation to increase productivity can have major benefits, including attracting foreign investment from transnational corporations and boosting economic growth. Government initiatives on accelerating technology development include:

- Made in China 2025
- Internet Plus
- Accelerating the Fostering and Development of Emerging Industries of Strategic Importance
- The National Medium- and Long-Term Programme for Science and Technology Development

Each programme has goals, specific priorities, a framework, a plan for implementation and financing. Furthermore, government investment drives the industries focused on – including the education, skills and labour required to achieve – goals within these industries.

The important point here is that government policies evolve to respond to changes in the global and local economy. China's strategies and initiatives to promote technology, and to sustain economic growth, remain competitive.

Second, it must be noted that changes in the labour market existed before technology became pervasive and it is instructive to contextualise one of the key contributors to this change – the demographic shift. The consequences of the one-child policy, together with an aging population, contributed to the shrinking labour force (Zaman, 2018; Reid et al, 2018). Furthermore, the working-age population is characterised by a pool of older workers that is large relative to the smaller young working-age group. Unsurprisingly, China's policy intervention on technology development is dominated by efforts to utilise industrial robots to respond to the challenge of an aging workforce and consequent labour shortage

Third, the diffusion of robots in manufacturing has been pushed by two interrelated facts: the rising costs of labour and China's leading international role as a producer of manufactured goods. On the one hand, in general, manufacturing labour costs have risen by 65 per cent since 2008, relative to other industries (Oxford Economics, 2019). For example, between 2006 and 2010 wages doubled in China's textile industry (Reid at al., 2018). In general, the adoption of technology for production in manufacturing means the upskilling of workers, which leads to increases in wages. In turn, this compels firms to relocate production to countries with low skills and a low-wage labour market or turn to automation as a cheaper option in the long-term, relative to rising labour costs. Lawrence (2019) found evidence of China's manufacturing employment in urban areas declining by 4 million jobs between 2014 and 2017.

On the other hand, the global demand for manufactured goods has accelerated the pace of robot densification in firms that manufacture household appliances and electronic devices. This is evident in China's acquisition of new industrial robots; in 2017 the country alone accounted for 36 per cent of the total sales of industrial robots worldwide and the upward trend is set to continue (Oxford Economics, 2019). If the trend for investing in and adopting robots continues, it could lead to much higher job displacement. In the long run, decisions to bring robots into production create winners and losers in the labour market; workers being the most affected since they are the ones substituted. This is the dilemma in China, as it promotes technology development and incorporates it into the economy. It has become a leader in the use of industrial robots, although the associated risks of displacing workers have adverse economic and social consequences.

However, the disruption can be positive. Reid at al. (2018) found evidence of increased labour productivity in almost 100 out of 200 firms that halved their workforce in favour of industrial robots. This means that while overall employment in manufacturing is likely to fall, increased productivity means that wage growth continues. They also noted the spill-over effects of wage growth spreading to the services sector, which is currently experiencing employment growth as China's economy evolves. This is an important observation; the new jobs emerging in the service sector offset those lost in manufacturing. The three interrelated factors listed above reveal that the impacts of automation on labour markets are not as straightforward as perceived. The evidence seems to suggest:

- The deployment of productivity-increasing technologies as part of governments' policy intervention can be influenced by factors such as an aging population and a shrinking labour force.
- Firms' aspiration to become internationally competitive and to meet the demands of global markets drives investment in industrial robots.
- The high cost of wages incentivises the move from labour to automation.
- Automation does have a positive effect on wages in manufacturing while at the same time jobs.
- Policy initiatives and design need to respond to technology-induced challenges such as labour market disruption.

Summary

The different country experiences demonstrate that technological change has divergent and complex effects in companies and industries and the labour market. In the context of this chapter, the main argument is that robotic technology and automation has negative effects on employment growth and on labour displacement. However, technological change and the densification of robots can lead to increases in wages for sectors that are dependent on it. In Germany and Japan, industrial robots' displacement effects have been limited.

In the US, automation technology encountered a manufacturing industry that was already declining due to internal and external factors. In short, although the displacing effect of robots was low, it added to the woes of an already struggling sector. China and Japan shed light on how robots in the labour market can cushion the weaknesses resulting from an aging population and the ensuing problem of a shrinking workforce. Although South Africa has the opposite, policymakers would do well to note that policy design needs to factor in demographics. Further, in China, for instance, as conditions of workers in automated sectors improved, new service sectors emerged to support their needs. All in all, these country-specific experiences indicate both negative and positive impacts on the labour market as the global production system moves from one technological revolution to another. Technological changes also have economic repercussions on countries and industries in ways that can alter the labour market to the detriment of workers.

In sum, robot-driven labour displacement affects countries differently and is an instructive lesson for South Africa's exploration of technology's ability to transform production and the labour market. This transformation will have far reaching consequences for the future of work. There is, however, another dimension to consider, namely that capitalism modifies relations of production, particularly with technology.

SETTING THE SCENE – SOUTH AFRICA

The previous section painted a picture of the detrimental effects of robots on overall employment. However, it showed that these effects have divergent effects in different countries; in developed economies the impact is low, but it is more pronounced in China's context of an emerging economy. Robots, though, are not the main cause of the employment decline.

Evidence suggests the automation technology has compositional effects that lead to job creation in economies that are evolving towards the services sector. Where wages are high, firms invest in industrial robots or move production offshore to countries where labour is cheap. Turning to South Africa, two factors are central to understanding the current impact of technology on banking and mining employment. These are 1) productivity in general; and 2) the synergies of current technology with 4IR's new disruptive technology.

Wittenberg (2014) discusses labour productivity and wages using employment and working hours from 1994 to 2011. He finds that trends show a 30 per cent aggregate increase in labour productivity. However, this increase accounts for the discrepancy between low productivity defined by lower growth and higher growth in the high productivity sector.

The fall of labour's income share in post-apartheid South Africa contributed to this decline. Strauss and Isaacs (2016) observe that technological changes, together with different policies, have resulted in labour income share declining by five per cent relative to other emerging economies. This evidence contradicts the touted narrative in South Africa's public discourse about wages affecting productivity and the concern raised when workers make their wage demands known. Wittenberg (2014) finds no evidence of average wages outstripping productivity or vice versa. Instead he stresses that while aggregate wages in South Africa appear to increase, it's the stagnant real median wages that tell the true story.

Mongale (2019) provides an innovative econometric study of the implications of labour productivity and costs in South Africa from 1998 to 2018, demonstrating the positive effects of productivity on economic growth and the negative effects of increasing labour costs in some industries, hence negative growth. However, this argument fails to take cognisance of other factors besides labour costs contributing to falling productivity levels. In the context of the mining case study in this chapter, Stoddard (2019) finds that aging mines, declining grades and deeper deposits contributed to a 0.3 per cent decline per year from 2013 to 2017 in South Africa's platinum, iron ore and base-metal mining. Mining in North America and Asia increased productivity by five per cent in the same period. In this context, a decline in productivity occurs for many reasons beyond labour costs. Stoddard observes that the country's mining is behind the disruptive technology curve relative to other countries; thus, sticking to the old ways of mining contributes to productivity decline.

Current technologies synergise with 4IR emerging innovations; artificial intelligence systems enhance labour capabilities, adding another dynamic to effects on jobs. Understanding this interplay is crucial to an understanding of implication of the 4IR for South Africa's future of work.

In South Africa, technology converges with the contentious politics intrinsic to the conflict between interests of capital and labour, and with government policies that are inadequate to tackle unemployment, poverty and inequality. Where does this leave organised labour and the response of unions?

According to Harvey (2019), the pervasiveness of technology in production has contributed to the decline of trade unions and has been the catalyst to the decline of South Africa's largest labour federation, the Congress of the South African Trade Unions (COSATU). Harvey writes that this is because organised labour has failed to reform itself to respond to globalisation and economic integration.

Mhene (2019) supports that view, suggesting that unions in agriculture and industry must be cognisant that traditional methods of bargaining cannot be used in an industry that is mechanising and automating. Training farmworkers, and improving their technical skills, must be one of the key demands unions put forth in wage negotiations.

In general, Harvey (2019) claims that South African trade unions have failed to produce any comprehensive analysis or literature on how they will deal with the implications of 4IR for 1) the industries they organise in and 2) union strategies for organising, bargaining and servicing their members. Kaggwa (2018) finds that trade union attempts at developing coherent strategies are likely to be undercut by the pervasiveness of shifting and unstable employer-employee relations: relations vary between subcontractors; owners of technology; traditional employers; and flexible, temporary and non-contract work.

The 4IR should not be misread as a magic wand that makes labour market challenges disappear (Molopyane, 2018). Earlier, this chapter pointed out the perils of looking at the 4IR as a process that is yet to touch South Africa and the rest of Africa, and as an external force with predetermined effects. The important point is that countries adopt automation technology to varying extents, at varying rates, and for different reasons. Any study of the impact of this technology will miss the full picture if it automatically assumes technology to be a prime driver of change.

TECHNOLOGY IN BANKING AND MINING

Banking and mining have been net employment creators in the past and are currently shedding jobs. Moreover, they now have to contend with operations, production and, ultimately, industries moving towards new technologies. This raises questions about which technologies are currently adopted. What is the impact of disruptive technology on employment in the two industries? Are jobs being displaced or created?

The mining industry

Mining is foundational to South Africa's economy. As Davenport (2013: 1) puts it, 'nowhere else in the world has a mineral revolution proved so influential in weaving the political, economic and social fabric of a society'. Colonialism facilitated extraction of mineral wealth, enabled the emergence of secondary industries and bestowed a capitalist economy (Magubane, 1979) on South Africa.

This chapter does not seek to interrogate the history of mining. It is mainly concerned with the apartheid capitalist nature of mining that founded the 'mode of exploitation' (Williams, 1975: 3) of African migrant workers by paying them wages lower than their fellow white mineworkers while relying on their labour power to extract profit. This factor contributed to black mineworkers' fight for better wages and working conditions and spurred trade union activity.

Exploitation founded the migrant labour system. Despite the reproach by some scholars (Lye, 1984; Harington et al., 2004), the migrant labour system founded on the 'principles of racial segregation, control and exploitation of Africans' (Gwatidzo and Benhura, 2013: 14) still persists. Furthermore, migrant labour systems have aided companies to sidestep traditional standard employment relations by outsourcing the recruitment of labour[1] of desperate migrant workers from neighbouring countries such as Lesotho and Mozambique, with adverse effects on union recruitment, wages and working conditions (Crush, 2001).

Back mineworkers were from the onset paid low wages. The cheap cost of their labour relative to the costs of production ensured and maintained profitability and growth. This is a defining feature of mining in South Africa and illustrates the extent to which exploitation has been central to the growth of the industry and companies.

Thus, in thinking about the impact of technology on mining, we must realise that mechanising and automating operations may lead to job displacement and declining direct employment. Hattingh et al. (2010) observed the complex impact of mechanisation and automation on overall employment. Currently, mining employs 454,861 people – the Platinum Group Metals (PGMs) employs 164,513 and together with coal and gold, they account for 78 per cent of employment in mining. The industry contributes at least 8.1 per cent to the gross domestic product (GDP); in 2019 it contributed R360.9 billion[2] to the GDP. The spill-over effects of mining sustain industries and businesses which service and create more jobs. These industries, which have grown around the formal mining sector, accounts for nearly five per cent of South Africa's workforce and supports more than 1.3 million indirect jobs (Arnoldi, 2020).

However, mining's contribution to the GDP has been declining. At its peak in the 1980s, it accounted for 21 per cent of the GDP (Arnoldi, 2020). According to the FTI report (2019), a number of global and local trends in production have contributed to South Africa's mining woes. These include changing demand patterns; varied commodity price cycles; falling ore bodies; policy uncertainty and lagging cost competitiveness; labour costs; electricity and water price increases and decreases. These developments had negative effects on employment in mining; Du Plessis

---

1   For a detailed analysis on sub-contracted labour in the platinum sector, see Buhlungu and Bezuidenhout (2008).

2   See the *Facts and Figures Pocketbook 2019* by the Minerals Council of South Africa: https://www.mineralscouncil.org.za/industry-news/publications/facts-and-figures, accessed 2 May 2020.

(2019) finds that full-time employment plummeted between 2013 and 2018 with more than 56,366 jobs cuts.

South Africa's mining future is further impacted by the increase in tariffs granted by the National Energy Regulator of South Africa (NERSA) to Eskom, the state-owned electricity supplier (Casey, 2019). The gold mining operations are going to be the most affected, requiring more electricity to process. Casey (2019) points out 95,000 jobs could be lost in gold mines alone with the higher tariffs having far-reaching consequences on the industry. South Africa's Mineral Council claims over 200,000 jobs are at risk in the near future.

This contextualisation is important because it foregrounds conditions that contribute to the industry's decline. However, South Africa is part of the global mining industry and its sustainability is linked to global trends. Unless companies are globally cost competitive, anticipate change in demands and adapt to technology currently being deployed in mining for operational efficiencies and output, South African mining is at risk of becoming an irrelevant player in the world mining industry. Logically, the domestic industry will use technology to increase production, remain competitive and meet demands. These technological changes along with other factors have unfavourable outcomes for employment. Across the world, mining companies are deploying autonomous and robot technologies in production and they are investing in the next generation of 4IR technologies for efficiency and to improve their competitive advantage (Cosbey et al., 2016). The following autonomous equipment are game changers:

- Driverless haul truck and loaders – already operated remotely by one person – with new algorithm programmes that improve their efficiency;
- Automated drilling and tunnel-boring systems – used in open-pit mines requiring one operator monitoring three or more machines from remote stations;
- Automated long-wall plough and shearers – currently being implemented in the coal mining sector.

The point therefore is that technology is speeding up the evolution of global mining as companies shift their strategies and adopt new operational models, which are increasingly becoming mechanised and automated.

What do these changes in technology mean for employment? The implications for the workforce are that the demand for labour will decline due to mining companies utilising emerging technologies in production (Deloitte, 2018a).

Cosbey et al. (2016) anticipate that automation will have a displacing effect on blue-collar, low-skilled and low-paid mineworkers. Machines – such as automated rock splitters – can replace workers in jobs that are repetitive, unchanging and labour intensive: truck and ship loaders, crushing and grinding machine operators and setters, excavating and loading machine operators, and dragline operators can all be substituted with automation (Cosbey et al., 2016; Clifford et al., 2018). Technological innovation can result in tasks being shifted from one occupation to another, thus displacing some workers (Bessen, 2015).

Literature on an automated mining sector (Marr, 2018; Ghebrihiwet, 2019; Clifford et al., 2018) indicates that remote controlling and supervision will require few workers, thus reducing employment and changing the future of work. What has been South Africa's mechanisation and automation experience? What implications has technology had on employment in mining?

South Africa's experience

Evidence relating to the diffusion of technology in mining production in South Africa is shared between mechanisation and automation. Webber et al. (2010) find that mechanised operation in Lonmin's Hossy shaft increased production, improved safety and upskilled supervisors for mechanised operation. Overall the net positive effect outweighed the implementation challenges. Moreover, the costs of training workers were far less than repairing damaged mechanised equipment.

In open-pit platinum mining, there is evidence that mechanisation has been a key driver of successfully achieving a record-breaking production target at Anglo American Platinum Mogalakwena mine in 2019; utilisation of machines reduced disruptions to production. McKay's (2018) evidence shows that the mine yielded much higher ounces per worker per month. This is in line with Autor's (2014) argument that machines and technology make workers more productive. A dedicated project to monitor and improve equipment efficiency through a tracking system enabled the mine to achieve its targets. Central to the success of Mogalakwena is the relationship between stakeholders and the people-centred approach to incorporating technology on the mine.

In gold mining, the experience has been different. The South Deep mechanisation project has been a nightmare for Gold Fields. The project has failed to yield positive return on a R22-billion acquisition that included transitioning from traditional methods of mining to modernised ones (Kotze, 2020). The underperformance adds to South Deep's economic concerns, as the mine already spent R100 million a month to operate.

Recently, autonomous trucks were successfully tested at coal mines in South Africa as proof of concept: this project revealed the potential to increase productivity since the trucks can haul 120 loads per day compared to 60 loads for current trucks (Moolman, 2018). However, autonomous trucks have not been implemented permanently; workers are still used for hauling.

Tassell (2018) finds that at the Finsch diamond mine in the Northern Cape a fleet of six autonomous trucks in operation for more than ten years has increased productivity, improved safety and reduced operational costs. He notes that South Africa currently has three mining projects that could mechanise and eventually adopt autonomous technology.

The evidence thus far suggests that mining companies in South Africa are increasingly using autonomous and mechanised technology for production. However, the decisions to automate or mechanise are mostly motivated by seeking to maintain and gain competitive advantage, improve safety and productivity and reduce costs. The Mineral Council of South Africa (2019) confirms this evidence, showing that companies believe technology can be a solution to accessing narrow reef resources; modernising from labour-intensive manual drilling activities; and reducing the cost of production, electricity and water. Technology potentially augments the capital-intensive character of mining, simultaneously increasing production output and breathing new life into mines.

Mining's different sub-sectors will always deploy technology to varied degrees, and it is important to recognise this when studying the impact of mechanisation and automation in the industry.

Implication on employment and jobs

Gumede (2018) draws attention to the socioeconomic effects of mechanising mines in South Africa, with evidence that employers and employees agree that mechanisation and subsequent automation will lead to job losses in the short-term but will create new jobs since it will extend the life of a mine. Trade unions consider mechanisation and autonomous technology to be a threat to current mineworkers by changing the profile and demographic of the workers required to young, technically skilled workers. Like Cosbey at al. (2016), Gumede stresses the displacing effects are likely to be felt most among lower-level, older employees in routine- and operator-type jobs with limited skills and education.

The effects of automation on employment and jobs in South Africa is context-specific, in that the country's history of discriminatory, apartheid labour market practices, together with mines' exploitative history, created a workforce characterised by low education and low skills levels. This means, as automation becomes part of operations and machines perform tasks previously carried out by workers, the value of labour declines along with their usefulness to production.

For example, if production becomes completely automated and the mineworker of the future is required to ensure that production continues by synching current technologies with 4IR innovations, then South Africa's current black, older mine workers with low education and skills levels cannot manage such processes or operations. They simply do not have the skills needed for mining that is technology reliant. This development is a great opportunity for young, skilled black women entering the workforce and points to a South Africa mining future that must take this shift into account. These new developments raise an important question for policymakers and companies alike: What will be the composition and defining features of a technology-reliant mining workforce?

Tassell (2018) finds the anxieties about job displacement due to automation to be overstated, as this automation enables access to areas previously deemed uneconomic and unsafe for mineworkers. In this sense, automation prolongs the life of a mine while safeguarding workers' safety. Tassell confirms the findings of Dauth et al. (2017) on the impact of the nature of German industrial relations by emphasising that automating decisions are best received and supported when they are taken as part of consultation processes that address the concerns of workers and unions.

Another point to make about decisions to automate is in line with the cautions voiced earlier, namely that the introduction of technology in to production processes is not predetermined. In South Africa's mining industry, currently rife with contentious politics and dissenting views, engagement is a better approach than top-down imposed decisions.

Earlier, the chapter discussed the diverse causes of falling productivity in South African mines, including evidence of declining grades, deeper deposits and aging mines, which have all contributed to influencing mining companies' decisions to automate. High wages are often assumed to be the reason for automation. But Strauss and Isaac (2016) have shown that in fact the median wage of miners has remained stagnant. Thus higher wages as reason for automating production makes no sense. There are more pronounced internal and external factors influencing decisions to automate than the need to displace wage-earning workers.

Summary

Multiple dimensions inform decisions to mechanise and automate production in mines. Despite its declining status, South Africa's mining industry is still part of the global one which is affected by changes in demands. In sum, mechanisation and automation have become part of mining. These technologies are going to be changed by the diffusion of new and more disruptive 4IR technologies, which will increasingly affect low-skilled, older workers. From this perspective, the future of work in mining, related to these dynamics, remains uncertain.

Furthermore, relative to other mining countries where low-skilled mineworkers are being trained to adapt to technology, in South

Africa reskilling raises a serious question for government, unions and companies. Considering the high levels of unemployment among graduates and youth, do they reskill older mineworkers with low education levels, or replace them with younger workers?

The digital revolution in banking

The banking sector's relationship with technology can be understood as one based on the need for survival, necessitated by innovations made in the information technology (IT) arena and the need to streamline operations and products in order to service customers. The digitalisation and synchronising of existing systems, with new ones currently unfolding, in South Africa's banking industry is not merely to keep up with the global trend. Rather, it is to establish systems that are compatible with 4IR disruptive technologies, which are changing the banking ecosystem – including a complete overhaul of the industry, its operations and the kind of workforce it requires.

External forces acting on the banking industry include the competitive pressure from financial technology (FinTech) start-ups and companies like PayPal, Facebook, Amazon and Apple entering the banking business (Jakšič and Marinč, 2019). These companies' ability to profitably scale the connection between technology and banking poses a threat to the competitive advantage banking has. Verhage and Surane (2019) confirm this with evidence showing that tech companies are considering incorporating payments, lending and insurance into their business models as they venture into the financial services.

The effect on employment as automation becomes a defining feature of banking is the displacement of workers due to digital technologies taking over tasks and performing work without the need for human input. Meena and Parimaralani (2019) report that this will lead to a decline in the workforce. However, they also point out that the role of staff is not outdated since, in this new banking ecosystem, employees can work along with technology to ensure efficiency.

As alluded to earlier on in this chapter, South Afrcan banking is currently in the throes of digital transformation occasioned by global trends, which include modernising old systems to better serve customers. PricewaterhouseCoopers (PwC) South Africa (2017) reports that banks are also responding to threats posed by non-traditional new players who are making inroads into financial services by offering customers banking solutions. This development is already evident with Discovery's recent opening of a banking division.

This evidence supports the findings of Verhage and Surane (2019), cited earlier, that banking needs to adopt new technology to find ways to maintain a competitive advantage and remain relevant to changing markets. Therefore, decisions to incorporate emerging technologies in operations and the digital transformation underway is a key driver of the ability to survive. For example, the digital service offered by banks reduces the relevance of going into a branch to get statements and deposit money, as these activities can be done remotely using online banking, cellphone banking apps and e-wallet. Therefore, companies that can take advantage of the digitisation process to redefine customer experiences will retain current clients and attract new ones.

The digital transformation is determining the future of work in banking, with adverse effects on employment. This is the paradox of technology disruption in banking. Its systems and tools give banks a competitive edge, improve customer-service and make new markets accessible. However, this same disruption has concurrent displacing effects on jobs, which lead to the decline of employment in the industry. Additionally, traditional banking branches are being replaced by flexible, interconnected AI systems that enable banking from anywhere. In a countermove to keep branches open and sustainable, some banks are increasing their over-the-counter services, such as smart ID documents and passports in collaboration with Home Affairs.

Implication on employment and jobs in South Africa

South Africa's banking sector is part of a global industry that is being redefined by technology. There are 158,000 people employed by banks in South Africa; between 2015–2018 banking was a net job creator (Haffajee, 2019: 1). According to the 2019 *Transformation Report*, the biggest growth in employment was at junior management level with black workers accounting for 84 per cent, or 54,000 workers in that category – a six per cent year-on-year increase. In contrast, senior management progression has been slow, with only 3,119 black senior managers in 2017 (BASA, 2019). While there has been progress in meeting employment equity targets, black workers in banking mostly remain in junior positions.

Furthermore, the labour market profile of banking is dominated by Absa, First National Bank (FNB), Standard Bank and Nedbank: they account for 67 per cent of the total employment in the sector, employing more females than males (61 per cent women and 39 per cent men).[3] The banking sector in South Africa has a young workforce, with 53 per cent being younger than 35 years old, and 37 per cent aged between 35 and 55. Moreover, most employees have a diploma or first-degree qualification. The profile of the banking sector can be characterised as being dominated by young black female workers with post-matric education levels. However, most are in junior management positions.

We can therefore see how banking has been a net job creator, especially for black women in contemporary South Africa. Are they safe from disruptive technologies? And what about the industry itself: what is it doing to keep up with the integrated global economy and the changes caused by technology?

The banking sector in South Africa operates within a stagnated domestic economy that was already in recession before COVID-19 and the ensuing stringent lockdown from 26 March 2020. As part of the global financial industry, South Africa's banking sector operates within, and is affected by, the macroeconomic context of the country and the global economy (PwC, 2019). The synergy between old and new technologies, such as machine-learning-powered AI systems, blockchain and cloud computing is one of the causes of disruption of work and employment in the banking industry.

There is evidence (PwC, 2017; Deloitte, 2018b; Deloitte, 2019) that digitalisation and modernisation were among the factors that contributed to the 2019 job cuts in the global banking industry. Comfort (2019) estimated that 75,700 jobs were cut by banks around the world with 83 per cent being in Europe. Three major factors reinforce the precarious state South Africa's banking sector will find

---

3   See BANKSETA (2018) Banking Sectors Skills Plan Report.

itself in moving forward. First, the downgrading of South Africa to junk status by Fitch and Moody's in March and April (Mnyanda, 2020). Second, the downgrade of the top four banks' credit ratings, along with the sovereign rating, to junk by the same agencies (Rumney and Reuters, 2020). Third, the subsequent crashing of the rand to historical lows against the dollar amidst these downgrades and the COVID-19 lockdown (Mnyanda, 2020).

Banking in South Africa employs 59,971 clerical support workers[4] whose occupations are at risk: as in mining, tasks that are routine, predictable and repetitive face automation. Considering that most bank employees are young, black, female workers in junior positions the tasks central to their occupation put them at risk of being substituted by smart machines powered by AI, notwithstanding their education levels.

One of the reasons South African 'lenders are cutting jobs' (Business Tech, 2020: 2) is for cost reduction purposes but also as a result of AI's ability to change old systems to establish new operations or means of production. PwC South Africa (2017) suggests the role of disruptive innovation has compelled all major banks to: 1) replace core systems; 2) digitally transform from manual systems; 3) implement new cyber security systems to minimise risk exposure; and 4) digitise front office-customer and back-office operations.

Bloomberg News (2019) reports that, in 2019, Absa restructured operations across its business units; Standard Bank planned to close 91 branches, while Nedbank was consulting more than 1,500 employees about redeployment or retrenchment. This is in line with the global trend, where close to 80,000 jobs, the highest in one year, since 2015, were lost. In 2019, the trade union South African Society of Bank Officials (SASBO) said plans by four major banks to restructure their operations towards digital transformation would lead to at least 10,000 job losses (Mungadze, 2019).

The changes to South Africa's banking, and to the sector's employment landscape, raise questions about the education and

---

4 See BANKSETA (2018) Banking Sector Skills Plan 2017/18 Report for a breakdown per province.

upskilling of workers in responding to technology. These issues have also been evident in the mining sector, where connections are being made between workers with low education levels, in routine and labour-intensive occupations, and susceptibility to displacement.

This discrepancy can again be explained by evidence found in Bessen (2015), which shows how technologies can shift tasks from one occupation to another, potentially displacing workers. In banking, new tools such as big data, AI, smart machines and blockchain mean that jobs shift from workers to machine systems. In this way, these new technologies, together with digital transformation, are redefining operating models and redesigning the workforce.

The banking case study provides evidence that differs from that provided by Kaggwa (2018). He writes of employees with high education attainment levels being are at risk of being displaced by the diffusion of technology in operations and the workplace.

Summary

The convergence of new technologies, the need to maintain a competitive edge over the threat posed by new challengers, non-traditional tech companies and the need to keep the consumer happy all drive the changes occurring in South Africa's banking industry. Moreover, decisions are taken with a top-down approach that is informed by digital technologies' transformational nature, which is indeed transforming the banking landscape.

The evidence presented suggests that the role of technology in banking is having negative effects on jobs. Furthermore, the marriage between current technologies and increasingly evolving new ones, such as operating systems powered by AI, is redefining the banking sector workforce. In South Africa black women in operator and routine occupations have a higher risk of being displaced by AI automation in customer-service and clerical jobs. However, the role of workers in banking is not obsolete; new technology may be reinventing the work but implementing it still requires humans to maintain and ensure its efficiency.

CONCLUSION: THE ROLE OF INSIGHTS
IN SHAPING POLICY DISCUSSION

An assessment of the displacing effects of current technologies on employment should be based on an understanding of how these differ from country to country and from sector to sector. For example, the insights from country experiences demonstrate the varied effects technologies have on employment in manufacturing industries in different countries. In Germany, no displacement effect was found when industrial automation occurred; furthermore, the industrial relations system – defined by workers' councils – engages in decisions on how technology is adopted. German unions prefer full employment to job loss and are therefore receptive to wage flexibility to retain jobs.

In Japan, workers did not see robots as a threat to their jobs because these new technologies augmented their current vocational skills and qualifications. The threat for Japan is not industrial automation or other technology; it is rather the aging working population and the resulting decline in the country's labour force. Therefore, technology has a positive effect because it fills the gap created by insufficient numbers of workers.

In the US, evidence suggests a low substitution of workers by robots, with the negative effect of industrial robot adoption in manufacturing offset by jobs created in other industries. There are, however, other already mentioned causal factors for this. The decline in the manufacturing industry in the US, which has been detrimental to employment, occurred for a number of reasons and not necessarily because of the deployment of automation technology.

China's experience draws attention to the upward shift of its position in the global economy, precipitated by the booming demand for goods and services in the global market. Therefore, decisions to promote technology-based production were part of a deliberate government policy of financially supporting domestic companies to meet global demands, en route to China becoming the world's leading manufacturer. China, like Japan, has a shrinking labour force problem owing to an aging population; and therefore the move towards technology replacing labour is logical. Another reason for the adoption of technology in production is that wages in the manufacturing industry continue to rise; this has incentivised companies to automate production because robots are cheaper than labour costs.

The country experiences clearly illustrate the usefulness of understanding what transpired in developed states. They offer government and policymakers in South Africa insight into the possible course of technological impact in the country: we are an emerging economy operating within a global one and within the context of a changing global production system.

The insights from banking and mining demonstrate how mechanisation, automation, IT and digital transformation have an impact on the employment of two different sectors of the workforce: 1) blue-collar, low skills and education, migrant, black, male mineworkers; and 2) medium to high education, white-collar, black, female employees. Furthermore, structural change in the banking sector, from legacy systems to a reconfigured, modernised one, will result in the movement of workers in and out of the labour market. For example, banking, an industry that has traditionally had a workforce with high education levels, is affected by labour-substituting technology.

Taken together, the experience of mining and the contentious politics inherent in the industry, and the role of the Chinese government in promoting technology-themed policy, offer lessons that can be useful for South African policymakers. Moreover, the banking and mining industries localise the impact, identifying new technologies and anticipating their ensuing disruption effects, including on employment. This can better enable the government to devise action plans that consider multiple outcomes and futures. One example would be conducting systematic research on the role of industrial robots and automation in Africa, rather than studying only how technology improves productivity and efficiency.

Consider the following scenario, which depicts an action plan for facilitating a sustainable future for the mining industry. I will call this, mining that adapts to technological shifts. In this scenario, policy design is approached from a collaborative perspective and includes business, labour and academic institutions. For example:

- Academic institutions could conceptualise an education system that is linked to skills required by business.
- Business could assist in the design of basic and higher education curricula suited for varied industries and could support lifelong learning for workers.
- Government could create policies and legislation that are adaptive to technological changes including humans working alongside robots.
- Labour unions' organisational strategies can shift to favour employment creation and the retention of work over wage structures that only improve the conditions of unionised workers and discourage employment for unskilled workers.

Past experience is a prologue and it bestows lessons for strategies South Africa could undertake before making 4IR central to the country's economic blueprint for tackling unemployment challenges – especially as there is evidence that 4IR will not solve the country's structural challenges. In developed countries, firms in different industries are deploying technology for survival and competitive reasons. In South Africa, an emerging economy, the same reasons for embracing technology apply. However, the country faces the fact that technology can entrench the challenges inherent in a labour market already characterised by high levels of unemployment.

REFERENCES

Acemoglu, D. and Restrepo, P. 2020. 'Robots and jobs: Evidence from US labour markets'. *Journal of Political Economy.* https://doi.org/10.1086/705716, accessed 2 May 2020.

Andes, S. and Muro, M. 29 April 2015. 'Don't blame the robots for lost manufacturing jobs'. *Brookings Institute.* https://www.brookings.edu/blog/the-avenue/2015/04/29/dont-blame-the-robots-for-lost-manufacturing-jobs/, accessed 9 July 2020.

Arnoldi, M. 4 February 2020. 'South African mining far from former glory; needs to focus on tech, ESG'. *Mining Weekly.* https://www.miningweekly.com/article/south-african-mining-far-from-former-glory-needs-to-focus-on-tech-esg-2020-02-04/rep_id:3650, accessed 11 July 2020.

Autor, D. 2014. 'Polanyi's Paradox and the shape of employment growth'.

MA: *National Bureau of Economic Research*. http://www.nber.org/papers/w20485.

Autor, D. and Salomons, A. 2018. 'Is automation labour-displacing'. BPEA Conference Drafts. https://conference.nber.org/conf_papers/f100969.pdf, accessed 28 June 2020.

Banking Association of South Africa. 2018. 'Annual Report 2018- 07/08/2019'. https://www.banking.org.za/wp-content/uploads/2019/08/BASA-Annual-Report-2018.pdf, accessed 2 April 2020.

Bessen, J. 2015. 'How computer automation affects occupations: Technology, jobs and skills'. Law and Economics Working Paper No. 15-49, Boston University School of Law, Boston, Mass.

Bessen, J., Goos, M., Salomons, A. and Van den Berge, W. 2020. 'Automation: A guide for policymakers'. *The Brookings Institution*. https://www.brookings.edu/wp-content/uploads/2020/01/Bessen-et-al_Full-report.pdf, accessed 25 February 2020.

Buhlungu, S. and Bezuidenhout, A. 2008. 'Union solidarity under stress: The case of the National Union of Mineworkers in South Africa'. *Labor Studies Journal* 33 (3), 262–287.

Business Tech. 10 February 2020. 'Over 9,000 planned job cuts have been announced for South Africa in 2020 – these are the companies affected'. *Business Tech*. https://businesstech.co.za/news/business/372434/over-9000-planned-job-cuts-have-been-announced-for-south-africa-in-2020-these-are-the-companies-affected/, accessed 26 February 2020.

Calitz, A.P., Poisat, P. and Cullen, M. 2017. 'The future African workplace: The use of collaborative robots in manufacturing'. *SA Journal of Human Resource Management*, 15(1), 1–11.

Casey, J.P. 26 March 2019. 'South Africa could lose 200,000 mining jobs, according to Minerals Council'. *Mining Technology*. https://www.mining-technology.com/news/south-africa-could-lose-200000-mining-jobs-according-to-minerals-council/, accessed 10 July 2020.

Clifford, M.J., Perrons, R.K., Ali, S.H. and Grice, T.A. 2018. *Extracting Innovations: Mining, Energy, and Technological Change in the Digital Age*. Boca Rotan: CRC Press.

Comfort, N. 2019. 3 December 2019. 'Global bank job cull tops 75,000 this year as UniCredit cuts'. *Bloomberg*. https://www.bloomberg.com/news/articles/2019-12-03/unicredit-pushes-global-bank-job-cuts-past-70-000-mark-this-year, accessed 2 April 2020.

Cosbey, A., Mann, H., Maennling, N., Toledano, P. et al. 2016. 'Mining a mirage? Reassessing the shared-value paradigm in light of the technological advances in the mining sector'. International Institute for Sustainable Development, Columbia Centre on Sustainable Investment. http://ccsi.columbia.edu/files/2015/07/mining-a-mirage-CCSI-IISD-EWB-2016.pdf.

Crush, J. 2001. 'The dark side of democracy: Migration, xenophobia, and human rights in South Africa'. *International Migration*, 38, 103–133.

Dauth, W., Findeisen, S., Südekum, J. and Woessner, N. 2017. 'German robots – the impact of industrial robots on workers'. CEPR Discussion Paper No. DP12306.

Davenport, J., 2013. *Digging Deep: A History of Mining in South Africa, 1852–2002.* Johannesburg: Jonathan Ball Publishers.

Deloitte. 2018a. 'The future of mining in Africa: Navigating a revolution'. *Deloitte.* https://www2.deloitte.com/content/dam/Deloitte/za/ Documents/energy-resources/za_Future_of_mining.pdf.

Deloitte. 2018b. '2019 Banking and Capital Markets Outlook: Reimagining transformation'. https://www2.deloitte.com/content/dam/Deloitte/ global/Documents/Financial-Services/gx-fsi-dcfs-2019-banking-cap-markets-outlook.pdf, accessed 2 May 2020.

Deloitte Development LLC (Firm). 2019. 'The Fourth Industrial Revolution: At the intersection of readiness and responsibility'. https://www2.deloitte. com/content/dam/Deloitte/de/Documents/human-capital/Deloitte_ Review_26_Fourth_Industrial_Revolution.pdf, accessed 12 May 2020.

Du Plessis, G. 5 June 2019. 'SA mining is facing catastrophe unless it can manage the effects of its shrinkage' *Miningmx.* https://www.miningmx. com/news/markets/37205-sa-mining-is-facing-catastrophe-unless-it-can-manage-the-effects-of-its-shrinkage/, accessed 12 July 2020.

FTI Consulting Inc. 2020. 'Mining Sustainability in South Africa Report'. https://www.fticonsulting-emea.com/~/media/Files/emea--files/insights/ reports/2020/feb/mining-sustainability-south-africa.pdf, accessed 11 July 2020.

Ghebrihiwet, N. 2019. 'FDI technology spillovers in the mining industry: Lessons from South Africa's mining sector'. *Resources Policy*, 62(C), 463–471.

Gillwald, A. 20 August 2019. 'South Africa is caught in the global hype of the fourth industrial revolution'. *The Conversation.* http://theconversation. com/south-africa-is-caught-in-the-global-hype-of-the-fourth-industrial-revolution-121189, accessed 23 January 2020.

Graetz, G. and Michaels, G. 2018. 'Robots at work'. *Review of Economics and Statistics*, 100(5), 753–768.

Grossman, D. 22 June 2017. 'Highly-automated Austrian steel mill only needs 14 People'. *Popular Mechanics,* https://www.popularmechanics.com/ technology/infrastructure/a27043/steel-mill-austria-automated/, accessed 27 March 2020.

Gumede, H. 2018. 'The socio-economic effects of mechanising and/or modernising hard rock mines in South Africa'. *South African Journal of Economic and Management Sciences*, 21(1), 11.

Gwatidzo, T. and Benhura, M. 2013. 'Mining sector wages in South Africa'.

Labour Market Intelligence Partnership (LMIP) Working Paper, 12013.

Haffajee, F. 25 September 2019. 'Banks still net job creators says industry body as strike looms'. *Financial24.* https://www.fin24.com/Companies/Financial-Services/banks-still-net-job-creators-says-industry-body-as-strike-looms-20190925-2, accessed 3 April 2020.

Harington, J.S., McGlashan, N.D. and Chelkowska, E.Z. 2004. 'A century of migrant labour in the gold mines of South Africa'. *The Journal of The South African Institute of Mining and Metallurgy.* http://www.saimm.co.za/Journal/ v104n02p065.pdf, accessed 25 May 2020.

Harvey, E. 17 October 2019. 'Trade unions are unprepared for the 4IR tsunami coming their way'. *Business Day.* https://www.businesslive.co.za/bd/opinion/2019-10-17-trade-unions-are-unprepared-for-the-4ir-tsunami-coming-their-way/, accessed 2 April 2020.

Hattingh, T.S., Sheer, T.J. and Du Plessis, A.G. 2010, 'Human factors in mine mechanisation'. The 4th International Platinum Conference: *Platinum in Transition 'Boom or Burst',* The Southern African Institute of Mining and Metallurgy, Johannesburg.

Hogg, A. 12 March 2019. 'Mining Charter III: Certainty, but at a cost – Peter Leon'. *BizNews.* https://www.biznews.com/briefs/2019/03/12/mining-charter-peter-leon, accessed 31 March 2020.

Houseman, S. 7 September 2018. 'Is automation really to blame for lost manufacturing jobs?'. *Foreign Affairs.* https://www.foreignaffairs.com/articles/2018-09-07/automation-really-blame-lost-manufacturing-jobs, accessed 30 March 2020.

Jakšič, M. and Marinč, M. 2019. 'Relationship banking and information technology: The role of artificial intelligence and FinTech'. *Risk Management,* 21, 1–18. https://doi.org/10.1057/s41283-018-0039-y.

Kaggwa, M. 6 May 2018. 'Technology threatens trade unions survival.' *News24.* https://www.news24.com/Columnists/GuestColumn/technology-threatens-trade-unions-survival-20180504, accessed 2 April 2020.

Kotze, C. 13 February 2020. 'South Deep gold mine: Gold Fields' South African mine back in the black'. *Mining Review Africa.* https://www.miningreview.com/gold/south-deep-gold-mine-gold-fields-south-african-mine-back-in-the-black/, accessed 9 July 2020.

Lawrence, R.Z. 2019. 'China, like the US, faces challenges in achieving inclusive growth through manufacturing'. *China & World Economy,* 28(2), 3–17.

Leduc, S. and Zheng L. 2020. 'Robots or workers? A macro analysis of automation and labor markets'. Federal Reserve Bank of San Francisco Working Paper 2019-17. https://doi.org/10.24148/wp2019-17.

Lye, W.F. 1984. *Three Views of the Migrant Labor System in South Africa.* Bloomington: Indiana University Press.

Magubane, B. 1979. 'The political economy of race and class in South Africa/New York/London'. *Monthly Review Press.*

Marr, B. 7 September 2018. 'The 4th Industrial Revolution: How mining companies are using AI, machine learning and robots'. *Forbes.* https://www.forbes.com/sites/bernardmarr/2018/09/07/the-4th-industrial-revolution-how-mining-companies-are-using-ai-machine-learning-and-robots/#376b205c497e, accessed 15 January 2020.

McKay, D. 15 August 2018. 'South Deep, a story of the big mine that couldn't'. *Miningmx.* https://www.miningmx.com/news/gold/33993-south-deep-the-mine-that-couldnt/, accessed 9 July 2020.

Meena, R. and Parimalarani, G. 2020. 'Impact of digital transformation on employment in banking sector'. *International Journal of Scientific & Technology Research*, 9, 491–4016.

Mhene, N. 18 October 2019. 'Farmworkers and 4IR'. *Mail & Guardian.* https://mg.co.za/article/2019-10-18-00-farmworkers-and-4ir/, accessed 2 April 2020.

Miller, C. 21 December 2016. 'The long-term jobs killer Is not China. It's automation'. *New York Times.* https://www.nytimes.com/2016/12/21/upshot/the-long-term-jobs-killer-is-not-china-its-automation.htm, accessed 27 March 2020.

Mineral Council of South Africa. 27 March 2020a. 'Mineral Council files application for the review of Mining Charter 2018'. *Minerals Council of South Africa.* www.2019-03-27-minerals-council-files-application-for-review-of-mining-charter_2018.pdf, accessed 1 April 2020.

Mineral Council of South Africa. 2020b. 'Facts and Figures Pocketbook 2019'. https://www.mineralscouncil.org.za/industry-news/publications/facts-and-figures, accessed 2 May 2020.

Mnyanda, L. 3 April 2020. Fitch downgrades SA further into junk in another blow for local markets'. *Business Day.* https://www.businesslive.co.za/bd/economy/2020-04-03-fitch-downgrades-sa-further-into-junk-in-another-blow-for-local-markets/, accessed 4 April 2020.

Molopyane, M. 19 March 2018. 'Fourth Industrial Revolution is not a silver bullet for an ill-prepared nation'. *Business Day.* https://www.businesslive.co.za/bd/opinion/2018-03-19-fourth-industrial-revolution-is-not-a-silver-bullet-for-an-ill-prepared-nation, accessed 8 January 2020.

Mongale, I. 2019. 'The implications of labour productivity and labour costs on the South African economy'. *Journal of Reviews on Global Economics*, 8, 1298–1307. 10.6000/1929-7092.2019.08.113.

Moolman, V. 11 May 2018. 'Autonomous mining vehicle test at SA coal mine successfully completed'. *Mining Weekly.* https://m.miningweekly.com/article/autonomous-mining-vehicle-test-at-sa-coal-mine-successfully-completed-2018-05-11/rep_id:3861, accessed 11 July 2020.

Morikawa, M., 2018. 'Assessing the impact of AI and robotics on job expectations using Japanese survey data'. *RIETI Highlight*, 68, 21–23.

Mungadze, S. 25 September 2019. 'Sasbo vows complete blackout of digital banking'. *iTWeb*, https://www.itweb.co.za/content/mYZRXM9PjXb7OgA8, accessed 25 February 2020.

Ndung'u, N.S. and Signé, L. 2020. 'Capturing the Fourth Industrial Revolution: A regional and national agenda'. *Brookings Institution*, https://www.brookings.edu/wp-content/uploads/2020/01/ForesightAfrica2020, accessed 8 January 2020.

Oxford Economics. 2019. 'How robots are changing the World Report'. *Oxford Economics.* https://cdn2.hubspot.net/hubfs/2240363/Report%20-%20How%20Robots%20Change%20the%20World.pdf?utm_medium=email&_hsenc=p2ANqtz-9zijSKYffbkzW-_DaKCLPsc4WcEcExW0ZeP_KmbujqBzH9ZK4ui_7rUev862O4vxvlh7I2JPwLPS-pVxq1lxeyDTfumQ&_hsmi=74013545&utm_content=74013545&ut, accessed 25 May 2020

PwC South Africa. 2017. 'The future of banking: A South African perspective'. https://www.pwc.co.za/en/assets/pdf/strategyand-future-of-banking.pdf, accessed 12 January 2020.

PwC South Africa. 2019. 'South Africa Major Banks Analysis'. https://www.pwc.co.za/en/assets/pdf/major-banks-analysis-nov-2019.pdf, accessed 2 May 2020.

Rumney, E. and Reuters. 1 April 2020. 'Fitch, Moody's downgrade SA banks to junk'. *Moneyweb*. https://www.moneyweb.co.za/news/south-africa/fitch-moodys-downgrade-sa-banks-to-junk/, accessed 3 April 2020.

Reid, J., Templeman, L. and Mahtani, S. 2018. 'Automation – not a job killer'. Deutsche Bank Research. https://www.dbresearch.com/PROD/RPS_EN-PROD/PROD0000000000469843/Automation_-_not_a_job_killer.PDF, accessed 11 July 2020.

Schneider, T., Hong, G.H. and Le, A.V. 2018. 'Land of the rising robots.' *Finance and Development,* 55(2), 28–31.

Stoddard, E. 6 February 2019. 'BCG, McKinsey & Co say no reboot for South African mining without productivity boost'. *Miningmx*. https://www.miningmx.com/special-reports/mining-indaba/mining-indaba-2019/35874-bcg-mckinsey-co-say-no-reboot-for-south-african-mining-without-productivity-boost/, accessed 11 July 2020.

Strauss, I. and Isaacs, G. 2016. 'Labour compensation growth in the South African economy: Assessing its impact through the labour share using the Global Policy Model'. University of the Witwatersrand, Working Paper Series No. 4.

Tassell, A. 24 April 2018. 'Mining ready to embrace automation'. https://www.crown.co.za/modern-mining-featured-news/6882-mining-industry-ready-to-embrace-automation, accessed 10 July 2020.

Verhage, J. and Surane, J. 23 December 2019. 'Big tech is coming for banking: Experts predict Fintech's 2020'. *Bloomberg*. https://www.bloomberg.com/news/articles/2019-12-23/big-tech-is-coming-for-banking-experts-predict-fintech-s-2020, accessed 9 July 2020.

Webber, G., Van Den Berg, A., Le Roux, G. and Hudson, J. 2010. 'Review of mechanization within Lonmin'. Proceedings of the 4th International Platinum Conference: *Platinum in Transition 'Boom or Bust'*, Southern African Institute of Mining and Metallurgy, Johannesburg, 277–284.

Williams, M. 1975. 'An analysis of South African capitalism – Neo-Ricardianism or Marxism?'. *Bulletin of the Conference of Socialist Economists*, 4(1), 1–38.

Wittenberg, M. 2014. 'Analysis of employment, real wage, and productivity trends in South Africa since 1994'. ILO Conditions of Work and Employment Series, Working Paper 45. Geneva: ILO. http://nationalminimumwage.co.za/wp-content/uploads/2015/09/0038Wages-etc-Wittenberg_ILO_2014.pdf, accessed 12 July 2020.

Zaman, R. 2018. 'Do robots protect and create human employment in China?'. *Tech Policy View.* https://www.techpolicyviews.com/do-robots-protect-and-create-human-employment-in-china/, accessed 11 July 2020.

EIGHT

The impact of new technologies on labour relations and market structures in the economy: The case of Uber

KHWEZI MABASA AND MZUKISI QOBO

INTRODUCTION

The Fourth Industrial Revolution (4IR) is no longer something that is speculated about. It is very much a part of how we live, work and play. It is evident in the digital products we use for leisure and work. Technological adoption has accelerated across the different sectors of the economy, including mining, manufacturing and services. This trend is likely to accelerate as companies increase the utilisation of labour-saving technologies as part of their risk-mitigating strategies.

For this reason, we should look at the precise ways in which 4IR manifests itself. Many countries around the world are putting in place processes and measures to harness the positive effects of 4IR while countering its adverse effects, especially those felt by blue-collar workers and vulnerable groups. This chapter argues against passive automaticity towards 4IR, and that there is scope for new regulatory forms to manage distributional consequences. It is our contention that the 4IR, and the various forms in which it expresses itself, creates winners and losers. It is important that we consider the impact of new technologies on labour relations and market structures in the economy. In the analysis, we develop a deeper understanding of the impact of these technologies, especially for those workers that are governed by Artificial Intelligence (AI)-based algorithms in the transport sector.

The 4IR has features that are vastly different from what we have seen over the centuries in the evolution of the capitalist mode of production. While the First to the Third Industrial Revolutions were largely characterised by labour-intensive processes, with systemic automation emerging as a feature of the Third Industrial Revolution, the 4IR has come to place greater emphasis on comprehensive substitution of technology for human labour with higher returns to capital. It was not conceivable in the old era to experience economic growth without increased employment, yet in the 4IR we see de-coupling of growth from employment (Ford, 2015). It is not the first time that technology has displaced labour; automation in the previous era performed a similar function, but not at the scale and depth at which AI-powered technologies today displace human labour.

The 4IR can thus be viewed as a fundamental shift in the way production is organised and how society consumes what it produces. The 4IR is a trend that is associated with the introduction of AI and machine learning in production processes or, in short, robotics on a vast scale. The field of AI, which is a critical part of the 4IR, has been described as 'the science and engineering of machines that act intelligently' (Norvig, 2012).

This chapter will look at how these trends reconstitute employment relations for a certain category of workers – the taxi drivers who are 'driving partners' of Uber. The chapter also examines Uber's broader effects on the market structure of the public passenger taxi industry. We explore the following essential questions: how does the transition towards digital AI-driven economic models shift capital, state and labour relations? What is the distributive socioeconomic impact of the digital transition on labour markets and passenger taxi market structures? What forms of regulations are required to ensure an economically inclusive transition to a digital economy in South Africa?

The chapter addresses these questions using a heterodox political economy method. This approach is useful for grappling with the underlying socioeconomic contradictions and policy contestations developing in the digital economy transition. The chapter draws on data from academic journals, media reports, book chapters, Competition Commission reports and documented sources on state legislation. It commences with a discussion that locates the 4IR, digital economies and AI in South Africa's political economy context. It then proceeds to examine the Uber case study within debates on the shared political economy model. The case study is discussed using three essential themes: state regulation, labour market and technology, and pricing. The chapter concludes with some recommendations on state regulations.

SOUTH AFRICA AND THE 4IR

Changes and technological innovations heralded by the 4IR have left the regulators way behind. Industries and countries are faced with tough political, regulatory and moral choices. South Africa's economy is constantly evolving because of digital transformation and new technologies that redefine the landscape of commerce. The arrival of Uber in 2013 was a significant development. Google opened its new offices during 2015 in Johannesburg to promote a shift to the digital economy. In 2016, the digital streaming platform, Netflix, made its debut in South Africa. The country has been moving with speed in integrating its real economy into a dynamic digital economy. This chapter argues that we need to explore new avenues of regulation that are fit for purpose, especially to define and set out clear rules of the game in the digital economy and to protect workers from exploitative conduct. Additionally, the new regulations should address economic exclusion and advance the country's economic transformation policy imperatives.

Digital markets are an important source of competitiveness and innovation, but they can also exploit weak regulations to limit tax exposure or to opt out from obligations companies have towards labour. This can be damaging to social capital. As we argue here, new regulatory designs are needed to tame the social costs of AI such as labour being replaced by technologies and market structure disruptions. It is, therefore, essential to direct digital markets to engender a legitimate social purpose and not just to serve goal of profit maximisation. As Schwab (2017: 92) has pointed out, 'robots and algorithms increasingly substitute capital for labour, while investing (or, more precisely, building a business in the digital economy) becomes less capital intensive'.

The bulk of South Africa's future growth may likely come on the back of activities related to the digital economy. Thus, it is important to assess the precise impacts of these digital trends on the structure of work, labour relations and the structure of the taxi transport market. Digital platforms such as Uber should be viewed through the lens of regulation and social dynamics. South Africa is a country that has a dualistic economy, where modernity and affluence are juxtaposed with high levels of inequality and unemployment. The digital platforms are mostly accessible to those with secure incomes, in the middle-class and upper layers of society who have access to banking services and technology whereas the majority of the population lack connectivity or access to ICT products and services consumable within their social contexts.

The interactions that take place on the Uber platform mirror the socioeconomic structure of society somewhat. Many of the Uber drivers live in communities that are on the margins of urban centres. Some have been retrenched from the real economy and have little prospects of finding another job given the skills-bias nature of the economy; others are immigrants who have come to South Africa to eke out an income for their families.

Yet, some Uber drivers are professionals who moonlight in the gig economy while simultaneously running parallel businesses in IT, restaurants or trading activities. Many of these drivers have socioeconomic vulnerabilities due to the economic strain the country is going through, and this will be more so in the wake of the COVID-19 pandemic. What is common among Uber drivers is that they are at a relative disadvantage vis-à-vis the company they work for and have limited agency to shape their circumstances differently or influence the way they are treated by Uber. Digital platforms can play an important role in broadening opportunities for the unemployed and in cushioning the effects of structural change in the economy.

The key factors for a successful, thriving digital economy include the right policy setting and clear regulation. Building a digital economy for greater inclusion, innovation and competitiveness is a critical mission for South Africa. But this needs to be realised within a normative framework of inclusion and equity. While we may need to dispense with the lenses of traditional policy and regulation, there is still a need for co-designing appropriate regulatory frameworks for those companies that are offering services through digital platforms. The next section discusses the interplay between technological trends and social dynamics through the prism of the Uber platform.

SHARING OR TAKING ECONOMY?

The development of Uber is best understood when situated in a broader debate on the digital-driven shared economy model (Allen and Berg, 2014; Collier et al., 2017; Geitung, 2017). Allen and Berg (2014: 4) define this political economy as one based on:

> sharing the knowledge of goods and services to better exchange them. These exchanges are leveraged by ubiquitous and cheap knowledge made available through disruptive technology.

A shared economy operates on alternative development principles, with a particular emphasis on 'decentralised exchange, access over ownership of resources, and firms becoming the facilitator of exchange (rather than acting as a producer), and mechanisms of self-governance'. As Mazzucato (2018: 217) points out, the sharing economy 'works by reducing the frictions between the two sides of the market: connecting buyers to sellers, potential customers to advertisers, in more efficient ways.' There are contrasting views in contemporary political economy literature on whether shared economies ameliorate the negative externalities associated with conventional capitalism.

Companies such as Uber are part of what some economists refer to as 'platform capitalism', and operate in two-sided markets, developing and connecting the demand- and the supply-side of the market (Mazzucato, 2018: 216).

Shared-economy advocates argue that it decentralises ownership patterns in the political economy. They also point out how this economic model addresses market-entry barriers for enterprises and labour. This competitiveness is developed through decentralised market structures that differ from existing consolidated, rigid product markets. Flexibility in shared economies is presented as a unique feature, which affords workers or independent entrepreneurs more autonomy in market transactions. Proponents also highlight how digital platforms enhance economic efficiencies that lead to more sustainable resource management and usage (Allen and Berg, 2014; Pollio, 2019). All these factors, according to supporters, produce better service and lower costs for all market players, including consumers. Companies such as Uber, Airbnb and Kickstart are perceived to epitomise the shared economy business model (Allen and Berg, 2014; Matzler et al., 2015). However, what is forgotten are the constrained margins for the drivers and other employees in these companies. And the negative externalities experienced by customers.

The shared economy detractors cite shortcomings in deregulated markets that operate on *laissez-faire* principles. They highlight how minimal regulation exposes entrepreneurs, employees and customers to market failures (Stemler, 2016). A prominent example in the literature is the dominance of a firm that allows it to unilaterally determine transactions and pricing mechanisms (Collier et al., 2017; Geitung, 2017). This subsequently leads to socioeconomic exploitation and unequal distribution of income in the shared economy. Critics also believe that virtues of autonomy for employees or micro-enterprises are over-stated. Several case studies are presented in the critical literature to illustrate market dominance by large firms, which unilaterally determine how workers and micro-enterprises operate (Clarke and Urata, 2016; Rinne, 2018; Robinson, 2017).

There is also vast literature and reporting on the poor safety and product quality that develop as a result of weak regulatory oversight

(Stemler, 2016; Rinne, 2018). As Stemler (2016: 43) explains, 'the sharing economy puts an enormous number of people in inherently vulnerable positions – getting into someone's car, inviting someone into your home, etc.' The debate about Uber, state regulation and managing social conflict is related to these contrasting views on the shared economy. This chapter examines how Uber operates using a political economy lens that considers these contradictions.

UBER'S ORIGINS AND ITS FORAY INTO SOUTH AFRICA

The Uber app launched in South Africa in 2013 as part of its global expansion. The company was founded by two entrepreneurs, Travis Kalanick and Garrett Camp in California (2009), where its headquarters are based. In 2014 it was valued at $17 billion, yet it has no tangible assets or 'workforce' contracted in the traditional sense. It is what Ismail (2014) refers to as an exponential organisation. These are companies that use new business models and employ technology to drive rapid shifts in industry by leveraging limited assets and staying lean. More than technology, Uber relies on network effects built on the back of a diffuse network hub of users and drivers who are connected by an algorithm. Network effects are built on the back of market leadership by a first mover or a company that has quickly established dominance through augmenting its extensive network (McAfee and Brynjolfsson, 2017: 213). This network becomes irresistible for those who are outside of it and attracts them using varied marketing tools. A perfect illustration of a platform with network effect is Facebook.

Exponential organisations rely on network effects, generated by a growing number of people (including relatives, friends and work colleagues) who are using digital platforms for their enhanced user experience and driven by a sense of being connected to a larger community of 'smart' users. Uber has also relied on a 'carefully cultivated network of drivers', without whom it would not be possible to successfully run Uber, especially in the absence of an infrastructure for autonomous vehicles (Hoffman and Yeh, 2018: 42). It is not just technology that matters for the success of Uber, but a business model that is centred on a network of drivers and effective distribution of service to customers in a way that enhances the customer experience.

Without a doubt, Uber has upended the old taxi system. An aspect of its business model has been to take the taxi market away from the conventional meter taxis to a platform that is run on an algorithm. Algorithms became the core of many AI-driven processes in the 2000s and they are based on statistical models that are trained on data to discover preferred patterns. In some instances, the pattern formation is structured to approximate how humans would think or behave. So, machine learning is about developing algorithm patterns based on data training. In other words, it is a form of programming big data in a way that allows for autonomous learning after certain patterns are established. This is the underpinning of Uber platforms. Its ability to modulate prices based on demand and developing knowledge of how to allocate drivers to customers based on algorithms. Uber epitomises the disruptive business model associated with a data-driven or AI-powered sharing economy. As Samit (2015: 17) puts it, 'disruption causes vast sums of money to flow from existing businesses and business models to new entrants.'

The following sections explore how Uber's emergence disrupted traditional taxi market structures. This discussion raises salient points about the relationship between competition policy, state regulatory frameworks and the transition towards an inclusive digital economy. The discussion commences with an analysis of state regulation in the market, and then proceeds to examine labour relations within Uber's operations. Then it ends with a discussion on pricing and the underlying dynamics influencing fare costs.

THE DIGITAL ECONOMY, MARKET DISRUPTION AND SOCIAL DYNAMICS

There is little doubt that the digital economy upends existing regulatory forms and accentuates social risks. This requires that policymakers and regulators consider an array of social factors when designing regulations aimed at new digital platforms. In this section, we take a closer look at state regulation, labour relations issues and competitiveness.

State regulations

Structural factors cause South Africa's high inequality levels and socioeconomic exclusion. Several literature sources identify rigid market structures as one of the main inequality drivers in society (MISTRA, 2014; World Bank, 2018; DPME, 2019). This dominance is prevalent throughout the economy, with a few companies accounting for a large portion of the market share in each sector. The World Bank (2018: 15) report expresses this trend in the following words:

> South Africa inherited an economy that made it difficult for historically disadvantaged entrepreneurs to accumulate wealth by competing in the country's rigid product markets. South African cartels and dominant firms are partly a symptom of this.

There are two main outcomes of this market failure: high prices and increased barriers to entry (World Bank, 2018). This market concentration challenge is also discussed in the Treasury's 2019 economic development strategy document. It specifically identifies lowering market barriers as an essential enabler for inclusive growth and SMME development (Treasury, 2019: 16).

Improved competition requires an integrated policy approach that considers different aspects of micro- and macroeconomic planning. Policymakers must also consider the state's role in shaping competition in varied market structures (Makhaya and Roberts, 2013: 567). State regulation and pricing are at the heart of the competition rivalry between Uber and meter taxis (Henama and Sifolo, 2017; Thosago, 2017; Competition Commission, 2020). Meter taxi driver associations and companies argue that they face unfair competition from Uber because it does not have to meet the regulatory requirements imposed on meter taxis. Representatives of meter taxi associations allege that Uber engages in predatory pricing, which allows the large corporation to access more market share (Competition Commission, 2020: 11). Both these issues cause social conflict and violence in the sector. It is important to note that social conflict over pricing and inconsistent regulatory regimes is not unique or peculiar to South Africa. Other countries such as France, Spain, England and Kenya had similar experiences when Uber emerged in transport markets (Dube, 2016; Competition Commission, 2020).

Meter taxi drivers have highlighted that the regulatory regime is unjust and has negative effects on their competitiveness. These operators navigate a complex maze of regulation, which includes operating permits, area restrictions and specialised driver licenses. They point out that most e-hailing drivers are operating illegally because they do not have operating permits. This is substantiated in the Competition Commission Market Inquiry report on transport. The report states that 'overall 79 per cent of e-hailing operators do not have valid licences for the major cities, including Cape Town, Durban, Port Elizabeth and the metros in the Gauteng City Region (Competition Commission, 2020: 25). However, e-hailing operators say that a major part of the reason for non-compliance is related to state capacity. Government authorities have failed to process operating permits efficiently and there is a significant backlog (Davis, 2015; Henama and Sifolo, 2017). The processing of permits is complex and involves various regulatory agencies at both provincial and municipal levels. These agencies, especially in municipalities, do not have the necessary capacity to make the process more efficient.

A major grievance of the meter taxi drivers is the area restrictions imposed on their companies and on individual taxi entrepreneurs (SAMTA, 2016; Competition Commission, 2020). These are prescribed in legislation and have significant effects on drivers' abilities to access markets and generate income. However, in contrast, e-hailing drivers are not obliged to comply with area restrictions like their meter taxi counterparts (SAMTA, 2016; Thosago, 2017).

The debate on area restrictions is more complicated because it is related to general regulations in the transport sector. These restrictions are imposed formally and informally through state and non-state institutions. For example, several taxi associations participate in overseeing transport area restrictions in different cities. There are varied economic, political and security issues to consider when discussing the most appropriate response for these policy inconsistencies. We cannot address all of them in this chapter. But it is important to highlight that the National Land Transport Amendment Bill (2016) does prescribe some area restrictions for e-hailing transport services. Both meter taxi and e-hailing associations oppose area restrictions for themselves on the grounds that they create inefficiencies and competitive disadvantages. The Competition Commission report (2020) advocates for their removal citing international case studies. Several states such as Finland have withdrawn these geographic restrictions so market players can compete on an equal footing.

All the factors discussed above raise pertinent points about the sharing economy model described earlier. The state has to respond to these shifts because it manages the social costs and benefits associated with shared or platform economy business models. These social costs and benefits have implications for South Africa's socioeconomic development. It is, therefore, important to propose some measures relating to how the state should adapt to the changes in the market structure. The last sections of the chapter provide some recommendations on possible policy responses. These recommendations include the labour relations issues explored below.

LABOUR RELATIONS AND TECHNOLOGY

The nature of work and employment across the globe has been shifting away from the Standard Employment Relationship (SER) since the late 1970s (Collier et al., 2017). South Africa's integration into the global political economy after democratisation accelerated this shift in the country. The impact of the transition on the post-apartheid labour market is documented elsewhere (Von Holdt, 2003; Buhlungu, 2010). In this section, we emphasise that technology and innovation play a significant role in restructuring the SER. Work restructuring is one of the essential features of shared and platform economies' enterprise development strategies. These business models operate on flexibility and self-regulation principles, which permeate all aspects of enterprise development (Allen and Berg, 2014; Pollio, 2019). This flexibility even extends to labour and necessitates creating non-traditional forms of work (Collier et al., 2017: 3).

Uber's investment is in the once-off algorithm-driven system, which has established the enterprise as one of the dominant e-hailing companies, if not a near monopoly. Uber is part of what Brynjolfsson and McAfee (2016: 148) refer to as 'the winner-take-all market' which redistributes incomes to the top, while compressing them at the bottom, thereby contributing to income inequalities. The company Uber Technologies Inc reached US$120 billion market value in 2019, and its CEO, Dara Khosrowshahi, received US$45.3 million compensation in 2018. He is entitled to further bonuses once the company hits US$120 billion valuation mark, unlike the drivers who are behind the wheels and face multiple social risks (Melin, 2019).

Uber is exploring a future of driverless cars to cut out the commission it pays to the driver. This conclusion is surprising because Uber does not assume the risk of car ownership; its drivers do. It has no employment contract with the drivers and regards them as independent contractors. Their relationship with Uber is cast at arms-length: Uber makes the application available to drivers, which allows those drivers to find customers; Uber then levies its commission on the earnings of the driver. The company has also attempted, with minimal success, to challenge its categorisation as a transport company in many jurisdictions to circumvent labour legislation (Collier et al., 2017; Nemusimbori, 2017; Robinson, 2017).

There are two primary responses to this practice in the literature on the sharing economy and in debates on public policy. The first view welcomes this innovative business model of managing labour, with proponents arguing that it provides more income benefits and autonomy than traditional employment. Advocates further state that it increases the quality of service and productivity (Allen and Berg, 2014; Henama and Sifolo, 2017). The second perspective adopts a more sceptical stance and accuses Uber of exploiting the driver's labour power, using superficial notions of entrepreneurial economic autonomy and control over their work (Clarke and Urata, 2016; Collier et al., 2017; Pollio, 2019).

The drivers who are part of Uber's network are human beings who are susceptible to various infirmities and have limited protection from the company. They enjoy neither a minimum wage nor benefits related to paid leave that apply in the old industrial paradigm. This is understandable since there is no company that manages them. In the old taxi industry, drivers working for riding companies are entitled to basic conditions of work – whether these are applied or not is a different matter altogether. Uber drivers can – and do – get into accidents and if they are physically injured or hospitalised, they will not continue to earn an income and they and their families are likely to be exposed to economic vulnerabilities. At the time of writing, these vulnerabilities have become accentuated under COVID-19. Even though the company has sought to cushion its employees from income losses during the pandemic, drivers have high exposure to COVID-19 infections. An Uber driver in the UK succumbed to COVID-19 (Booth, 2020).

A question at the heart of Uber operations worldwide has been whether the drivers can be classified as employees or independent contractors. This has been a hotly disputed matter in many countries: In May 2018 there were various strikes organised by Uber drivers around major cities such as Los Angeles, Chicago, Philadelphia, Boston, New York City, Washington DC and London. Similarly, South Africa has experienced intermittent strikes between 2017 and 2019, with Uber drivers complaining about unfairness in the distribution of proceeds, poor working conditions, unilateral fare determination and unjust account deletions (IOL, 2018; Kahla, 2020; Malinga, 2020).

In South Africa, the debate about whether the drivers are employees or independent contractors was tested in both the Commission for Conciliation, Mediation and Arbitration (CCMA) and the Labour Court between 2017 and 2018. The CCMA ruled that Uber drivers are employees, but this was overturned by the Labour Court in 2018 because of a technicality. Applicants in the case cited Uber's South African subsidiary as the respondent instead of presenting the case against the Dutch-based parent company: Uber BV. The presiding judge concluded that the question on the employee status of Uber drivers remains unanswered even though he overturned the CCMA 2017 ruling (Chambers, 2018; Van Rensburg, 2018).

South African labour law sets out detailed criteria for identifying an employee in section 200 of the Labour Relations Act (Act 8 of 2018). It contains seven criteria to determine if there is an employer–employee relationship. There are two essential aspects to consider when making this determination: the status of the contract and the relationship between the two parties. The relationship issue is at the heart of the debate in South Africa and other countries where Uber operates.

Court rulings on the matter have based their conclusion on assessing the relationship between drivers and Uber. The drivers are regarded as independent (contractually), but Uber exercises significant control over their work and entry into the platform, just like a conventional employer (Clarke and Urata, 2016; Geitung, 2017). The company determines fare structures and how these structures are administered without consulting the drivers. This effectively translates to setting the driver's weekly wages or income. Findings in the Competition Commission transport inquiry reveal that some 'operator's earnings are below the minimum wage and they cannot quit because some have invested in cars' (Competition Commission, 2020: 57). The gross income of drivers for services below two kilometres declined from R85 in 2013 to R20 by 2017.

This top-down process is mediated through technology and offers no room for negotiation or co-determination. Many drivers who lease cars have to deal with another layer of complexity, which is the relationship they have with the owners of the vehicles. The owners of the vehicles set weekly targets for Uber drivers. This is roughly R3,000 a week, translating to about R12,000 a month, that Uber drivers have to pay an owner for renting the vehicle. In turn, the owner of the car takes care of all maintenance except operating expenses related to fuel and food for the driver. Many drivers find the rents they have to pay to owners very high. They are hemmed-in by owners of the vehicles who demand high rents on the one hand, and by Uber, which extracts 25 per cent commission upfront, on the other. In 2016 the company processed over US$26 billion of payments globally (Hoffman and Yeh, 2018: 55).

Uber drivers that are renting vehicles from owners net roughly R6,000 per month, minus operating expenses. Some drivers find this much higher than the wage they used to earn in the security industry or in driving trucks long distance. Some also see flexibility as a benefit since they can start other side businesses. Others see Uber as a holding job until they find some breakthrough in their next career or entrepreneurial idea. There are also students who use the proceeds to finance their studies. Uber drivers are an eclectic mass that are driven by a variety of interests. However, the structure of the relationship between Uber and its drivers needs regulation as this can lend itself to super-exploitative modes of work.

Uber has the right to remove drivers from the platform by unilaterally deleting their accounts (Clarke and Urata, 2016). It also sets out specifications for vehicles and detailed quality standards for the service. An additional form of control is the rating system, which allows Uber to assess the driver's performance through algorithms (Collier et al., 2017). The algorithm is part of what Zuboff (2019) characterises as 'surveillance capitalism', which mediates the relationship between the Uber driver and the rider through an anonymous rating system. This system is a disciplinary mechanism, used by the application's managers to shape the behaviour of the drivers and maintain professional standards. Uber drivers can also rate their customers.

Some have argued that this mutual rating system overcomes information asymmetry about the driver or the customer, and that it performs the role of regulation – so it is a peer regulatory system of sort (McAfee and Brynjolfsson, 2017: 208–209). Its positive value is that it keeps track of integrity or trust equity between a network of customers who use the platform and the drivers. We argue that the system serves a different purpose: the mutual rating process is aimed at maintaining a certain equilibrium in the behaviour of the rider and the customer. It may introduce biases or preconceptions that could lead to more negative ratings if a driver has had poor ratings.

Crucially, it is not that clear if Uber will use the ratings for other purposes, or whether they can be sold on to other organisations. The algorithm is trained on certain perceived behaviours of the customer and the driver; a practise that is part of layering the texture of the data on the behaviour and, possibly, the personalities of the participants in this digital ecosystem.

'Surveillance capitalism', Zuboff observes (2019: 11):

> operates through unprecedented asymmetries in knowledge and the power that accrues to knowledge. Surveillance capitalism knows everything *about us*, whereas their operations are designed to be unknowable *to us*. They accumulate vast domains of new knowledge *from us*, but not *for us*.'

This statement amplifies how surveillance capitalism, which is driven by algorithms, is an essential part of the AI-driven digital economy.

What really constitutes surveillance capitalism that is embedded in digital products and services is less about the technology itself, and more about the structure or logic of its thinking, in other words, the motive of capital and its ends, as well as how it repatterns human relations to serve its ultimate goals. The technology is instrumentalised to create a sense of inevitability – and omnipotence – so much so that whoever questions it would be deemed to be questioning the natural order of things. As we argued earlier, digital platforms and services such as Uber require regulation even if these are different from those that regulated old industries.

Nemusimbori (2017) and Geitung's (2017) case studies on the relationship between drivers and Uber in South Africa highlights some of the points in the preceding sections. They concur that the company exercises significant control over the driver's work. Geitung (2017: 59) explains this power relation in the following words:

> drivers did not feel they could influence Uber's decisions, and they did not feel that Uber listened to them. The feelings of being one's own boss as independent contractors were overshadowed by Uber's interference in the business.

Nemusimbori (2017) goes further and investigates whether the relationship meets the seven criteria set out in the LRA for defining an employee. He concludes that Uber drivers meet the minimum elements in the legislation. Courts in other countries such as the USA, Switzerland and the United Kingdom have ruled that Uber operates like an employer, and ordered the company to pay defined minimum wages, social security and unemployment benefits (Barainsky et al., 2016; Competition Commission, 2020).

These cases demonstrate that there is no neat distinction between the real economy and the digital economy. However, hard Uber tries to eliminate or suppress the human factor, the challenges of human welfare, or contention over distributive costs, will not disappear. Uber constantly seeks to upend the regime of accumulation with its regulatory structures, while expanding the surplus capital that it generates for its owners. The mode of regulation, as Amin (1994: 8) explains it, refers to 'institutional ensemble and the complex of cultural habits and norms which secure capitalist reproduction as such'. In the mould of an extreme version of post-Fordist production or service, Uber regards the driver as an appendage and facilitator of its service, which is distributed through a depersonalised and algorithm-driven application.

Uber tolerates its drivers because it does not have a better option to activate its business model. In 2015, the company announced that it was investing in a new facility in the US to test self-driving car models that could be part of its fleet of taxis in a project that was supported by Carnegie Mellon University. The then CEO, Travis Kalanick, decried the fact that Uber is perceived to be expensive, shifting the blame to the driver. In his words, 'The reason Uber could be expensive is because you're not just paying for the car – you're paying for the other dude in the car… When there is no other dude in the car, the cost of taking an Uber anywhere becomes cheaper than owning a vehicle' (Lowensohn, 2015). It may take some decades before autonomous vehicles arrive in developing economies such as South Africa and other high-density cities that are marked by lack of driver predictability. For the foreseeable future, the tensions between the human driver and the app, as well as the distributive contradictions that contribute to social inequalities, will play themselves out for a long time to come.

Uber merges human and digital labour in one: the driver has to interact with the technology (app) to accept customers. It is difficult for drivers to mount successful strikes since there is no front-line management to respond to the grievances of the drivers. Besides, if the drivers insist on protesting for an extended period, they might lose their place or suffer income loss. It is this vulnerability that the company exploits. The contractual arrangements between Uber and the driver remain shrouded in ambiguity – and it is not clear who determines the contractual framework. The role of regulatory bodies also remains ambiguous. This discussion on the status of Uber drivers illustrates how technology in the shared or platform economy transforms the SER. It supports alternative, flexible forms of work that test existing labour legislation frameworks.

The South African state, through the CCMA and Labour Court, has attempted to provide guidance on how Uber work should be regulated. These two institutions provided different rulings, and the question of whether South African Uber drivers are employees remains unanswered. This policy dilemma presents several challenges for stakeholders because it has fiscal, regulatory and income ramifications. For example, Uber could be required to pay employee tax and social security benefits if it is defined as an employer. Similarly, drivers might be subjected to the work and wage regulations of conventional transport sector agreements if they want to be hired as Uber employees. It is, therefore, important to consider all these dimensions before concluding on the employee status of drivers.

Another option for the drivers is to take advantage of a unique structural feature in Uber operations. Most forms of work in the digital-driven economy do not allow individuals to meet regularly and share work-related experiences (Collier et al., 2017; Robinson, 2017). Uber is an outlier because drivers normally meet and create platforms for discussing common challenges. This presents an opportunity for potential collective action around pertinent socioeconomic issues, which relate to their relationship with Uber. The collective action does not have to be structured on the basis of a traditional trade union movement. It should rather emerge from the existing networks of collaboration and coordination displayed in the recent protests against Uber. In the same measure that Uber uses technology as a platform for accumulation, the possibility exists for creative technology-based forms of organisation by its drivers.

This chapter argues that coordination and approaching Uber as a unified drivers' association is a possible solution for addressing the drivers' grievances. Already, Uber drivers coordinate via another digital platform – WhatsApp – which they use to alert each other of any dangerous spots or when coordination is needed for a strike. The problem is that this form of coordination has not evolved to position Uber drivers fully as an effective workers' right advocacy force. Collective organising and agency must suit the structural features of the digital-driven Uber business model. The regulatory framework must also shift to accommodate the structural changes imposed by the digital-based economy. This requires rethinking in various areas of regulation, especially labour and fiscal legislation. These regulations must appreciate the peculiarity of employment and market transactions in digital economies. Another essential space in the Uber regulation debate is the pricing of fares and the impact it has on income for drivers and their competitors. This issue is explored further in the next sections.

PRICING AND COMPETITIVENESS

The pricing issue is related to market structure dynamics and business model differences. According to the National Land Transport Act (5 of 2009), meter taxi pricing ought to be regulated by the minister or the provincial member of the executive council (MEC) responsible for transport. This legislation also permits customers and meter taxi drivers to negotiate fares before a trip commences (NLTA, 2009). However, evidence presented in the Competition Commission hearings illustrates that meter taxi companies and associations set their own prices. The only exception to this trend is the Western Cape, which prescribes a price ceiling for fares (Competition Commission, 2020: 30). This finding points to a governance and policy implementation failure, which has a significant influence on price determination in the sector. Another crucial factor that determines prices are the financial implications of regulatory differences, which are cited throughout public policy debates. Some meter taxi companies have to pay employee tax and wages because they hire drivers. These financial commitments place an additional cost on the company that is passed on to clients. Individual meter taxi drivers and associations also factor in regulatory costs when determining prices.

Prices charged by meter taxi drivers are also influenced by the traditional business model (Dube, 2016; Henama and Sifolo, 2017). These operators typically work from a base such as a hotel, taxi rank or a mall. So, they wait for customers and drive them to their destinations. Then the meter taxi driver returns to this base after dropping off the client and this increases service costs. Charges must cover driving to the customer's destination and coming back to the base of operations (Competition Commission, 2020).

The fixed price model used in meter taxi services limits their access to markets. Fixed prices impede drivers from gaining advantages from supply and demand changes. For example, meter taxi drivers cannot benefit from price surges when demand peaks (OECD, 2018: 6). This model also prohibits these operators from lowering prices to accommodate customers whose purchasing power is less than that of their traditional client base. Meter taxis in South Africa rely primarily on tourists and high-income earners.

It is evident that the meter taxi drivers and companies have limited market reach because of failures to introduce innovative technologies over the years (Barainsky et al., 2016: 7). The authors (Berainsky et al., 2016: 7) state:

> taxi companies in all parts of the world relied on a similar kind of business model for almost a century. Until a few years ago, the taxi sector was not showing any sign of evolving and users have to accept the service as it is.

The lack of innovation described in the preceding statements has a direct impact on market share, which inherently determines price competitiveness. A company's ability to lower prices to match demand fluctuations is dependent on scale and market share. The low levels of innovation in the traditional meter taxi business model limits its potential for market share expansion.

The regulation of Uber pricing is not catered for in the existing National Land Transport Act (2009) because it does not recognise e-hailing transport services (NLTA, 2009; Davis, 2015). Uber operators have been governed through a special Department of Transport (DoT) practice note that guides provincial and municipal authorities' transport regulation. The government noted this shortcoming and drafted a National Land Transport Amendment Bill in 2016 that recognises e-hailing transport operators. This piece of legislation has been discussed in both legislative houses and public consultation processes are complete.

The Minister of Transport tabled a final version for parliament ratification on 10 March 2020 (Onoja, 2020). Section 66 in the amended Bill describes a different process for Uber pricing, which establishes a limited role for the minister or MEC in price regulation (National Land and Transport Amendment Bill, 2016). The new section of the Bill states that e-hailing services must have the capacity to 'estimate fares and distances, taking into account distance and time, and must communicate the estimate to passengers in advance electronically'. The amended bill allows the minister or MEC to issue regulations that 'ensure accurate readings of the e-hailing or technology-enabled application' (National Land and Transport Amendment Bill, 2016). This process is different from the one described in earlier sections, in which political authorities have greater capacity to determine prices for meter taxi operators. The Competition Commission Report recommends that these powers be removed because price regulation in the meter taxi industry increases 'administrative burdens' for government regulatory agencies and impedes taxi operators from responding efficiently to shifts in market demand trends (Competition Commission Report, 2020: 39).

Another difference in the Uber pricing model is the usage of a market-based regime called dynamic pricing. This fare mechanism is determined through market factors such as distance, time spent on the trip, supply-demand dynamics and congestion on the roads. Transparency is enhanced for customers, as they receive upfront calculations of how fares are determined (Barainsky et al., 2016; Dube, 2016). The main component that makes the system competitive is the app which facilitates dynamic pricing in the following ways (Allen and Berg, 2014; Henama and Sifolo, 2017). First, it increases transparency in market transactions as consumers see varied estimates of fare calculations through accessing the app on their mobile phones. Second, Uber's application connects drivers (suppliers of services) with customers requesting specific trips (demand). This resolves demand and supply market failures, increases efficiencies and lowers 'market transactions costs' (PwC, 2014; Cramer and Krueger 2016; Dube, 2016). Third, Uber drivers can earn higher income when the prices surge because of supply-demand dynamics or preferential pricing for holidays (Manavhela and Henama, 2018; Competition Commission, 2020).

Dynamic pricing, which is powered by AI in Uber's operations, has several advantages for both consumers and Uber operators. Clients have more choice in determining how much they are willing to pay for a trip. They can assess price fluctuations and organise trips to suit their purchasing power and service needs (Allen and Berg, 2014). Uber's large network of drivers also guarantees access to more efficient supply and avoids the productivity losses experienced in traditional meter taxi services (Henama and Sifolo, 2017). Dynamic pricing decreases trip costs in some cases when it encourages drivers to enter areas experiencing price surges. This subsequently leads to price reduction because the supply exceeds the demand. However, there are some drawbacks or potential market failures in the dynamic pricing mechanism.

The first is the potential for overcharging, especially when prices increase as a result of demand and supply dynamics during concerts, sport events or other large group gatherings. An example in the South African context is the overcharging which took place during a musical festival in 2018. According to the Competition Commission (2019: 35), some customers 'were left stranded because of dynamic pricing. Some concert goers were charged R1000 compared to a typical cost of R200'. Customers raise concerns about exorbitant fares during price surge periods and the lack of transparency in determining these prices (Dholakia, 2015; Manavhela and Henama, 2018). The point on minimal transparency illustrates the information asymmetry between Uber and customers in price surging market transactions. There are additional complaints about the instability in the Uber price surge system, which makes it difficult for customers to budget and plan trips. This inevitably leads to customer dissatisfaction and, in some cases, exploitation (Dholakia, 2015).

Some Uber drivers raise serious concerns about the pricing model and its impact on their income. The first issue is sudden fare price decreases, which occur when supply outstrips demand in specific areas. These drivers attribute this fall in fares to Uber and state that it causes financial challenges. This point is important because the operating costs of the Uber business model are externalised on to drivers. They pay for fuel, car insurance, licensing permits, car maintenance and financing (Pollio, 2019). An additional concern is the lowering of prices because of promotions with partner organisations. Uber establishes commercial partnerships with several companies, and this allows clients to obtain price discounts. The drivers allege that the cost of running these promotions is externalised. In other words, they are subsiding these reductions through lower trip fares (IOL, 2018; Kahla, 2020; Malinga, 2020). They also argue that there is an oversupply of drivers in the market and this lowers their incomes. Drivers criticise Uber for expanding the driver network by taking in new driver partners. This has a huge impact on their earnings because it lowers prices drastically when there is an oversupply (Malinga, 2020).

CONCLUSION: LESSONS AND RECOMMENDATIONS

When Uber arrived in South Africa, it initially confronted regulatory challenges, but it was able to work around these through engagement with public affairs and stakeholders – capabilities that the old taxi industry could not muster. Regulations governing public transport and taxis require a detailed set of procedures. There is an onerous bureaucratic process that self-employed operators are required to go through, with long time lags that augment business costs. South Africa's Uber subsidiary managed to circumvent this regulatory maze claiming that it was not a transport provider and it should, therefore, be exempted from the regulatory procedures demanded of conventional taxi operators. This argument was based on the use of an algorithm-based application to match drivers with customers. The chapter has illustrated how the weak regulation of Uber operations produces negative socioeconomic outcomes such as income inequalities, distorted sector competition and driver exploitation. The evidence presented throughout the chapter proves that Uber operates as a business and an employer in the transport sector.

We propose that government authorities amend general transport taxi regulations, so they respond to the socioeconomic benefits and costs (externalities) associated with digital e-hailing services. This would be in line with international regulatory precedents because Uber has been legally demarcated as a transport sector company in Europe and other jurisdictions. There is no rational policy basis for applying a different regulatory regime in South Africa, especially if one employs an organisational status test that examines both formal registration and the economic relationship that Uber has with other transport market actors. This regulatory regime must resolve some of the policy inconsistencies cited in the chapter's discussion of state regulations. The protests in the industry are fuelled by policy incoherence and weak responses to digital business models. Local and provincial authorities are advised to introduce measures, which will improve efficiencies in permit and licensing systems. These systems have a huge bearing on sector competitiveness and the costs of doing business.

Furthermore, it is imperative to create regulations aimed at diversifying ownership patterns in the e-hailing sector and regulating these for the local context, as illustrated in the discussion on international experiences. This supports the argument we present regarding a just transition to a digital political economy. South African regulators must use this transition to address inequality and socioeconomic exclusion. We argue that industrial and competition policy interventions are essential for attaining this goal. Competition regulation in the taxi market structure should be coordinated, consistent and streamlined. It should prioritise addressing regulatory barriers that distort competition and fuel social conflict among various market players. This recommendation applies specifically to the policy areas governing pricing, licensing and area restrictions in the industry. Additionally, South Africa's state authorities, across different departments, should develop a national strategy for supporting local e-hailing operators that includes industrial and business operation incentives. This must be accompanied by competition regulations that address the market entry barriers identified in the Competition Commission market inquiry report (2020). These barriers include the following: finance; marketing/advertising costs; market access through networks and established business partnerships.

Uber has exposed that labour regulations are not adaptable enough to govern the digital economy organised around AI. The company could sidestep questions about the working conditions of drivers since it does not employ them in an SER but offers them a platform to generate an income. In this sense, it was not more exploitative than the conditions under which taxi drivers operate, especially those that have lease arrangements with owners of vehicles. Often the point of tension, between Uber and its drivers, is the pricing of the service: that this does not fully value the drivers' time or wear and tear on their vehicles, nor the cost of living they face, and thus makes the Uber driving experience a drudgery. In its defence, Uber invokes the dictatorship of the algorithm, which mediates price points depending on various market factors.

Uber performs what Karl Marx (1844) referred to as alienation of the worker from the production process and fellow human beings; or, put differently, the estrangement of the constituent parts of the production processes from one another. Drivers' existence in this ecosystem is imposed upon them since they have constrained choices outside of it, especially given the high levels of unemployment or precarious employment in the old industrial sectors in South Africa. They are, in reality, participating in what Marcuse (1964) identifies as 'a more progressive stage of alienation,' a false sense of progress that soon turns into unhappiness as rents harvested by the app become a source of pain for the driver. Terms such as 'partner', used to refer to the driver, that seek to soften the edges of exploitation are cloaking some insidious forms of it.

All these factors highlight that South Africa's labour regulation is based on the SER in a Fordist industrial relations model. We have pointed out the shortcomings of this labour regulatory framework, and how it presents opportunities for companies such as Uber to circumvent labour law while exercising employer control over drivers. This has subsequently led to the exploitation of drivers who experience an array of socioeconomic challenges. The end result is a social conflict between different actors in the transport taxi value chain, and resulting protest action. We urge drivers to organise and establish a collective drivers' association that engages both Uber and regulatory authorities on their challenges. This advocacy must be centred on addressing their unequal working relationship with Uber and on clarifying their status in labour regulatory frameworks. As argued earlier, there is international precedence set in other countries where Uber was ordered to pay minimum wages, employee benefits and employee tax because it operates like an employer. The 2018 Labour Court judgment did not provide a conclusive resolution and regulatory authorities need to address this policy gap. They must consider the structure of a digitised business model, and how to adapt or amend labour legislation so it protects employees in the transition towards a digital economy.

South Africa would do well to note and learn from Uber's regulatory challenges in markets around the world. A few years after Uber was established, California passed regulations that required transportation network companies such as Uber and Lyft to conduct criminal background checks on their drivers (McAfee and Brynjolfosson, 2017: 208). Authorities in California had earlier ordered the removal of driverless vehicles from the roads since they did not have the licence to operate. Uber had also deceived drivers in some major US cities about what they were likely to pay if they signed up to its platform and was forced to pay a US$20 million settlement for this chicanery. It had also been embroiled in IP theft when Waymo, the self-driving car owned by Google, filed a lawsuit against Uber (Levin, 2017).

In late 2019, Uber was stripped of its licence to operate in London. This demonstrated that the rider-hailing service is not invincible. This was the second time it had happened. Regulators deemed the service not 'fit and proper' to operate in London. It was criticised for various regulatory breaches and that it had put passengers' safety and security at risk (Bradshaw, 2019). Some of the most specific complaints about Uber related to driver fraud in London. There were also cases concerning distribution of indecent material to children. Elsewhere Uber was accused of running illegal spy programmes and that there were widespread incidents of sexual harassment and discrimination at its head office. All these cases show that regulators in South Africa need to either amend or introduce laws that address externalities created in the digital economy.

REFERENCES

Allen, D. and Berg, C. 2014. 'The sharing economy: How over-regulation could destroy an economic revolution'. Institute of Public Affairs. https://www. academia.edu/10374868/The_sharing_economy_How_over-regulation_ could_destroy_an_economic_revolution, accessed 20 May 2020.

Amin, A. 1994. *Post-Fordism*. Massachusetts: Blackwell Publishers.

Barainsky, L., Gumberidze, E. and Nurul, A.M. 2016. 'Uber and taxi regulations: Are member states preserving a legal monopoly to the detriment of consumers?' https://www.researchgate.net/publication/309679194, accessed 19 May 2020.

Booth, R. 17 April 2020. 'Uber driver dies from Covid-19 after hiding it over fear of eviction'. *The Guardian*. https://www.theguardian.com/ world/2020/apr/17/uber-driver-dies-from-covid-19-after-hiding-it-over-fear-of-eviction, accessed 19 May 2020.

Bradshaw, T. 25 November 2019. 'Uber loses licence to operate in London'. *Financial Times*. https://www.ft.com//content/78827b06-11ea-a225-db2f231cfeae, accessed 21 May 2020.

Brynjolfosson, E. and McAfee, A. 2016. *The Second Machine Age*. New York: W.W. Norton and Company.

Buhlungu, S. 2010. *A Paradox of Victory: COSATU and the Democratic Transition in South Africa*. Pietermaritzburg: University of KwaZulu-Natal Press.

Chambers, D. 17 January 2018. 'Uber and out for fired drivers in labour court fight'. *TimesLive*. https://www.timeslive.co.za/amp/news/south-africa/2018-01-17-uber-and-out-for-fired-drivers-in-labour-court-fight/, accessed 19 May 2020.

Clarke, C. and Utara, M. 2016. 'Uber: Don't take us for a ride'. *Global Labour Column*. http://column.global-labour-university.org/2016/06/uber-dont-take-us-for-ride.html?m=1, accessed 18 May 2020.

Collier, B.R., Dubai, V. and Carter, C. 2017. 'The regulation of Labor platforms: The politics of the Uber economy'. https://brie.berkeley.edu/ sites/default/files/reg-of-labor-platforms.pdf, accessed 19 May 2020.

Competition Commission. 2020. 'Market inquiry into land based public passenger transport: Metered taxis and e-hailing services report'. http:// www.compcom.co.za/wp-content/uploads/2020/02/PROVISIONAL-REPORT-ON-E-HAILING-AND-METERED-TAXIS-19FEBRUARY2020-NON-CONFIDENTIAL-VERSION1.pdf, accessed 19 May 2020.

Cramer, J. and Krueger, A.B. 2016. 'Disruptive change in the taxi business: The case of Uber'. NBER Working Paper 22083, The National Bureau of Economic Research. https://www.nber.org/papers/w22083.pdf, accessed 19 May 2020.

Davis, R. 1 July 2015. 'Uber vs. Authorities: Regulatory headaches abound for popular taxi service'. *Daily Maverick*. https://www.dailymaverick.co.za/article/2015-07-01-uber-vs-authorities-regulatory-headaches-abound-for-popular-taxi-service/, accessed 18 May 2020.

Department of Planning, Monitoring and Evaluation (DPME). 2019. 'Towards a 25-year review 1994-2019'. https://www.dpme.gov.za/news/SiteAssets/Pages/25-Year-Review-Launch/Towards%20A%2025%20Year%20Review.pdf.

Dholakia, U.M. 2015. 'Everyone hates Uber's surge pricing – here's how to fix it'. *Harvard Business Review*. https://hbr.org/omp/2015/12/everyone-hates-ubers-surge-pricing-heres-how-to-fix-it, accessed 18 May 2020.

Dube, S.C. 2016. 'Uber: a game-changer in passenger transport in South Africa?'. *CCRED Quarterly Review*. https://static1.squarespace.com/static/52246331e4b0a46e5f1b8ce5/t/56521e01e4b0e332af41071b/1448222209348/CCRED+Review7.4-6.pdf, accessed 19 May 2020.

Ford, M. 2015. *Rise of the Robots: Technology and The Threat of a Jobless Future*. New York: Basic Books.

Geitung, I. 2017. 'Uber drivers in Cape Town: Working conditions and worker agency in the sharing economy'. Master Thesis. Department of Sociology and Human Geography, University of Oslo.

Hanusch, M. 2018. 'An Incomplete Transition: Overcoming the legacy of exclusion in South Africa'. World Bank. https://blogs.worldbank.org/nasikiliza/an-incomplete-transition-overcoming-the-legacy-of-exclusion-in-south-africa, accessed 22 July 2020.

Henama, U.S. and Sifolo, P.P.S. 2017. 'Uber: The South Africa experience'. *African Journal of Hospitality, Tourism and Leisure*, 6(2), pp 1–10.

Hoffman, R and Yeh, C. 2018. *Blitzscaling: The Lightning-fast Path to Building Massively Valuable Businesses*. New York: Currency Publishers.

IOL. 13 November 2018. 'Uber, taxify drivers strike over 'slavery-like' conditions'. https://www.iol.co.za/amp/motoring/industry-news/uber-taxify-drivers-strike-over-slavery-like-conditions-18094851, accessed 21 May 2020.

Ismail, S. 2014. *Exponential Organization: Why new organizations are ten times better, faster, and cheaper than yours (and what to do about it)*. New York: Diversion Books.

Kahla, C. 19 February 2020. 'Bolt and Uber drivers protest over safety concerns and high commission'. *The South African*. https://www.thesouthafrican.com/news/bolt-uber-drivers-protest-safety-concerns-commission/, accessed 19 May 2020.

Kai-fu, L. 2019. *AI Superpowers*. Boston, Massachusetts: Houghton Mifflin Harcourt.

Levin, S. 10 March 2017. 'Google's self-driving car group tries to block Uber from using allegedly stolen tech'. *The Guardian*. https://www.theguardian.

com/technology/2017/mar/10/uber-google-self-driving-car-technology-waymo-lawsuit, accessed 21 May 2020.

Lowensohn, J. 2 February 2016. 'Uber just announced its own self-driving car project'. *The Verge*. https://www.google.com/amp/s/www.theverge.com/platform/amp/2015/2/2/7966527/uber-just-announced-its-own-self-driving-car-project, accessed 22 July 2020.

Makhaya, G. and Roberts, S. 2013. 'Expectations and outcomes: Considering competition and corporate power in South Africa under democracy'. *Review of the African Political Economy*, 40(138), 537–555.

Malinga, S. 20 February 2020. 'Uber, bolt drivers plan shutdown of ride-hailing services'. *ITWeb*, https://www.itweb.co.za/content/8OKdWMDYgygqbznQ, accessed 18 May 2020.

Manavhela, P. and Hemana, U.S. 2018. 'Surge pricing as a new pricing model for transport services: The case of Uber in South Africa'. Working paper presented at a conference on 8th advances in hospitality and tourism marketing and management (AHTMM), Bangkok, 599.

Mapungubwe Institute for Strategic Reflection (MISTRA). 2014. 'National Planning Commission Social Compact Report.' Johannesburg: MISTRA. http://www.dac.gov.za/sites/default/files/social-compact-comparative-analysis.pdf, accessed 22 July 2020.

Marcuse, H. 1964. *One-Dimensional Man*. Boston: Beacon Press

Marx, K. 1844. 'Economic and philosophical manuscripts of 1844. Estranged labour'. https://www.marxists.org/archive/marx/works/1844/manuscripts/labour.htm, accessed 20 April 2020.

Matzler, K., Veider, V. and Kathan, W. 2015. 'Adapting to the sharing economy'. *MIT Sloan Management Review*, 56(2), pp 71–77.

Mazzucato, M. 2018. *The Value of Everything: Making and Taking in the Global Economy*. New York: Allen Lane.

McAfee, A. and Brynjolfsson, E. 2017. *Machine, Platform and Crowd: Harnessing our Digital Future*. New York: W.W. Norton and Company.

Melin, A. 11 April 2019. 'Uber 's executives have big payouts riding on a $120 billion value target'. *Bloomberg*. https://www.bloomberg.com/news/articles/2019-04-11/uber-s-executives-have-big-payouts-riding-on-120-billion-target, accessed 21 May 2020.

National Land Transport Act (NLTA) No. 5 of 2009. 8 April 2009. https://www.gov.za/sites/default/files/gcis_document/201409/32110413.pdf.

National Land Transport Amendment Bill. 2016. https://www.gov.za/sites/default/files/gcis_document/201904/national-land-transport-amendment-bill-b7d-2016.pdf.

Nemusimbori, N.E. 2017. 'An appraisal of the status of Uber drivers in South African labour law'. Masters Dissertation, Faculty of Law, University of Pretoria.

Norvig, P. 31 October 2012. 'Artificial intelligence: Early ambitions'. *New*

Scientist. https://www.newscientist.com/article/mg21628892-600-artificial-intelligence-early-ambitions/, accessed 23 July 2020.

Onoja, S. 10 March 2020. 'South Africa to further regulate Uber and other e-hailing services'. *Business Elite Africa.* https://businesselitesafrica.com/auto/south-africa-to-further-regulate-uber-and-other-e-hailing-services/, accessed 21 May 2020.

Organisation for Economic Co-operation and Development (OECD). 2018. 'Taxi, ride-sourcing and ride-sharing services'. Directorate for Financial and Enterprise Affairs Competition Committee Working Party No. 2 on Competition and Regulation. https://www.oecd.org/daf/competition/taxis-and-ride-sharing-services.htm.

Pollio, A. 2019. 'Forefronts of the sharing economy: Uber in Cape Town'. *International Journal of Urban andRregional Research*, 43(4), 760–775.

PricewaterhouseCoopers (PwC). 2014. 'The sharing of economy: How will it disrupt your business?'. https://pwc.blogs.com/files/sharing-economy-final_0814.pdf, accessed 19 May 2020.

Rinne, A. 2018. 'The dark side of the sharing economy'. World Economic Forum. https://www.weforum.org/agenda/2018/01/the-dark-side-of-the-sharing-economy/, accessed 21 May 2020.

Robinson, H.C. 2017. 'Making a digital working class: Uber drivers in Boston, 2016–2017'. PhD Thesis, Massachusetts Institute of Technology, Boston.

Samit, J. 2015. *Disrupt You: Master Personal Transformation, Seize Opportunity, and Thrive in the Era of Endless Innovation*. New York: Flatiron Books.

Schwab, K. 2017. *The Fourth Industrial Revolution*. London: Penguin.

South African Metered Taxi Association (SAMTA). 2016. 'CC Submissions'. https://static.pmg.org.za/160921samta.pdf.

Stemler, A. 2016. 'The myth of the sharing economy and its implications for regulating innovation'. *Emory Law Journal*, 67, 197–241.

Thosago, M. 2017. 'South Africa's public transport market inquiry: Integrating modes'. *Quarterly Competition Review*, 13–14.

Treasury. 2019. *Economic Transformation, Inclusive Growth, and Competitiveness: Towards an Economic Strategy for South Africa*. Pretoria: National Treasury. http://www.treasury.gov.za/comm_media/press/2019/Towards%20an%20Economic%20Strategy%20for%20SA.pdf.

Van Rensburg, D. 21 January 2018. 'False start for Uber drivers' employee claim'. *Fin24*. https://www.google.com/amp/s/www.news24.com/amp/fin24/tech/companies/false-start-for-uber-drivers-employee-claim-20180119, accessed 18 May 2020.

Von Holdt, K. 2003. *Transition from Below: Forging Trade Unionism and Workplace Change in South Africa*. Pietermaritzburg: University of Natal Press.

Zuboff, S. 2019. *The Age of Surveillance Capitalism: The Fight for a Human Future at the New Frontier of Power*. New York: Public Affairs.

Section Three

Application of Advanced Technologies in Sectoral Developments

The potential of Fourth Industrial Revolution technologies to transform healthcare: The question of access for the marginalised

Zamanzima Mazibuko-Makena

INTRODUCTION

Throughout the years, medical research, technology and innovation in their divergent forms have introduced noteworthy changes to medicine and healthcare globally. Countless advancements in critical knowledge of medicine and healthcare have allowed for several improvements to be made in diagnosis, treatment and care. With the onset of the COVID-19 pandemic, the need for these improvements became ever more urgent and the race began to develop technological innovations and scientific insights that would counter the pandemic. This chapter is being written at the onset of the COVID-19 outbreak, when governments, doctors and scientist worldwide are still grappling with what COVID-19 is, and what it will mean for our future. It is too soon to take stock of the role of technological innovations in tackling the pandemic. However, what is clear is that the world is turning to the technologies of the Fourth Industrial Revolution (4IR) to solve some of its most pressing healthcare issues, and questions about the role of these technologies in transforming health practices have never been more pressing.

Researchers have turned to 4IR technologies during the pandemic as these are the latest in the cutting-edge innovations said to be transforming medicine, due to rapid advances in genomics, genetic engineering, nanotechnology, Artificial Intelligence (AI), Internet of Things (IoT), robotics and much else (Thuemmler & Bai 2017; WEF, 2019). These scientific and technological advances promise to provide more accurate and personalised healthcare, giving patients autonomy and the chance to participate in their treatment, as well as improving health outcomes (WEF, 2019). Essentially, 4IR technologies have the potential to answer the pressing challenges we have today in healthcare. These potential benefits are undoubtedly desirable. However, despite these potential benefits, these innovations and technologies raise important questions about the multifaceted medical, economic and social implications of their implementation in an unequal, resource-strained society like South Africa's, with a weak healthcare system.

In many countries, citizens are entitled to the provision of healthcare and it is commonly their constitutional right. Globally, the most critical issue in healthcare is providing widespread and effective treatment options that improve quality of life. Advances in medicine and the emergence of new concepts of healthcare should thus not necessarily lead to reduced access to basic services or exacerbated disparity of access for marginalised communities compared to the elite. Several health-related interventions have been implemented in Africa to improve health conditions and outcomes, yet the continent still carries the greatest burden of disease (MISTRA, 2019). In some measure, access to these health-related interventions has not been extensive nor sustainable, leaving the most vulnerable population groups with the persistent burden of disease. This chapter, thus, explores how we can ensure the benefits provided by 4IR technologies in healthcare are distributed equitably.

Part of what this chapter aims to do is to position 4IR technologies in healthcare within context – exploring the realities of introducing technologies associated with the 4IR into a country like South Africa with its extreme inequalities, where internet penetration is uneven due to inadequate infrastructure; where data costs are high especially for poorer communities; where there are scarce human resources and skills; deficient legislation; and, sometimes, unreliable power supply.

The chapter attempts to identify the features of the 4IR that could pragmatically be implemented in South Africa's overburdened and under-resourced healthcare system and that could lead to a transformative impact on health and medicine. The analysis draws on secondary research in South African as well as other African and international literature to evaluate key challenges and opportunities for 4IR technology in South Africa's health sector. Using the conceptual framework for value co-creation (Prahalad and Ramaswamy, 2000; Voorberg et al., 2014), the chapter argues for inclusive innovation in order to ensure equitable access to new technologies. Citizens should actively participate in the development of technological and public service innovations meant for them. Innovations have traditionally been imposed on citizens, without in-depth understanding of their daily needs and without cognisance taken of their level of access to, and understanding of, the new innovations. Thus, the chapter puts emphasis on the importance of ensuring that patients from all socioeconomic backgrounds are able to access new technologies and are educated on how to use them.

Furthermore, through a description and brief analysis of South Africa's attempts to implement an electronic health records (EHR) system, the chapter explores the barriers to adoption of technologies that the government has attempted to implement previously. Technology adoption is an intrinsically social and developmental process, therefore involving users of technology in the process of developing technologies is crucial. The chapter concludes by reflecting on 4IR-related policies that can potentially address issues in the healthcare system and argues for complementary health policies and precautionary and realistic implementation of 4IR technologies in South Africa.

HEALTHCARE SYSTEMS AND TECHNOLOGICAL ADVANCEMENTS: GLOBAL TRENDS

Healthcare systems across the globe are grappling with an incessantly changing health landscape that comes with countless uncertainties. The increase in the number of people with chronic illnesses; the growing population (in many countries this includes an exponential increase in aging populations); expensive medical technologies and infrastructure; and a decreasing health workforce are common challenges the world over (Allen, 2019).

As we find ourselves in an era where ground-breaking technology is at the forefront of several industries, the pressing search for sustainable solutions to healthcare has seen increasing advocacy for the use of advanced technologies for improved healthcare to meet this need. So-called 4IR technologies have the potential to improve healthcare in providing both more efficient diagnoses and therapies (treatment and vaccines) as well as in the reliable and timely delivery of healthcare services (WEF, 2019). High-income countries are already enjoying the benefits of incorporating 4IR technologies (ICT, AI and high-performance computing technology) into their healthcare systems and continue to advance their healthcare using these technologies. Examples include an AI-based symptom and cure checker that uses algorithms to diagnose and treat illness; a digital pathology platform that uses AI to detect patterns in cancer cells, and AI software that notifies doctors when a patient's health deteriorates (Daley, 2020). Additionally, for innovative and transformative diagnostics and therapies, current and emerging technologies,[1] such as biotechnology and nanotechnology, have been extensively researched. Some have already been integrated into medical practice while others are still in the research and development (R&D) phase (MISTRA, 2018; WEF, 2019).

The opportunities presented in the initial stages of the global COVID-19 pandemic for these technologies were utilised to showcase

---

1   Emerging technology is a term generally used to describe a new technology, but it may also refer to the continuing development of an existing technology. The term is commonly reserved for technologies that are creating, or are expected to create, significant social or economic effects.

their value. The pandemic highlighted the importance of providing sufficient diagnostic kits in the initial stages of the spread of infectious diseases. Countries with well-established ICT and AI systems, such as South Korea, were able to develop COVID-19 diagnostic kits within a short period of time (Government of the Republic of Korea, 2020). The diagnostic kits swiftly became widely accessible and played a major role in reducing uncertainties in the early stages of the viral spread in South Korea. South Korea's ability to rapidly develop diagnostic kits is due to companies investing in nurturing an R&D environment based on ICT such as big data and AI.

To fortify their contact tracing process, the government in South Korea is using surveillance technology that had already been in place, to support a public health response (Fendos, 2020). Through electronic transaction data, mobile phone location logs, and surveillance camera footage, information on a patient's contacts and whereabouts can easily be accessed and used to rapidly quarantine affected individuals and decontaminate implicated locations (Fendos, 2020).

Furthermore, in South Korea AI is being used to detect lung conditions such as pneumonia, a major symptom of COVID-19 patients, and to identify expeditiously those who need intensive care (Government of the Republic of Korea, 2020). This technology has been valuable in regions where there is an insufficient number of healthcare workers to perform diagnostic tests and swiftly screen patients in order to prioritise patients with severe symptoms.

The unparalleled advancements in digital health solutions has stimulated the development of a range of digital medical technologies, such as the ones used in South Korea, for new care delivery models (Bartlett et al. 2017) that establish analytical, preventive and personalised healthcare (Reddy, 2019). These advancements also lead to large-scale collaboration among stakeholders in healthcare, which brings about rapid, more precise and less invasive diagnostics, treatments and therapies (Reddy, 2019). Consumers, particularly in high-income countries, have become accustomed to convenience, access and personalised products and services in other aspects of their lives (for example, shopping; food orders; banking; education) and are demanding the same for healthcare. There is growing pressure for healthcare to be more proactive and focused on wellbeing with greater involvement of patients (more engagement and empowerment) (Allen, 2019). In this regard, advanced technologies are being utilised to address issues of simpler access; improved and early diagnosis; increased consumer engagement; and an emphasis on prevention and wellbeing and less on treatment (Allen, 2019). Furthermore, transformational technologies in healthcare are generally being used to expedite decision-making and reduce routine administrative duties, enabling healthcare workers to perform optimally (Reddy, 2019) thereby prioritising their time, attention and effort on patients.

While many applications of these technologies have been implemented in contexts where there are sufficient resources and relatively strong healthcare systems, their use in under-resourced settings remains comparatively limited. Nonetheless, governments globally (whether well resourced or not) have adopted the concept of 4IR and there is optimism that these technologies could have a transformative impact on public health in developing countries. The discourse, however, needs to be more nuanced. There needs to be a more critical look at the landscape in developing countries, particularly in Africa, where these technologies are purported to have a transformative impact. The introduction of imported policies or strategies from countries with an extremely different context deprives developing countries of an approach that is tailored to their milieu and one that could have a better chance of being successfully implemented.

4IR IN THE SOUTH AFRICAN HEALTH SECTOR

The benefits that 4IR technologies are providing in high-income countries could completely alter the healthcare landscape in South Africa. However, the South African healthcare system is facing challenges that require more than simply integrating AI, precision medicine and big data analytics into the healthcare ecosystem. The notion of the 4IR necessitates different considerations in South Africa, with its under-resourced and overburdened healthcare system. The country needs to satisfy foundational requirements first in order for its healthcare system to fully benefit from the implementation of 4IR technologies, including socioeconomic redress; upskilling of its health workforce, and upgraded infrastructure. This next section of the chapter thus presents what 4IR technologies have to confront in the South African healthcare system. It paints a different picture to the one in high-income, well-resourced countries, where the notion of 4IR was first conceptualised and advanced. In locating the 4IR within this context, one can provide a realistic analysis of the possibility of widespread and successful implementation of the 4IR in the South African health sector.

Context: The healthcare system in South Africa

The South African healthcare system and health policies have experienced many changes over decades, which have been closely linked to the country's political history (Katuu, 2018). Widespread inequalities entrenched by colonialism and apartheid have had calamitous and enduring effects on the health outcomes of the mainly black population. Notwithstanding considerable progress and continuous endeavours by the government to remedy past disparities, the South African healthcare system is still unable to meet the health demands of the majority adequately and efficiently. Not only does the South African healthcare system have to contend with a quadruple burden of disease (HIV/AIDS, TB, non-communicable diseases, and violent injuries) it has to do so with limited resources exacerbated by underfunding, mismanagement, shortages of health personnel and deteriorating infrastructure (Delobelle, 2013; Katuu, 2018; MISTRA, 2019).

The World Health Organization (WHO) describes healthcare systems as being made up of 'all the people and actions whose primary purpose is to improve health. They may be integrated and centrally directed, but often they are not' (WHO, 2000). The healthcare system in South Africa still carries the legacy of exclusion and fragmentation and this is apparent in how it is divided into two parallel sectors – a highly developed private sector and a disintegrated, dysfunctional public sector (Belli et al., 2018; Katuu, 2018). The public sector, which caters for 84 per cent of the population, is unduly underfunded and there is disproportionate distribution of healthcare workers between private and public institutions, as well as between urban and rural settings (Belli et al., 2018; Di Paola and Vale, 2019). The conditions highlighted above have left the public healthcare system on very shaky ground.

Different studies conducted in public healthcare institutions across the country have confirmed the unfavourable conditions patients and healthcare workers have to grapple with (Di Paola and Vale, 2019; Malakoane et al., 2020; Maphumulo and Bhengu, 2019; Mutshatshi et al., 2018). Prolonged waiting time for patients is among the most widespread problems. In South Africa, visits to public health facilities commonly take a long time, sometimes resulting in patients missing appointments or finding alternative health facilities with a shorter waiting time (Maphumulo and Bhengu, 2019).

Another common issue is that of poor record-keeping in public health facilities. A study at particular public hospitals in Vhembe district, Limpopo Province, South Africa found that nurses, who are responsible for keeping records of every intervention they administer to patients, had challenges with keeping consistent and accurate records. This was due to 1) time needed to complete recording forms; 2) increased admission of patients; and 3) insufficient supply of recording materials resulting in incomplete recording (Mutshatshi et al., 2018). This is in the face of pressures resulting from management reforms to healthcare aimed at improving productivity and financial security over the need for patient care (Di Paola and Vale, 2019). Under the new public management,[2] healthcare workers have been assigned responsibility for stringent administrative measures, which have been prioritised in the running of the hospital in order to achieve cost-cutting and efficiency goals, neglecting the relationship between patient and healthcare provider (Di Paola and Vale, 2019).

The shortages of medicine and equipment have also been reported to be a major stumbling block in public health facilities (Maphumulo and Bhengu, 2019). A study conducted across healthcare facilities in the Free State, South Africa, to determine healthcare system challenges associated with poor public healthcare delivery, revealed several issues,

---

2   The new public management is part of management reforms in healthcare under which public services have been increasingly encouraged to behave like businesses, aimed at efficiency and performance management measured in terms of administrative rather than clinical goals (Di Paola and Vale, 2019).

including 1) 'non-functional ICT system'; 2) 'shortage of resources/ equipment'; 3) 'pharmaceutical system challenges'; and 4) 'supply chain problems' (Malakoane et al., 2020). It was found on one hand that patients experienced poor availability of healthcare workers, medicine and diagnostic devices, which translated into a lack of access to quality healthcare. On the other hand, healthcare workers were dealing with shortages in medicines, vaccines and technologies due to lack of access and high costs, as well as procurement and distribution challenges (Malakoane et al., 2020).

It thus becomes clear that the South African healthcare system requires extensive reform and strengthening. In its goal to rehabilitate the country's healthcare system, the South African government has pinned the healthcare system on the values of primary healthcare and the district health system (Katuu, 2018). According to the National Development Plan Vision 2030 (NDP), primary health care is anchored in 'values such as universal access, equity, participation, and an integrated approach' (NPC, 2013). Furthermore, some of the most crucial components of primary healthcare embrace preventive healthcare and the use of suitable technology (NPC, 2013). The NDP further states that other components entail 'better access to and use of first-contact care, a patient-focused (rather than a disease-focused) approach, a long-term perspective, comprehensive and timely services, and home-based care when necessary' (NPC, 2013).

In 2011, South Africa's National Department of Health (NDoH) launched the model for re-engineering primary healthcare (PHC) in the continuous efforts to transform healthcare in South Africa, followed by an audit on PHC clinics in the public sector (Health Systems Trust, 2013). In July 2013, subsequent to the audit, the NDoH established the 'Ideal Clinic' programme as a means to systematically reform and rectify the deficits in PHC clinics (NDoH, 2018). According to the NDoH (2018):

an Ideal Clinic is a clinic with good infrastructure, adequate staff, adequate medicine and supplies, good administrative processes and adequate bulk supplies that use applicable clinical policies, protocols, guidelines as well as partner and stakeholder support, to ensure the provision of quality healthcare services to the community. An Ideal Clinic cooperates with other government departments as well as with the private sector and non-governmental organisations to address the social determinants of health.

The introduction of a National Health Insurance (NHI), which is still under consideration by parliament, is also anticipated to address some of the shortages and inequalities in the healthcare system. The NHI Bill (NDoH, 2019) aims to provide universal access to quality healthcare for all South Africans regardless of socioeconomic status. Certainly, the issues prevalent in South Africa's healthcare system require radical, but also practical, solutions. Reforms should not leave the poor and vulnerable in worse positions than they are already in. Poverty, low socioeconomic status, and inequality have been linked to vulnerability to diseases (Farmer, 2004) and it has been shown that Africa has some of the widest gender income disparities in the world, which contributes to African women being among the poorest people. This makes African women even more vulnerable to the negative impacts of a weakened healthcare system (UNESCO, 2015; UNESCO, 2017). Furthermore, the challenges in the healthcare domain are exacerbated in rural areas in the country, something that is common throughout the continent. Attempting to uncritically implement imported strategies focused purely on advanced technologies, in a healthcare system with complex and clearly entrenched issues, will only serve to deepen those issues and leave the vulnerable in a worse state.

THE CASE OF ELECTRONIC HEALTH RECORDS

The discourse on 4IR technologies, particularly the AI and ICT-based technologies, has not adequately addressed the inefficient ICT systems that have been implemented in the healthcare system. Health information systems that the government has previously attempted to apply have not been effective. The ICT system in healthcare is like a fragmented healthcare system and it generates the same undesirable outcomes (Botha et al., 2016). In order for advanced technological innovations to be valuable, a foundational ICT system has to be in place, as well as skilled personnel to operate the system. A brief analysis of the electronic health records implemented in South Africa's healthcare system reveals the challenges faced in this domain.

Prior to 1994, South Africa had several information systems in public healthcare which were not interoperable (Katuu, 2016; NDoH, 2019). These systems were fragmented and failed to deliver appropriate information to make logical design decisions. Thus, part of transforming South Africa's public healthcare system was to establish an integrated management system that would prioritise the patient (Katuu, 2016). However, fragmentation in health information systems used in South African public health facilities is still evident and this was detailed in a 2013 review of hospital information systems (NDoH and CSIR, 2014). The study revealed that several different systems from different suppliers were implemented and up to 31 per cent of these systems were unable to exchange patient information.

The National Department of Health together with the Department of Science and Technology (now Department of Science and Innovation) as well as the Council for Scientific and Industrial Research (CSIR) established a patient registration system compatible with an electronic health record (EHR) (NDoH and CSIR, 2014). This system aimed to provide an efficient process for integrating data about patient care throughout the public and private healthcare domains. This pooled data was intended to be used by health services for reporting and planning at provincial and national levels, and concurrently for addressing challenges in the absence of interoperability, disintegration and the lack of a National Patient Master Index.[3] However, in order to make these systems interoperable there is a need for further development of their structural design, and an integrated platform. In 2017, five out of the nine provinces in South Africa had some version of an EHR system operational in public hospitals. However, over 50 per cent of public health facilities in South Africa still utilise a paper-based filing system, years after EHRs were introduced (Cilliers and

---

3 An electronic database that holds demographic information on every patient who receives healthcare services. The MPI aims to accurately match and link records by uniquely identifying individuals. This is done by storing information such as name, date of birth, gender, etc., and assigning everyone a unique identifier.

Wright, 2017; Katurura and Cilliers, 2018).

A study by Katurura and Cilliers (2018) to determine the barriers to implementation of EHR systems found the following main issues from various countries in Africa:

1. insufficient user training, skills and commitment, meaning that healthcare workers are not equipped to use EHR systems resulting in a lack of adoption;
2. absence of information about the technology infrastructure and a high cost of implementation; and
3. strategies for the implementation of EHRs that lack coordination and synchronisation.

Additionally, there is a lack of specificity in national strategies.

These findings are supported by research conducted by Cilliers and Wright (2017) who established that because of unreliable electricity supply, poor internet connectivity and limited bandwidth, healthcare workers in South Africa found EHR systems to be undependable and inaccessible. Cilliers and Wright (2017) further report on a failed Google Health application implemented in 2008 and discontinued in 2012. Users of the application were not provided with adequate information on its use and benefits and there was thus a lack of trust and concerns about privacy for users. Healthcare workers were also not recommending it to patients as a useful product to receive health information. Deficiencies in EHR systems have further been attributed to limited computer availability, slow sign-in processes and convoluted methods of recording or retrieving information (Bardach et al., 2017).

Cohen et al. (2015) found that there can be unintentional consequences of an efficient EHR system such as a decrease in time spent with physicians but an increase in waiting time. Cohen et al. (2015) point to users' possible lack of experience in computer usage which reveals a lack of training and support to develop competences in overall healthcare.

Adequate engagement with end users of new technologies – whether patients or healthcare personnel – and the lack of sufficient and reliable infrastructure among other issues have led to the failed adoption of technologies with great potential to improve healthcare. Therefore, it is important to acknowledge that there have been many dynamic initiatives that have been unsustainable due to the barriers mentioned. Resources have been directed towards establishing and advancing various ICT systems, without provision of an enabling environment for the systems to succeed. The application of technological innovations requires complementary strategies that will ensure the acceptance and optimal utilisation of the technology.

ENSURING EQUITABLE ACCESS TO TECHNOLOGIES IN HEALTHCARE

For the most part, technological innovations are enjoyed by those who already have good access to healthcare and who have a good understanding of the basic operations of current and new technologies (Van Winkle et al., 2017). It has been shown that the benefits provided by innovative healthcare technologies are unevenly distributed towards those from privileged socioeconomic backgrounds (Cozzens and Thakur, 2014). This population group is usually familiar with digital technologies and is able to navigate them easily to search for pertinent health information. This is usually followed by using the information acquired to attend to health issues (Nambisan and Nambisan, 2017). This means that interventions in healthcare that are focused exclusively on technology (particularly digital technologies) are likely to benefit the privileged with higher levels of e-health literacy while excluding marginalised communities (senior citizens; those who are illiterate and/or have restricted access to the internet and other basic digital technologies).

An example of inequality in health technologies is aptly illustrated by Cozzens and Thakur (2014) in an anecdote about a diabetic Mozambican man who lives in a small house in an impoverished area without electricity and with limited access to clean water. The Mozambican government has provided access to free recombinant insulin – a refined, highly efficient medical treatment produced through genetic engineering. The use of this treatment entails testing for blood sugar levels, storing and injecting insulin with a syringe, which all require a strict routine, a level of literacy and household infrastructure that the man does not have. His doctor makes the decision to rather provide him with a simpler treatment that is not as efficient, but that will be more successful in the long run, given the man's circumstances. The doctor then sees another patient who is a professional and prescribes the recombinant insulin for the patient.

Thus, as much as the intention of the Mozambican government in this narrative was to provide an efficient health technology that would benefit all citizens, it ended up benefitting a middle-class patient while excluding an impoverished man who needs it the most. The value of health technologies is undisputed and, more so, the advantages of digital technologies ranging from simpler organisational systems such as EHRs and e-prescribing systems to more intricate digital healthcare systems such as personal health records, are widely accepted. However, in a country as unequal as South Africa and many others, funds directed towards technology-centred solutions (such as the ones promoted by 4IR strategies) would result in the majority of the 84 per cent of people that rely on the public healthcare system being excluded from the benefits. Therefore, mechanisms need to be put in place to ensure that the advantages of digital technologies and their associated innovations reach the intended end-users who need them the most.

The question of access to medicines and other technologies has been ongoing for years. Attention has been given to drugs and vaccines in contentious deliberations about poor access in developing countries; but, this also applies to other healthcare technologies (Frost and Reich, 2009). Furthermore, the debates on access have focused on barriers such as pricing and patents and less on distribution, delivery and technology adoption challenges. Inadequate access to healthcare technologies in developing countries is a multifaceted problem consisting of a mixture of market failures and an inefficient government (Cozzens and Thakur, 2014). The solutions should encompass strategies involving socioeconomic, political and environmental approaches. Moreover, it is important to determine people's perceptions of different technologies, illnesses and treatments when crafting solutions.

Case studies by Frost and Reich (2009) show that it is difficult to formulate strategies that produce unanimous acceptance of a technology at either a national or global level. Creating access therefore requires the time and resources to develop carefully designed strategies, together with end-users, for dealing with the manifold impediments along the pathways to access. With South Africa's NHI Bill still being deliberated, the unequal structure of the nation's healthcare system and issues such as access to and adoption of healthcare technologies are essential for both the implementation of an NHI as well as of a 4IR strategy.

The next section of the chapter explores interventions that can be implemented to ensure that the benefits from healthcare technological innovations are equitably disseminated among all population groups, particularly poor and marginalised communities. The chapter draws on the conceptual framework for value co-creation introduced by Prahalad and Ramaswamy (2000), which discusses the *active participation* of consumers in the many phases of the invention process as opposed to the generic concept of partaking, or *passive participation*. Voorberg et al. (2014) speak of the importance of citizen engagement in social innovation, which requires that the conditions under which citizens co-create be well understood.

Co-creation for healthcare technologies

Co-creation entails a paradigm shift away from simply consulting end-users to actively working together with them in order to determine the issues at hand and create solutions. This concept was initially applied to the private sector (businesses working together with customers). However, it has gradually been explored in the public sector where citizens have predominantly been categorised as passive users of public services, to thinking of them as co-creators (Voorberg et al., 2017). In this setting, opportunities are created for citizens to partake in identifying which are the services and outcomes that are important to them, and how these should be coordinated. The co-creation concept encourages governments and public officials to take into account alternative knowledge systems, different sources of information and unique experiences (Voorberg et al., 2017).

The advent of the internet and other digital technologies has made it viable for consumers to participate at various points of innovation (Sawhney et al., 2005), which results in an enhanced product and consumers being satisfied with the quality of the innovation. In healthcare, the concept of co-creation speaks to the inclusion of patients as collaborators in innovation and value creation. This results in healthcare providers being able to offer patients enhanced experiences through the provision of innovative healthcare services at more affordable costs.

Nambisan and Nambisan (2017) identify three themes that can be adopted to ensure equitable access to technological innovations in healthcare: educate, engage, evolve.

First, under 'educate' they recognise that what is known about healthcare challenges and solutions is substantial, always developing and shared among various stakeholders, including diverse patients. Patients may have unique knowledge about what they need from their digital technology, such as more detailed information on what the data it produces means, and how potential solutions may work (or not) in their exceptional context (Nambisan and Nambisan, 2017). Therefore, knowledge must be bidirectional – healthcare institutions need to educate patients on the technological innovations, and, at the same time, patients need to inform and educate healthcare institutions on the context of their routine experience of the technological innovation. This is crucial for improving the development and adoption of technological innovation.

Second, 'engage' entails patients being active players in the redesigning of sociotechnical systems[4] that involve infusion and adoption of new technological innovations. Nambisan and Nambisan (2017) provide an analysis of how healthcare consumers have established online forums to review issues related to the use of wearable devices. These reviews and discussions among healthcare consumers help advance and provide unrestricted solutions to some of these challenges. In a setting like one in rural South Africa, this engagement would need to take a different form. Perhaps through a communication platform easily accessible to most or, in communities where internet and data is an issue, meetings or focus groups could be utilised. Dynamic and effective patient participation in the multiple stages of innovation may result in healthcare providers being more active in creating space for improved innovation.

---

4 Sociotechnical theory has at its core the idea that the design and performance of any organisational system can only be understood and improved if both 'social' and 'technical' aspects are brought together and treated as interdependent parts of a complex system.

Third, under 'evolve', the authors examine wearable devices and how the health data they provide has been used to create data portals. Furthermore, these devices have fuelled an analysis by healthcare institutions of how healthcare practitioners can use such consumer-owned data in diagnosis and treatment. This is an indication of how new technological innovations generate changes in sociotechnical structures and how these changes sequentially alter how new technologies are advanced, understood or applied by healthcare institutions and patients. As such, co-creation could persist through the lifespan of the innovation; healthcare institutions need to be mindful of these subtleties and adjust their approaches and procedures accordingly.

eHealth interventions in African countries

eHealth is the use of ICT to improve a flow of information, using electronic devices, thus supporting the delivery of health services and the management of health systems (WHO and ITU, 2012). eHealth has been lauded as a viable solution to equitable distribution of healthcare to the marginalised and vulnerable population groups, particularly across African healthcare systems. What is encompassed in eHealth has been debated for a few years as the term has been used to describe ICT technologies used for healthcare in similar but subtly different ways. A qualitative study on what constitutes eHealth was conducted to establish a conceptual framework of eHealth that sufficiently describes its intricacy and possible areas of intersection (Shaw et al., 2017). The study, which entailed obtaining perspectives from professionals in healthcare delivery, research, education, practice, governance and policy, revealed that there are three distinct but intersecting domains of eHealth (see Table 9.1). The researchers in this study argued that there was a need for a model of eHealth that provided an opportunity for shared discourse and for eHealth implementation and operationalisation (Shaw et al., 2017).

These domains are respectively defined as 1) the use of digital technologies to monitor, track, and inform; 2) the use of digital technologies to facilitate communicative encounters between health stakeholders; and 3) the use of data to improve health and health services. In a systematic review of published research, Cohen et al. (2015) examined the use and

Table 9.1: eHealth domains and subcategories

Domain	Subcategories
Health in our hands: the use of eHealth technologies to monitor, track and inform health	• Health, not just healthcare • Consumer-driven and -controlled health • Health via social media and the internet
Interacting for health: the use of technologies to communicate between stakeholders in health	• Connecting for real-time health • Social discourses and storytelling • New ways of interacting to personalise care • Supporting health professionals
Data enabling health: the collection, management and use of health data sources	• Data management systems and data repositories • Data for precision health • Data enabling quality

Source: Shaw et al. (2017)

impacts of eHealth in community-based health facilities in developing countries. The review found numerous studies referring to eHealth initiatives. However, upon further investigation the authors concluded that there may be limited diffusion of these eHealth technologies within community-based facilities in developing country contexts. The authors also established that several eHealth technologies are aimed at data collection for reporting purposes rather than for patient care.

One of the most common eHealth platforms used across Africa is mobile health, or mHealth in short, which is the use of mobile devices for healthcare delivery. Mobile connectivity has been a driver of transformation for millions of people across Africa, through facilitating the distribution of basic resources and services, including education, healthcare and financial inclusion. Mobile devices have allowed for the circumvention of some of the challenges brought on by considerable gaps in infrastructure and funding. The number of smartphone connections in the sub-Saharan African region reached 302 million in 2018, and this number is forecast to reach approximately 700 million by 2025, an adoption rate of 66 per cent (GSMA, 2019). Populations in developing or low-income countries have been able to take advantage of the availability of low-cost smartphones that have overcome the barrier to affordable access (Wood et al., 2019). These low-cost smartphones provide data sensing and processing capabilities comparable to more expensive mobile devices (Wood et al., 2019).

The proliferation of mobile technologies across the African continent has given rise to interventions that use mobile devices for healthcare delivery. Approximately 98 per cent of all internet experiences in Africa happen on mobile smartphones (GSMA, 2019). In Kenya, the accessibility of smartphones has been used advantageously in several eHealth projects implemented in an attempt to offer cost-effective solutions to health and healthcare system challenges (Njoroge et al., 2017). Although there has been a mixture of the types of eHealth platforms used, the majority of these were facilitated by mobile devices. This finding is supported by a report that revealed that most of the East African eHealth start-ups use mobile devices, with 64 per cent of Kenyan start-ups relying on mobile devices (High-Tech Health, 2017).

The use of mHealth in South Africa is also expanding. GSMA indicates there were 83 mHealth services in South Africa in 2013, with the majority focusing on HIV/AIDs and women and children (Cargo, 2013). In 2016, the most common mHealth applications were the registration of users and vital event tracking, data collection and reporting, and the creation and updating of electronic health records (Botha et al., 2016). As the technical barriers to using mHealth applications need to be as low as possible to increase reach, these services mostly utilise SMS (Botha et al., 2016).

Table 9.2 shows a few mHealth programmes that have been implemented in countries on the continent using different platforms.

The sustainability of many mHealth interventions over long periods of time, however, is a cause for concern. The reliance on donor funding means the services are discontinued after some time. Thus, there is a need for a more viable solution to mHealth implementation. Botha et al. (2016: 145) argue for:

an integrated mHealth & Wellness Innovation Ecosystem as opposed to the development of additional free-standing

Table 9.2: Health programmes in a few African countries

Project	Platform used	Purpose
MomConnect[5] Since 2014 (South Africa)	• SMS • Unstructured Supplementary Services Data (USSD) • Mobile website	• Platform allows pregnant women to reach out with pressing questions and get feedback. • Was used as a platform to improve service at healthcare facilities.
Project[6] Masilukeke Since 2008 (South Africa)	• SMS	• Health education and awareness programme (one million messages per day to subscribers to encourage HIV/AIDS testing).
MSos[7] Since 2013 (Kenya)	• SMS • Web portal	• Text message facility that allows real-time communication between health facility workers and disease surveillance coordinators at sub-county, county and national levels.
Safer Mom[8] Since 2014 (Nigeria)	• Web App • Mobile App • SMS • Voice service	• Platform that monitors and tracks pregnancy and babies' development.
MTrac[9] (Uganda)	• Toll free MTrac SMS hotline 8200	• SMS based technology connecting hospitals to the national drug chain.

Source: Njoroge et al. (2017)

---

5   MomConnect http://www.health.gov.za/index.php/mom-connect
6   Project Masilukeke https://www.innovations.harvard.edu/project-masiluleke
7   MSos http://www.africanstrategies4health.org/uploads/1/3/5/3/13538666/msos.pdf
8   Safer Mom https://www.changemakers.com/makingmorehealth/entries/safermom
9   MTrac http://www.mtrac.ug/

mobile applications and services. mHealth in South Africa is not only about improving the availability, access and delivery of healthcare services, but is also about enhancing the country's strategic capabilities to create, adapt and implement novel mHealth solutions within, and by, the public and private sector, towards enhancing the country's overall innovation capacity.

Importantly, collaboration between stakeholders in the public sector, the private sector and broader civil society is required to overcome barriers to successful mHealth implementation and so to unlock its full potential. The capacity for mHealth to contribute towards improved health equality and ultimately to pave the way for successful implementation of even more advanced ICT systems should be harnessed. The lack of readily available and sufficient evidence on the impacts of eHealth on healthcare delivery and patient outcomes (Cohen et al., 2015) is a missed opportunity to determine the usefulness of mHealth applications and to establish areas for improvement. It also means there is lack of sufficient evidence to conduct studies examining readiness for emerging technologies.

This section has highlighted that, in a developing context, mHealth is a more readily available technology to improve healthcare and make it more equitable. The importance of taking cognisance of the context in which technological innovations are intended to operate cannot be overstated. Essentially, the design and development of healthcare systems and technologies should be mindful of a nation's needs and available resources. Furthermore, co-creation should be at the centre of the development of mHealth solutions to increase the chances of uptake.

CURRENT ADVANCES IN 4IR TECHNOLOGIES IN THE SOUTH AFRICAN HEALTH SECTOR

Even with the country's unfavourable state of healthcare and sometimes ineffective implementation of health technologies, there are small pockets where advanced 4IR technologies are being employed in South Africa's healthcare sector. A few examples are provided below in Table 9.3.

Table 9.3: Examples of 4IR technology advancement in South Africa

4IR Technology	Application	Example
Artificial Intelligence	Diagnosis and triage of patients	To streamline and improve medical imaging diagnosis for radiologists. With this technology, radiologists are empowered to prioritise urgent cases and, in that way, present a more patient-centric model of health service delivery (Envisionit, 2020)
Artificial Intelligence	Automated triaging	Reduces patient waiting time at clinics, and ensures records are always accessible (Phulukisa, 2020)
Artificial Intelligence	Enhanced communication with patients	Multilingual chatbot service for self-diagnosis (doctor on your smartphone). Additionally, the app connects healthcare workers and patients in real-time (Datawizzards, 2020).
Artificial Intelligence	Counselling services	An app providing counselling and tackling GBV and mental health (AI for Good, 2018)
Big Data	Clinical decision-making support	Vantage technology integrates data immediately from a large range of sources to analyse service delivery. It then makes recommendations for step-by-step action, directing staff to take the right action at the right time, rather than make decisions based on gut instinct, habit or anecdotal evidence (BroadReach, 2019).

4IR Technology	Application	Example
Biotechnology	Advanced genetic engineering and synthetic biology to develop human therapeutic proteins	Using plant cells to express human recombinant proteins with increased safety, production speed, clinical efficacy, cost and scalability (Arzagen, 2019)
Drones	Deliver blood and medical supplies	The South African National Blood Service (SANBS) partnered with the Western Cape Blood Service to introduce a nationwide blood delivery drone service (Malinga, 2020).

Furthermore, COVID-19 has led to various private-sector and academic institutions producing face shields for frontline health workers using 3D printing (3D community, 2020). 3D printing requires little human interaction and can even be done in the manufacturer's home. Additionally, algorithms have been used for data modelling (Covid19sa, 2020) to assist in decision-making during the COVID-19 pandemic. The models are able to show and predict factors such as hotspots, numbers of infections and deaths, and available functional hospital beds. They have been valuable in navigating through the uncertainty presented by this novel virus.

While these exciting technological innovations are impressive and have the capacity to transform healthcare, there is a low probability of their benefits being widespread and inclusive without a concurrent strengthening of the healthcare system and socioeconomic redress.

WAY FORWARD: A MULTI-LAYERED, SYNDEMICS APPROACH TO HEALTHCARE

As described above, the issues in the South African healthcare system are multifaceted, historical and systemic. It then follows that efforts to strengthen the healthcare system should take a comprehensive approach. The use of AI and other digital technologies in under-resourced settings requires a firm understanding of socioeconomic, healthcare and political conditions, infrastructure constraints and further associated infrastructure requirements, such as IT, communications systems and programmes for delivering primary health services. In addition, several AI technologies rely on robust electronic health record systems.

Technological innovations are a crucial aspect of healthcare reforms, but they require complementary and supportive interventions for their benefit to be experienced by all. Some areas that policymakers could consider include the following:

- Systemic issues in the country that need to be addressed. The South African government has drafted numerous policies that aim to redress inequalities, but many of those challenges persist. These systemic challenges are a barrier to realising the full potential of 4IR technologies, particularly for communities that bear the brunt of a defunct system.
- The South African healthcare system relies heavily on community healthcare workers (CHW). They play a pivotal role in the management of diseases in healthcare delivery and can be used extensively in the process of technology co-creation and/or diffusion. They have unique knowledge of communities, and this can be used to forge partnerships between healthcare providers and patients. However, CHWs need to be fairly compensated for the work they do – and their employment formalised. In this way, they will be incentivised and will have an enabling environment to work optimally.
- A national, widespread intervention for healthcare infrastructure development is required, including ensuring water and sanitation facilities are up to standard. Such an infrastructure plays a major role in health outcomes.
- A healthcare-focused 4IR strategy tailored to the South African healthcare system should be drafted.
- Priority should be given to the training and upskilling of healthcare workers in technologies introduced into the healthcare system.

- Electronic health records (and any other ICT systems) need to be maintained by skilled personnel.
- Lastly, adequate resources are required to plan for and effectively meet the health needs of the whole population. The NHI should, therefore, be prioritised to aid in redressing the fragmented healthcare system, while providing quality healthcare and access to health technologies to all, regardless of socioeconomic factors, geographical location or ability to pay.

CONCLUSION

This chapter has explored the benefits of 4IR technologies and shown how these are currently being enjoyed by high-income countries. The issue of access to technological innovations for marginalised communities was shown to be a possible barrier to equitable distribution of the benefits of 4IR technologies. Solutions to healthcare in the South African context, and more broadly the African context, require a syndemic approach (MISTRA, 2019). Technology alone will not be the panacea to healthcare system challenges in the country, and on the continent at large. Rather, parallel strategies and policies need to be in place, such as programmes that ensure that healthcare workers are upskilled, and that patients are included in the development of technologies meant for them. These are some of the various factors that impact on health outcomes, including socioeconomic, environmental and political factors, which need to be addressed, with technological advances as enablers of solutions. Health inequalities will continue being entrenched and the majority of the population will be left out of the so-called revolution if 4IR technologies are treated as a blanket solution, and a comprehensive, transdisciplinary approach is neglected.

These concerns have been brought to the fore with the race for a COVID-19 vaccine. There is a danger that in the urgency to have a vaccine, questions about access to and distribution of the vaccine will be neglected. Will developed, wealthier countries be the first to gain access to the vaccine? And within both developed and developing countries, will privileged groups have better and quicker access? The need for an equitable, fair strategy for a potential vaccine underscores the importance of the central issue raised in this chapter: The technologies of the Fourth Industrial Revolution hold exciting potential for transformation but their benefits are greatly dependant on the context in which they are implemented. In South Africa, one of the most unequal countries in the world, failure to discuss what 4IR technologies will mean for its marginalised groups, and their access to healthcare, runs the risk of entrenching inequalities. In exploring 4IR technologies and their potential for transformation, researchers need to keep at the forefront the question of who these innovations are for.

REFERENCES

AI for Good. 2018. 'rAInbow'. https://www.hirainbow.org/, accessed 10 June 2020.

Allen, S. 2019. '2020 global health care outlook: Laying a foundation for the future'. *Deloitte Insights*. www.deloitte.com/insights, accessed 12 June 2020.

Arzagen. 2019. 'Plant-based expression technology.' ArzaGen. http://azargen. com/#plant-based-expression-technology, accessed 2 August 2020.

Bardach, S.H., Real, K. and Bardach, D.R. 2017. 'Perspectives of healthcare practitioners: An exploration of interprofessional communication using electronic medical records.' *Journal of Interprofessional Care*, 31(3), 300–306.

Bartlett, R., Dash, P., Markus, M., McKenna, S. and Streicher, S. 2017. 'New models of healthcare'. Mckinsey and Company. https://www.mckinsey. com/industries/healthcare-systems-and-services/our-insights/new-models-of-healthcare, accessed 27 July 2020.

Belli, P., Matsebula, T., Ndhlalambi, M. and Ngarachu, M. 2018. 'A brief profile of the status of health and the health system in South Africa: Background note for the South Africa systematic country diagnostic'. World Bank. https://elibrary.worldbank.org/doi/abs/10.1596/30036, accessed 18 June 2020.

Botha, A., Herselman, M. and Kotze, D. 2016. 'mHealth & wellness innovation ecosystem'. In: Herselman, M. and Botha, A. (eds.). *Strategies, Approaches and Experiences: Towards Building a South African Digital Health Innovation Ecosystem*. Pretoria: CSIR Meraka.

BroadReach Consulting. 2019. '4IR technology holds key to better healthcare in Africa'. https://www.broadreachcorporation.com/4ir-technology-holds-key-to-better-healthcare-in-africa/#:~:text=Fourth%20Industrial%20 Revolution%20technology%2C%20called,in%20tackling%20the%20 HIV%20epidemic, accessed 21 July 2020.

Cargo, M. 2013. 'South Africa mHealth landscape.' GSMA mHealth. https://www.gsma.com/mobilefordevelopment/ programme/mhealth/south-africa-mhealth-landscape/, accessed 7 May 2020.

Cilliers, L. and Wright, G. 2017. 'Electronic health records in the cloud: Improving primary health care delivery in South Africa'. *Studies in Health Technology and Informatics*, 245, 35–39.

Cohen, J.F., Coleman, E. and Abrahams, L. 2015. 'Use and impacts of e-health within community health facilities in developing countries: A systematic literature review.' *ECIS Proceedings*. https://aisel.aisnet.org/ecis2015_cr/33/, accessed 5 August 2020.

Cozzens, S. and Thakur, D. 2014. *Innovation and Inequality: Emerging Technologies in an Unequal World*. Cheltenham, UK: Edward Elgar.

Dahiya, A.S., Thireau, J., Boudaden, J., Lal, S., Gulzar, U., Zhang, Y., Gil, T. et al. 2020. 'Energy Autonomous Wearable Sensors for Smart Healthcare: A Review'. *Journal of The Electrochemical Society*, 167(3), 037516.

Daley, S. 2020. '32 examples of AI in healthcare that will make you feel better about the future.' *Built in*, https://builtin.com/artificial-intelligence/artificial-intelligence-healthcare, accessed 3 August 2020.

Datawizzards. 2020. http://datawizzards.io/, accessed 10 June 2020.

Delobelle, P. 2013. 'The health system in South Africa: Historical perspectives and current challenges'. In :Wolhuter C.C. (ed.). *South Africa in Focus: Economic, Political and Social Issues*. Editor: Nova Science Publishers, Inc.

Department of Health (DoH). 2018. *Ideal Clinic Manual Version 18*. Pretoria: DoH.

Di Paola, M. and Vale, B. 2019. 'Knowledge, power and the role of frontline health workers for South Africa's epidemic preparedness.' In: Mapungubwe Institute for Strategic Reflection (MISTRA). *Epidemics and the Health of African Nations*. Mazibuko, Z (ed). Johannesburg: MISTRA, 229–269.

Envisionit Deep AI. 2020. https://www.edai.africa/, accessed 10 June 2020.

Farmer, P. 2004. 'On suffering and structural violence: Social and economic rights in the global era'. In: Farmer, P. (ed.). *Pathologies of Power: Health, Human Rights, and the New War on the Poor*. Los Angeles: University of California Press.

Frost, L.J. and Reich, M.R. 2009. 'Creating access to health technologies in poor countries'. *Health Affairs*, 28(4), 962–973.

Gatouillat, A., Badr, Y., Massot, B. and Sejdić, E. 2018. 'Internet of Medical Things: A review of recent contributions dealing with cyber-physical systems in medicine'. *IEEE Internet of Things Journal*, 5(5), 3810–3822.

Gordon, W.J. and Catalini, C. 2018. 'Blockchain technology for healthcare: Facilitating the transition to patient-driven interoperability'. *Computational and Structural Biotechnology Journal*, 16, 224–230.

Government of the Republic of Korea. 2020. 'Flattening the curve on

COVID-19: How Korea responded to a pandemic using ICT'. https://extranet.who.int/goarn/flattening-curve-covid-19-how-korea-responded-pandemic-using-ict-information-communication, accessed 13 June 2020.

GSMA. 2019. 'The mobile economy sub-Saharan Africa 2019'. https://www.gsma.com/mobileeconomy/wp-content/uploads/2020/03/GSMA_MobileEconomy2020_SSA_Eng.pdf, accessed 23 June 2020

Health Systems Trust. 2013. 'The national health care facilities baseline audit: national summary report'. Health Systems Trust. http://www.hst.org.za/publications/national-health-care-facilitiesbaseline-audit-national-summary-report, accessed 13 July 2020.

High-Tech Health. 2017. 'Exploring the African E-health Startup Ecosystem Report'. https://disrupt-africa.com/high-tech-health/, accessed 18 July 2020.

Johnson, J.L. and Manion, S. 2019. 'Blockchain in healthcare, research, and scientific publishing'. *Medical Writing*, 28(4), 10–13.

Katurura, M.C. and Cilliers, L. 2018. 'Electronic health record system in the public health care sector of South Africa: A systematic literature review'. *African Journal of Primary Health & Family Medicine*, 10(1), a1746. https://doi.org/10.4102/phcfm.v10i1.1746.

Katuu, S. 2018. 'Healthcare systems: Typologies, framework models, and South Africa's health sector'. *International Journal of Health Governance*. 23(2), 134–148.

Limaye, A. and Adegbija T. 2017. 'A workload characterization for the Internet of Medical Things (IoMT)'. *IEEE Computer Society Annual Symposium on VLSI (ISVLSI)*, 302–307

Mackey, T.K., Kuo, T, Gummadi, B., Clauson, K.A., Church, G. et al. 2019. '"Fit-for-purpose?" – challenges and opportunities for applications of blockchain technology in the future of healthcare'. *BMC Medicine*, 17, 68.

Malakoane, B., Heunis, J.C., Chikobvu, P., Kigozi, N.G. and Kruger, W.H. 2020. 'Public health system challenges in the Free State, South Africa: a situation appraisal to inform health system strengthening'. *BMC Health Services Research*, 20(58). https://doi.org/10.1186/s12913-019-4862-y.

Maphumulo, W.T. and Bhengu, B.R., 2019, 'Challenges of quality improvement in the healthcare of South Africa post-apartheid: A critical review', *Curationis* 42(1), a1901. https://doi.org/10.4102/curationis. v42i1.1901.

Mapungubwe Institute for Strategic Reflection (MISTRA), 2018. *Beyond Imagination: The Ethics and Applications of Nanotechnology and Bio-Economics in South Africa*. Mazibuko, Z (ed). Johannesburg: MISTRA

Mapungubwe Institute for Strategic Reflection (MISTRA), 2019. *Epidemics and the Health of African Nations*. Mazibuko, Z (ed). Johannesburg: MISTRA

Mujawar, M.A., Gohel, H., Bhardwaj, S.K., Srinivasan, S., Hickman, N. and Kaushik, A. 2020. 'Aspects of nano-enabling biosensing systems for intelligent healthcare; towards COVID-19 management'. *Materials Today Chemistry.* https://doi.org/10.1016/j.mtchem.2020.100306.

Mutshatshi, T.E., Mothiba, T.M., Mamogobo, P.M. and Mbombi, M.O. 2018. 'Record-keeping: Challenges experienced by nurses in selected public hospitals'. *Curationis* 41(1), 1931. https://doi.org/10.4102/curationis. v41i1.1931.

Nambisan, S. and Nambisan, P. 2017. 'How should organizations promote equitable distribution of benefits from technological hnnovation in Health care?'. *AMA Journal of Ethics,* 19(11), 1106–1115.

National Department of Health (NDoH). 2018. 'Ideal Clinic Manual Version 18.' NDoH. https://www.idealhealthfacility.org.za/docs/v18/Final%20 Ideal%20Clinic%20Manual%20-%20version%2018%20(26%20 July%202018).pdf, accessed 12 July 2020.

National Department of Health (NDoH). 2019a. 'National Health Insurance Bill'. NDoH. https://www.gov.za/sites/default/files/gcis_ document/201908/national-health-insurancebill-b-11-2019.pdf, accessed 12 July 2020.

National Department of Health (NDoH). 2019b. 'National Digital Health Strategy for South Africa 2019 – 2024.' NDoH. http://www.health.gov.za/ index.php/2014-08-15-12-54-26?download=3651:national-digital-health-strategy-for-south-africa-2019-2024, accessed 26 July 2020.

National Department of Health (NDoH) and Council for Scientific and Industrial Research (CSIR). 2014. 'South African national health normative standards framework for interoperability in eHealth'. https://www.gov. za/sites/default/files/gcis_document/201409/37583gen314.pdf, accessed 10 June 2020.

Njoroge, M., Zurovac, D., Ogara, E.A.A., Chuma, J. and Kirigia, D. 2017. 'Assessing the feasibility of eHealth and mHealth: A systematic review and analysis of initiatives implemented in Kenya'. *BMC Res Notes,* 10 (90), DOI 10.1186/s13104-017-2416-0.

Phulukisa Health Solutions. 2020. https://phulukisa.co.za/e-triage/#scroll, accessed 10 June 2020.

Prahalad, C.K. and Ramaswamy, V. 2000. 'Co-opting customer competence'. *Harvard Business Review,* 78(1), 79–90.

Reddy, M. 2019. 'Digital transformation in healthcare in 2020: 7 Key Trends. Digital Authority Partners'. https://www.digitalauthority.me/resources/ state-of-digital-transformation-healthcare/, accessed 13 May 2020.

Sawhney, M., Verona, G. and Prandelli, E. 2005. 'Collaborating to create: The internet as a Ppatform for customer engagement in product innovation'. *Journal of Interactive Marketing,* 19, 4–17.

Shaw, T., McGregor, D., Brunner, M., Keep, M., Janssen, A. and Barnet, S. 2017. 'What is eHealth (6)? Development of a Conceptual Model for eHealth: Qualitative Study with Key Informants'. *Journal of Medical*

Internet Research, 19(10), e324. https://doi.org/10.2196/jmir.8106.

Thuemmler, C. and Bai, C. 2017. 'Health 4.0: How virtualization and big data are revolutionizing healthcare'. New York: Springer.

United Nations Economic, Scientific and Cultural Organization (UNESCO). 2015. 'EFA Global Monitoring Report'. www.unesco.org, accessed11 December 2018.

United Nations Economic, Scientific and Cultural Organization (UNESCO). 2017.'Gender equality in SSA. Innovative Programmes, visible results'. www.unesco.org, accessed 11 December 2018.

Van Winkle, B., Carpenter, N. and Moscucci, M. 2017. 'Why aren't our digital solutions working for Eeeryone?'. *AMA Journal of Ethics*, 19(11), 1116–1124.

Voorberg, W.H., Bekkers, V.J. and Tummers, L.G. 2014. 'A systematic review of co-creation and co-production: Embarking on the social innovation journey'. *Public Management Review*, DOI: 10.1080/14719037.2014.930505.

Voorberg, W., Bekkers, V., Timeus, K., Tonurist, P. and Tummers, L. 2017. 'Changing public service delivery: Learning in co-creation'. *Policy and Society*, 36(2), 178–194.

Wahl, B., Cossy-Gantner, A., Germann, S. and Schwalbe, N.R. 2018. 'Artificial intelligence (AI) and global health: How can AI contribute to health in resource-poor settings?' *BMJ Glob Health*, 3, e000798. doi:10.1136/bmjgh-2018-000798.

Wood, C.S., Thomas, M.R., Budd, J., Mashamba-Thompson, T.P., Herbst, K. et al. 2019. 'Taking connected mobile-health diagnostics of infectious diseases to the field'. *Nature*, 566, 467–474.

World Economic Forum (WEF). 2019. 'Health and healthcare in the Fourth Industrial Revolution: Global future council on the future of health and healthcare 2016–2018'. Cologny/Geneva Switzerland.

World Health Organization (WHO) and International Telecommunication Union (ITU). 2012. 'National eHealth Strategy Toolkit'. International Telecommunication Union. https://apps.who.int/iris/handle/10665/75211, accessed 23 June 2020.

World Health Organization (WHO). 2000. 'The World Health Report 2000: Health Systems – Improving Performance'. World Health Organization, Geneva.

TEN

The potential benefits and risks of technologies of the Fourth Industrial Revolution in addressing environmental transgression in South Africa

ANELILE GIBIXEGO AND NELSON ODUME

INTRODUCTION

The interaction between society and the environment has greatly been influenced by technological advancements. During the Stone Age, and into the agrarian economy, society depended largely on the natural environment for sustenance and livelihood (Amhez, 2019). The Second Industrial Revolution, which began in the 19th century, represented a disruptive leap into the proliferation of industries and led to massive production (Mokyr, 1998; Allen, 2012; Petrillo et al., 2018). Inventions in the Second Industrial Revolution were mainly in the oil and chemical sectors, the steel industry and the electricity and transportation sectors (Mokyr, 1998). However, much of the socioeconomic development witnessed in the Second Industrial Revolution has come at great cost to the natural environment, particularly freshwater, marine, land, air and forest ecosystems (Rockstrom et al., 2009). These environmental externalities continue to haunt humanity as the industries of the Second Industrial Revolution are among the top global emitters of greenhouse gases (Atkeson and Kehoe, 2001). This is also true for South Africa, as the mining sector is among the top emitter of greenhouse gases in the country (Casey, 2019). For example, diamond mines in the Northern Cape, the Witwatersrand gold mines and coal mines in eMalahleni were all pivotal for South African economic growth and job creation, but these have also caused serious environmental externalities such as water, land and air pollution (Akcil and Koldas, 2006).

The Third Industrial Revolution, which began following the Second World War, was characterised by rapid industrial progress aligned with information technologies, advances in computer hardware and software, communication technologies, and the internet (Snudden, 2019). These advances in information and computer technologies led to automation of most of the production processes from the Second Industrial Revolution. However, the combined advances made in the Second and Third Industrial Revolutions, coupled with a growing human population, led to changes in the earth's biosphere due to human activities, and significant increases in global temperatures, heralding a new geological epoch called the Anthropocene (Lewis and Maslin, 2015; Laurance, 2019). Specifically, advances in both the Second and Third Industrial Revolutions have led to many ecological transgressions such as escalating environmental pollution, land system changes, climate change, ocean acidification, over-exploitation of fisheries resources, deforestation and biodiversity loss (Lewis and Maslin, 2015). The Second Industrial Revolution can be considered as being resource intensive; the Third Industrial Revolution signalled a shift to a services-oriented economy.

The potential relationship between environmental perturbation and the mainstay of each revolution is shown by the Kuznets curve in Figure 10.1. The diagram shows that in post-industrial economies, which are characterised by more service-based economies, there is a decrease in environmental degradation as countries shift away from heavy polluting industries. This shift is accompanied by an increase in income per capita (Figure 10.1).

Figure 10.1: The Environmental Kuznets curve, showing the decrease in environmental degradation in the post-industrial economies

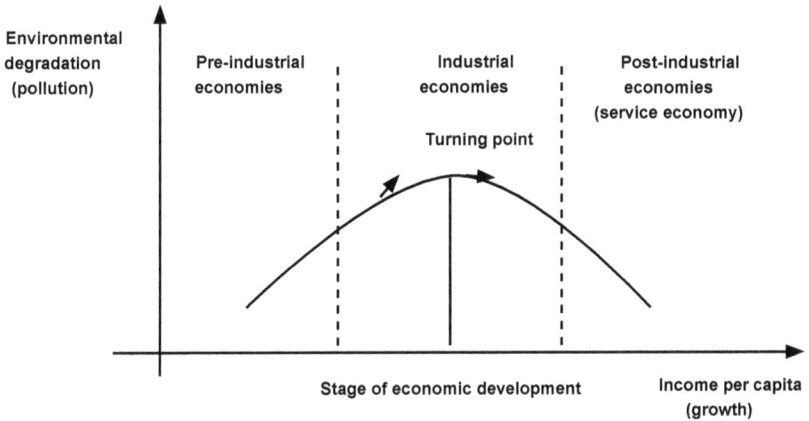

Source: Tanger et al. (2011: 1110)

The current era heralds the Fourth Industrial Revolution (4IR), as named by Schwab (2017), which represents a seamless integration of the physical environment, the biosphere and the digital environment. Key features of the 4IR include the Internet of Things (IoT), Artificial Intelligence (AI), bio- and nano-technological innovation, green technology, big data and fast computing, blockchain and governance innovation such as polycentricity (Schwab, 2017; Petrillo et al., 2018). It has been framed as the industrial revolution that will restore ecological integrity using advanced treatments and technologies.

The key question we seek to explore in this chapter is: To what extent can 4IR technologies limit South Africa's transgression of key ecological boundaries, systems and features and, in so doing, improve social-ecological interactions and relationships? We take a systems view, embedding our argument within a socioecological paradigm, which includes complexity thinking. We explore this question in relation to five important features of South Africa's economies and ecologies: 1) freshwater quantity, quality and ecosystem health; 2) land system change; 3) mining; 4) carbon emission and climate change; and 5) biodiversity.

These five features have been selected for the analysis because interventions across these domains are likely to make the greatest contributions to South Africa's strides towards sustainable development, including social, political and economic imperatives. For example, South Africa is a water-scarce country, and freshwater pollution and inadequate water supply present some of the greatest risks to the country (Stats SA, 2017). We, therefore, explore the role of 4IR technologies in addressing some of the pressing water challenges in South Africa. The mining industry is a large contributor to South Africa's economy and is strategically important in relation to the future of renewable energies. Therefore, we argue how its operations can continue in a more sustainable way. Land, biodiversity and climate areas have been contravened, largely due to human activities. We, thus, investigate how 4IR technologies can be utilised to address some of the ecological transgressions made and possibly to reverse them. We also assess the risks associated with an uncritical adoption of 4IR technologies in South Africa.

This chapter shows that 4IR technologies hold great potential to reverse some historical environmental transgressions in South Africa, but attention must be paid to potential risks and unintended consequences as well. It shows the importance of 4IR technologies in the just transition to a low-carbon economy. To minimise the risk posed through an uncritical adoption of 4IR technologies, the chapter argues that an environmental governance framework, embedded in complex social-ecological system (SES) theory, would be required (Folke, 2006). Such a framework would require serious consideration of the transformational imperatives of equity, social justice, and environmental and ecological justice in the adoption of any 4IR technologies. This consideration would need to take into account whether adoption of a technology would further the realisation of any of those transformational imperatives or not. Without such a framework, the chances of unintended consequences for either ecological or social systems may become heightened.

The chapter concludes with a critical reflection on possible environmental policies and governance implications for South Africa. The chapter shows that 4IR technologies hold great potential to reverse some historical environmental transgressions in South Africa, but attention must be paid to potential risks and unintended consequences as South Africa strives towards a more humane society with equal opportunities for all.

A JUST TRANSITION INTO A LOW-CARBON ECONOMY USING THE 4IR AS A CATALYST

South Africa is one of the most unequal societies in the world (Stats SA, 2019). Evidence suggests that the poor within any society are most at risk of environmental disasters occasioned by climate change and extreme weather events (NPC, 2013; Ziervogel et al., 2014). In South Africa, for example, while the 2014-2016 drought affected nearly all sectors of the economy and population, the effects were most pronounced for the poor because of their low adaption and resilience capacities (Archer, 2019; Vogel and Olivier, 2019).

The adjective 'just' in the concept 'just transition' refers to justice and fairness in the transition for all parties involved. It takes into cognisance that the poor and vulnerable are disproportionately affected by climate change (NPC, 2013). According to Heffron and McCauley (2018) the 'just transition' concept emanated from the labour movement with the realisation of the intricate relationship between the environment, employment and livelihoods. A just transition into a low-carbon economy is one which simultaneously addresses South Africa's imperatives for social and ecological justice (Sengupta et al., 2019), while reducing reliance on carbon-intensive industries (NPC, 2013). In South Africa, millions of people are employed in energy-intensive industries (such as mining and transportation) (Stats SA, 2020); therefore, a just transition is non-negotiable.

By social justice we imply a transition that actively addresses historical inequalities and equity imperatives within the social fabric of the country. Ecological justice implies a developmental path that seeks to protect ecosystems, while ensuring the judicious use of ecosystem services for socioeconomic development in an efficient, equitable and sustainable manner (NPC, 2013). The argument for a transition to a low-carbon economy is advanced in chapter 5 of the National Development Plan (NDP) Vision 2030.

The hypothesised magnitude of technological advancement for a low-carbon economy and institutional change is at a scale previously experienced in former industrial revolutions (Pearson and Foxon, 2012). The productive gains and economic benefits of a transition to a low-carbon economy would be at an even bigger scale than of previous revolutions, making the low-carbon economy more economically viable and environmentally desirable (Pearson and Foxon, 2012). While the successes of the previous revolutions relied on price incentives, the invention of new technologies and enhanced human capital (Allen, 2012), a transition to a low-carbon economy would be driven by social good and ecosystem wellbeing (Pearson and Foxon, 2012). This then implies that as 4IR technologies cause social disruption, decision-making around the sharing of costs and benefits associated with such technologies should be guided by considerations for distributive justice. This would ensure that the poor and the most vulnerable are protected from carrying the burden of the transition and disruptions. To achieve such desirable goals, carefully thought-through policy and implementation strategies would be needed, as well as consideration of the best possible social outcomes for South Africa. The next section examines specific sectors and the potential contributions of 4IR technologies in these sectors.

FRESHWATER QUANTITY, QUALITY AND ECOSYSTEM HEALTH

The majority of South African riverine, wetlands and reservoir ecosystems are being impacted by human activities, threatening their functionality and the services they provide to society (Thirion and Jafta, 2019; Bega, 2019). Effluent discharges from wastewater systems (Bwapwa, 2019) are major culprits as are run-offs from agricultural farmlands, stormwater return flows and industrial discharge. Owing to the vast pollution being witnessed in some of South Africa's major rivers, we may have transgressed ecological boundaries of these systems from a water quality perspective (Tempelhoff, 2009; Scherman and Palmer, 2013). 4IR technologies may help in addressing water-related challenges as discussed below.

Wastewater Treatment Works

The deteriorating condition of Wastewater Treatment Works (WWTW) in South Africa is a major contributor to environmental and human health problems (Herbig and Meissner, 2019). Pollution of water resources arising from untreated or poorly treated wastewater effluents has both ecological and social implications. For example, evidence indicates that water resources subject to influences of poorly treated wastewater effluents, such as the lower section of the Upper Vaal, have high incidences of pathogenic microorganisms. This type of pollution raises health and hygiene concerns. Therefore, deploying appropriate 4IR technologies in the wastewater sector to improve effluent quality and overall system efficiency would not only contribute to addressing environmental imperatives but also public health, safety and cultural issues, thus aiding the just transition.

An Artificial Intelligence (AI) algorithm has been developed in the United Kingdom that can be embedded into the operating system of wastewater treatment facilities. This AI algorithm allows for the automatic deployment of appropriate filtration systems for the treatment of wastewater (WEF, 2018). The algorithm considers the overall system performance to deploy the correct protocol, thereby saving energy and removing the need for mechanical interferences that are prone to human errors (WEF, 2018). A combination of smart sensors, AI and robots are deployed to detect and predict leaks in sewer networks and, once proactively detected, such potential leaks are addressed thereby avoiding contamination of water resources from discharges of untreated wastewater. In South Africa, the deployment of a similar system could contribute to reducing the widespread discharge of poorly treated or untreated wastewater effluent into receiving water bodies. As earlier mentioned, this would have both environmental and social benefits as it would reduce ecological degradation, and incidences of pathogenic microorganisms.

Remarkable progress has been made in the field of nanotechnology regarding wastewater treatment in South Africa (Bwapwa, 2019). Nanomaterials that selectively remove organics and novel pollutants such as pharmaceuticals and synthetic dyes (particles size from 0.01 to 0.001) are being developed (Bwapwa, 2019). Organics such as endocrine disrupting compounds (EDCs), which are a public health concern (Manickum and John, 2014), are being removed through nanotechnological innovations. For example, nano-adsorbents such as graphene and carbon nanotubes, nano-photocatalysts such as metals and metal oxide, and nano-filtration are some of the nanotechnology innovations used for the removal of novel pollutants from waste and drinking water (Chauhan et al., 2019). Membrane distillation (MD) is a thermally driven process that uses low-grade energy to operate and has been extensively explored as an alternative, cost-effective, efficient water treatment process compared to conventional membrane processes (Nthunya et al., 2019). Modification of MD membranes with nanoscale materials significantly improves their performance, preventing wetting and fouling (Nthunya et al., 2019).

Wastewater treatment entails utilisation of large amounts of energy, and processes that release carbon emissions (Wang et al., 2016). Utilising on-site renewable energy options would significantly transform this industry and make it more carbon neutral. Following the integrated resource plan, which aims to 'pursue a diversified energy mix that reduces reliance on a single or a few primary energy sources' (DMR, 2019: 11), water management bodies should include the onsite renewable energy options. This is another area where 4IR technologies can contribute to the just transition in the wastewater sector, by contributing to carbon neutrality. Changing how electricity is generated on site can also make a significant contribution to energy consumption and ease the financial burden on municipalities. Furthermore, utilising new renewable energies has new employment possibilities, according to Kiersz and Akhtar (2019).

Agricultural water withdrawal and precision irrigation

Data from the World Bank collection of development indicators shows that agricultural water withdrawal, which represent the volume of water taken from the natural environment for agricultural purposes, represented 63 per cent of water consumption in South Africa in 2014 (Trading Economics, no date). The implication is that addressing water use efficiency in the agricultural sector is likely to have a significant impact on water availability in South Africa. In other countries, precision irrigation that combines the use of smart sensors, AI technologies and drones that selectively irrigate crops in need of water are being deployed in the agricultural sector (WEF, 2018). These technologies can be adopted in South Africa to contribute to both food and water security. However, attention would need to be paid to the costs of their deployment, and measures would need to be in place to minimise their environmental and social risks, such as noise pollution by drones.

Water distribution and usage

In South Africa, leaks in water distribution networks are responsible for enormous water losses, constituting about 25.4 per cent of water that is produced and is lost before it reaches the consumer (McKenzie et al., 2012). These water losses have detrimental social, economic and environmental effects. For example, R5 billion is lost per annum in South Africa as a result of non-revenue water (McKenzie et al., 2012). Smart deployment of 4IR technologies can help address this loss by providing water management systems that can monitor water networks and resources in real time (Alabi et al., 2019). For example, in response to the challenges posed during the protracted Cape Town drought from 2015–2019, 4IR technologies were developed by a local start-up to detect leaks and monitor water usage (see Box 10.1).

Smart water systems

Dong et al. (2015) reviewed the enabling technologies for a smart water quality monitoring system (SWQMS), and explored three major subsystems within this SWQMS: 1) data collection subsystem, 2) data transmission subsystem/network, and 3) data management subsystem. The SWQMS was developed and deployed in Fiji to curb water quality problems (Mamun et al., 2019). The data collection subsystem is comprised of a network of smart wireless sensors and/or remote-sensing technologies that collect water quality information. Typical water quality variables that can be collected using these smart sensors and remote sensing techniques include total organic carbon (TOC), nutrients, temperature, electrical conductivity, water clarity and turbidity, residual chlorine, chlorophyll and metals (Dong et al., 2015).

While much research still needs to be done to improve quality

Box 10.1: 4IR technologies used in monitoring water use in Cape Town

In South Africa, during the peak of the Cape Town drought, a start-up called BRIGIOT developed a smart water meter system called Dropula, using 4IR technologies. The smart meter can monitor the volume and flow of water in real time, reporting water consumption every minute instead of once an hour for conventional meters. The meter also detects leaks and areas of high usage. Through Internet of Things connectivity, the meter sends users an SMS notifying them of their water consumption and possible leakages, and the user can then take immediate action to address water consumption or fix the leaks. Apart from sending SMS notifications to users, the data from the meter is stored in the Cloud and the user can have access to it at any time from any web browser. By installing these meters in some Western Cape schools, together with behavioural change practices, schools were able to save about 550 million litres of water in 17 months, translating to monetary savings of about R41.3 million (Reynolds, 2019).

control and quality assurance, as well as to expand the range of water quality variables that can be monitored, the ability of these sensors and remote sensing techniques to provide real-time continuous data is an important advance from traditional monitoring systems that rely on human effort and prolong analysis time. This SWQMS tool can empower citizens to track, investigate and observe activities and report back. The SWQMS management subsystem allows for the storage, analysis, manipulation and evaluation of water quality data. However, the contribution of these smart sensors to a just transition, in a social-ecological context, will depend on 'where' and 'how' they are deployed. The UN Data Revolution Group (2014) and the UN Global Pulse (2012) have cautioned that data inequality may open up another dimension of social inequality. For example, if these smart sensors for water monitoring are not equitably distributed this could exacerbate inequalities in South Africa as robust data for decision-making would be available for some areas but scarce for other parts of the country. This is where a governance framework and institutional settings that explicitly consider the interconnectedness between social and ecological systems is important.

In South Africa, water quality models that integrate both flow and quality components are now being developed. For example, the Water Quality System Assessment Model (WQSAM) (Slaughter and Hughes, 2013), which has been developed to support decision-making processes regarding water quality management in South Africa. Citizen science tools for the smart monitoring of water have been developed and these are also an important advance in the water sector in South Africa. The miniSASS online platform has been developed, enabling citizen data collection for river health (Box 10.2). Although miniSASS is a low technology tool, it increases citizen awareness, environmental education and democratisation of the science of water quality.

Water 4.0 – building an integrated water business

Closely aligned with smart water systems in South Africa is the emerging Water 4.0 initiative, which forms part of ongoing research on the use of wireless network sensors and IoT technologies to help address challenges to the public water supply (Poljak, 2018). Within this Water 4.0 initiative, Alabi et al. (2019) argue for the development of a business model that is integrated with the environment around it – a so-called 'integrated business model' for water in South Africa in the context of the 4IR. The model would have three components: the technologies aspect, the business stack of the systems and the business strategy aspect. However, there has been very little uptake of Water 4.0's appealing vision of a digitised water management system and integrated business model, as the 4IR technologies themselves have not been adopted into the water sector's operations. More research is required, specifically for the implementation of such an initiative. Fourth Industrial Revolution technologies must be able to improve the control of water reliability, improve monitoring systems and aid in the maintenance of water and wastewater infrastructures (Alabi et al., 2019).

Box 10.2: Internet of Things supported environmental innovation allowing citizens to monitor river health in South Africa

The miniSASS online platform allows South African citizens to use aquatic insects in rivers in their neighbourhood to assess the health of the rivers, and to upload their results onto the platform through a web portal. The platform is freely accessible and open source. A web-based visualisation tool allows anyone to interrogate the health of any river of interest (Graham et al., 2004). The platform is also accessible through an app; therefore, citizens can log their findings instantly. The project, which is enabled by IoT and funded by the Water Research Commission, also serves as a citizen's environmental education tool.

LAND SYSTEM CHANGE

Over the past five decades, South Africa has witnessed significant land system change (Niedertscheider et al., 2012). Grasslands are gradually being turned into cultivated lands, and forest plantations are growing. Land system change is often gradual but can have significant effects on natural and anthropogenic processes such as sediment entrenchment in surface water resources, the hydrological cycle, carbon sequestration, agricultural productivity, and livelihood stability and diversification (Niedertscheider et al., 2012). Without adequate attention to land system change, and its complex relationship to key drivers of change in the Anthropocene, the goal of ecologically sustainable socioeconomic development may elude South Africa (Niedertscheider et al., 2012).

In South Africa, key drivers of land system change include environmental drivers such as climate change and carbon emissions; socioeconomic drivers such as a growing human population, industrialisation and changing agricultural patterns; political and institutional drivers such as redressing the apartheid legacies of inequitable distribution of land; as well as foreign investors seeking arable lands (Gillson et al., 2012). These key drivers mean that addressing land system change in South Africa is complex, as are its effects on the country's ecology and economy. At the global scale, it is estimated that humanity still has some room to manoeuvre (three per cent expansion) regarding land system change (Rockstrom et al., 2009), but under the business-as-usual model, the land system change safe operating space worldwide may be transgressed within the next few decades (Rockstrom et al., 2009).

Accurate, reliable and efficient monitoring of land system changes, such as deforestation, erosion or grazing intensity, is necessary for effective decision-making. Developments in advanced satellite imagery, combined with smart sensors aided by big data analytics and cloud computing, hold enormous potential for monitoring land system change such as deforestation, cropland systems and even grazing patterns (WEF, 2018). Further, these advances in the future will democratise access to land system change data, enabling and strengthening accountability. Already the open-source web application, Global Forest Watch, is able to monitor global forest change in near real time, and access to the data is free, enabling citizens to track where forest change is taking place and the extent of it. Additionally, drones equipped with sensors can monitor activities happening in remote areas that were previously inaccessible to humans, making data available for decision-making purposes.

Practising sustainable intensification in agriculture can help to address pressure on land system change as agriculture accounts for the largest portion of land under use in South Africa (Niedertscheider et al., 2012). Precision agriculture, coupled with blockchain enabled water trading and allocation, as well as the use of drones and AI, can accelerate South Africa's strides towards sustainable agricultural intensification (AfDB, 2019). In addition, owing to the rural–urban migration in South Africa, 4IR technologies can assist in smart urban planning and efficient land utilisation (PwC, 2017).

MINING

Mining is a key sector in South Africa both from a political economy and an environmental perspective. For example, in 2019, the mining sector contributed about R361 billion to the GDP (Minerals Council,

2019: 8). Nevertheless, there is an urgent need for the sector to transform into an eco-sustainable production system. The mining sector is a significant contributor to the greenhouse gas emission levels (Brent et al., 2009; Carels et al., 2013). South Africa, being a signatory to the Paris Agreement (UN, 2015), is committed to reducing greenhouse gases emission, in other words, to a transition into a low-carbon economy. Such a transition would also facilitate the goal of equity as climate change effects are felt most by the most vulnerable in society.

Much mining is required to source the minerals and metals required for creating the renewable energy technologies needed to achieve a low-carbon economy. The way in which this further mining is done would determine whether South Africa can reduce its carbon footprint (Harvey, 2019). Fortunately, South Africa mines many of the minerals or metals used to produce solar panels, wind tunnels, and so on (Harvey, 2019). Thus, if South Africa is to achieve its quota of the climatic global agreement, the mining sector must play a pivotal role. Mining processes and operations can be transformed into cleaner production systems with fewer carbon emissions and a smaller ecological footprint.

Machine learning, robotics and Artificial Intelligence hold the potential to completely transform mining operations by increasing precision and efficiency, and creating a safer working environment. The adoption of 4IR features and advanced technologies like machine learning, robotics and hydrogen fuel cells can facilitate a transition to 'smart mines' and 'zero entry production zones' (Harvey, 2018). Overall, mines in South Africa could operate according to the vision of 'zero entry production areas' (Nikolakopoulos et al, 2015) and become smart mines. This would allow for unstaffed aerial vehicles that can detect where metal resources are located without excavation. Drones are being used to scope geographical areas for ore-bearing land (Harvey, 2018). Drones are also able to contribute to human safety as they have been used to identify dangerous and unsafe spots by scanning a mining site (Freire and Cota, 2017). In South Africa, this is an important social imperative because it has the potential to reduce deaths related to mining operations, which mostly affect the working class. The zero-entry production vision calls for more research aimed at high speed and high precision localisation in 6D, assisting robotic technology, and robotising and automating the manual work in mines (Rylnikova et al., 2017).

In South Africa, Hydrogen South Africa (HySA) has done much work on developing hydrogen fuel-cell-driven mining equipment. The fuel-cell vehicles are advantageous because the mining process becomes economical and environmentally benign as the operation does not emit noxious gases (Miller et al., 2012). The use of fuel-cell technologies will also increase sustainable employment opportunities through the introduction of new occupations that will offer both low- and high-skill jobs; these will be in technical development and research, engineering and administrative support (Bezdek, 2019).

These technologies have been trialled in selected mines and applied in industry, underground mining trucks and loaders; some fuel-cell vehicles have engines with power ratings as high as 750 kW (Hysa, 2020). Another piece of fuel-cell equipment, the Dok-Ing ULP mining dozer, is described in Box 10.3, and is an example of the effective use of advanced technology in South Africa to render mining safer.

Box 10.3: Green underground mining solutions developed by Hydrogen South Africa and Anglo-American

The Dok-Ing ULP mining dozer is a remotely controlled machine, under 50 centimetres in height. Its traction power and energy are provided solely by polymer electrolyte membrane (PEM) fuel cells and hydrogen stored in a reversible metal-hydride bed. Its low height is advantageous because it can access stopes that humans cannot reach. It can be run without any people at the stope. This dozer has been retrofitted with a fuel cell stack. It is a product of South Africa, designed and built in South Africa. The key advantages of this innovation include increased safety at mines and reductions in energy consumption and greenhouse gas emissions (Valicek and Fourie, 2014).

As the use of 4IR technologies begins to grow, the social justice question that needs to be addressed is whether these technologies will replace jobs in the mining sector. While the 4IR is the future of mining, it is also important that workers in the sector are trained and re-skilled in preparation for the new opportunities that the introduction of new technologies will bring. Paying attention to fast-tracking technological innovation, without adequate attention to upskilling the working class, would only exacerbate the social inequalities embedded in South Africa, thereby defeating the purpose of the just transition.

The South African Mining, Extraction, Research, Development and Innovation (SAMERDI) strategy was drafted by the CSIR in 2014. The focus of the strategy is 1) increasing the efficiency of extraction and improving health and safety within operations; 2) developing fully mechanised mining systems for hard rock mine drilling; and 3) developing non-explosive rock breaking technologies (Singh, 2017). Fourth Industrial Revolution technologies have been identified as critical to achieving the key objectives of the strategy, particularly in the areas of ore processing; increasing connectivity in mines; making mining a complete system; and reducing energy consumption. This is a guiding strategy for the future of South African mining.

CARBON EMISSION AND CLIMATE CHANGE

South Africa is the 14th largest global emitter of greenhouse gases (GHG) and the biggest in Africa (UCSUSA, 2019). The heavy reliance on coal, energy intensive industrial processes, land use changes and transportation are the biggest contributors to GHG emission in the country. In 2017, 88 per cent of South Africa's electricity was generated through coal-fired power plants, contributing significantly to carbon emissions (Mcsweeney and Timperley, 2018). The transport sector contributed about 11 per cent of the South African GHG emissions in 2015 (Mcsweeney and Timperley, 2018). At a global scale, humanity has already exceeded some of the markers established by scientists for measuring climatic turning points.[1] This is true locally in South Africa too, with the effects of climate variability becoming more pronounced (Ziervogel et al., 2014). For example, the prolonged drought in parts of South Africa such as Cape Town, the Eastern Cape and KwaZulu-Natal continue to impact on the economy and environment (Jordaan et al., 2019; Vetter et al, 2020).

While South Africa has made commendable strides towards reducing its global share of GHG emission – such as a diversified energy mix (DMR, 2019), international commitments, policy and institutional reforms like the Carbon Tax Act – these efforts can be accelerated through the deployment of 4IR technologies. The costs of solar PV and wind, for instance, have fallen significantly in South Africa due to advances in renewable energy technologies. The social implication of reducing the costs of solar PV is that the technology becomes accessible to many more people in South Africa. Advances in material technology are revolutionising the design of batteries in electric cars, which may outperform the combustion engine (WEF, 2018). Concerted efforts to accelerate widespread use of electric vehicles, supported by the electric vehicle roadmap of South Africa (Dane, no date), coupled with a well-coordinated Bus Rapid Transit System for public transport would reduce the GHG emissions attributable to the transport sector by 11 per cent (Dane, no date).

Future solutions to reducing GHG emissions may include technological advancements in satellite imagery coupled with those in drone development, AI, IoT as well as cloud computing and services.

---

1   In Rockström et al., 2009, seven planetary boundaries are identified that form a part of climatic planetary boundaries. The concentration of CO_2 in the atmosphere (<350 ppm and/or a maximum change of +1 W m-2 in radiative forcing); the ocean acidification (mean surface seawater saturation state with respect to aragonite ³ 80% of pre-industrial levels); stratospheric ozone (<5% reduction in O_3 concentration from pre-industrial level of 290 Dobson Units); biogeochemical nitrogen (N) cycle (limit industrial and agricultural fixation of N_2 to 35 Tg N yr-1) and phosphorus (P) cycle (annual P inflow to oceans not to exceed 10 times the natural background weathering of P); global freshwater use (<4000 km3 yr-1 of consumptive use of runoff resources); land system change (<15% of the ice-free land surface under cropland); and the rate at which biological diversity is lost (annual rate of <10 extinctions per million species).

These would all enable real-time monitoring of GHG transmission and the storage of, and access to, relevant data. These technologies would also allow for the mapping of landscapes vulnerable to climate change which could be overlaid with inequality maps. This would inform decisions on critical spots where investment can be made for greatest environmental sustainability and equity impact.

BIODIVERSITY MANAGEMENT AND ENVIRONMENTAL DECISION SUPPORT SYSTEMS

Nature conservation and biodiversity is a priority in South Africa (DEA, 2020). This is a strategic focus for government in order to fulfil citizens' constitutional rights to 'an environment that is not harmful to their health or wellbeing, and to have the environment protected for the benefit of present and future generations' (RSA Constitution, 1996: 9). The increasing rate of biodiversity loss is a key planetary transgression and concern that needs to be addressed (Rockstrom et al., 2009). Therefore, this section highlights the use of 4IR technologies that can contribute to protecting biodiversity through developing management, tracking and monitoring systems. For many years already, 4IR technologies, such as satellite imagery, have been deployed to map critical ecosystems and vulnerable and endangered species (Cortés et al., 2000). Remote sensing technologies are also being used to assess risk and identify hazards in order to protect biodiversity (Duan et al., 2020; Pham et al., 2020).

The focus area for 4IR technologies is in the transformation of environmental decision support systems (EDSS), which combine considerations of the complex natural environment with community engagement and concerns (Cortés et al., 2000; Reiter et al., 2018). Natural resource managers are challenged to make decisions in an increasingly complex and uncertain environment (Reiter et al., 2018). Fourth Industrial Revolution technologies have already been utilised for hazard identification, risk assessment, and evaluation, intervention and decision-making in nature (Thessen, 2016). These environmental informatic tools are important to combine Artificial Intelligence, geographical information systems (GIS), environmental ontology and traditional numerical methods in order to support environmental systems. Information technologies are playing an increasing role in the planning, prediction, supervision and control of environmental processes at many different scales and within various time spans (Cortés et al., 2000).

Innovation is allowing researchers to map, sequence and replicate biological systems on Earth in new ways that have not been imagined before. The Earth BioGenome Project (EBP) is one such project; it is sequencing and classifying all of Earth's eukaryotic diversity (±1.5 million species) (Exposito-Alonso et al., 2020). The project is aimed at providing data to inform a broad range of humanities issues, like the impact of climate change on biodiversity; the conservation of endangered species and ecosystems; and the preservation and enhancement of ecosystem services (Lewin et al., 2018). The EBP is one key decision-making tool that is continuously being developed and will have utility in biotechnology, medicine, conservation and sustaining human societies (Lewin et al., 2018; Pennis, 2007).

In the context of South Africa, through existing community-based conservation initiatives such as those in the Mzimvubu catchment (Madikizela and Dye, 2013; Bester et al., 2019) and others led by Conservation South Africa, citizens can be mobilised to support conservation efforts by supplying eukaryote species in their domains. The mobilisation of citizens through existing conservation programmes and initiatives can accelerate the protection of critical and endangered species.

HAVE WE LEARNT ENVIRONMENTAL LESSONS FROM PREVIOUS INDUSTRIAL REVOLUTIONS?

Much of the environmental degradation humanity has witnessed in the past has been due to the unintended consequences of the Second and Third Industrial Revolutions. While 4IR technologies hold enormous potential for reversing some of the environmental transgressions we have made in South Africa, there is a need to think carefully about potential unintended consequences of 4IR technologies (Herweijer et al., 2018). This section reflects briefly on the potential environmental risks posed by the 4IR.

Lag between pace of technological advancement and shift in South Africa skills base

The World Economic Forum (WEF, 2017b) predicts that 44 per cent of all work activities in South Africa are susceptible to automation, and the average information and communications technology (ICT) intensity of jobs in the country has grown by 26 per cent in the past decade. Approximately nine per cent of South African employers are already constrained by a scarcity of the required skills and this situation may get worse unless the education system is transformed (WEF, 2017b). This suggests that the transformation of the South African skills base lags behind 4IR technological advancement, and as jobs become automated, the lag may lead to joblessness for previously employed individuals. Individuals who may have been employed in the services sector may turn to land-dependent occupations such as small-scale farming. Increased dependence on land and resource-intensive occupations may further increase pressure on the environment (see Figure 10.1 above), leading to degradation while the workforce re-skill. For South Africa to address these challenges, its entire education system would need to respond adequately to the demand of the 4IR era, supported by a coordinated policy of re-skilling the workforce. If this is not done, there is a risk of escalating the already alarming inequalities in the country.

Environmental inequalities may be exacerbated

The UN Data Revolution Group (2014) and the UN Global Pulse (2012) caution that if action is not taken, data inequality is opening up a new world between data haves and have-nots, splitting the world between those who stand to benefit from data coverage, access and usage and those who stand to lose. The implication of this assertion for South Africa is that if 4IR technologies for environmental monitoring are not evenly deployed across the country, there is a possibility of favouring certain parts of the country over others. Data on exposure to environmental threats such as subtle water quality impairment, leakages from wastewater sewer systems and air pollution may not be readily available for areas lacking enough 4IR deployment, such as rural parts of the country. Since data guides effective decision-making processes (Cinnamon, 2020), such unequal deployment of 4IR may aggravate social and environmental inequalities, particularly in a country where apartheid legacies still impact negatively on the majority of the population. To address these challenges, the imperative of addressing social and ecological justice should guide the deployment of environmental monitoring technologies supported by a well thought-through policy and strategy.

High energy consumption and water scarcity

Fourth Industrial Revolution technologies such as blockchain, IoT and Cloud services are energy intensive (Herweijer et al., 2018). In a country like South Africa where coal constitutes the major energy source (Mcsweeney and Timperley, 2018), this means that heavy deployments of these technologies in South Africa are likely to increase energy consumption. This, in turn, is likely to increase South Africa's contribution to CO_2 emission, unless green energy sources are urgently mainstreamed. Further, without increasing energy production in South Africa, heavy deployment of energy-intensive 4IR technologies is likely to escalate the problems associated with power rationing and may have unintended social and economic consequences (Atebe et al., 2019). Under a business-as-usual model, pressure on energy production in South Africa may also negatively impact on water and other environmental resources such as land and forestry (Mcsweeney and Timperley, 2018).

Noise pollution

The proliferation of drones may lead to noise pollution (Herweijer et al., 2018). Wildlife sensitive to noise pollution may be impacted, and the distribution and diversity of these animals may be affected. To reduce this risk and minimise the noise from drones, consideration would need to be given to their design, and to their deployment in sensitive ecosystems.

Data security and the poaching of protected species

Data security will increasingly become an issue of concern as more and more devices are connected through IoT (UN Global Pulse, 2012;

UN Data Revolution Group, 2014). Security breaches of data about important habitats and protected species may give poachers clues about the location and numbers of protected species (WEF, 2018). Poachers may then use such data to trace the habitats of protected species. This is a risk in South Africa, where it is envisaged that digital data about protected species such as elephants and rhino, may be collected and stored through IoT. A well thought-through cyber security environmental governance system would need to be in place to guard against and mitigate the effects of an environmental security breach.

Environmental data governance
Without an appropriate governance framework for data in the 4IR, issues of data ownership, collection, transmission, storage and access may all compromise the competitive advantages of the 4IR technologies (WEF 2017a; Cinnamon, 2020). This poses perhaps the greatest risk in the area of environmental governance. It is crucial that South Africa sets up an environmental governance framework for 4IR-generated data. This chapter argues that such a framework needs to be holistic, seeing environmental governance in the context of social-ecological systems.

Environmental risks and toxicity to non-target species
Progress in nanotechnology, particularly in engineered nanomaterials, promises to address a range of environmental challenges. However, the toxicity of these materials, and the environmental risk they pose, remain a concern (Sharifi et al., 2012). The nano-scale size of these materials mean that they would be liable to absorption by non-target species, and that toxicity could therefore have dire consequences, with significant implications for the ecosystems (Blasco and Corsi, 2019). Research needs to be directed towards gaining a better understanding of the ecological and human health risks posed by these materials.

POLICY RECOMMENDATIONS AND ENVIRONMENTAL GOVERNANCE IMPLICATIONS

Appropriate environmental governance frameworks, coupled with agile institutions, are necessary to regulate the proliferation of 4IR technologies to ensure the equitable distribution of benefits. A just low-carbon economy, achieved through the use of 4IR technologies, will require a comprehensive policy on environmental governance that would address issues of access, control, costs and benefits of 4IR technologies and data. These new policies should speak to the imperative for responsive institutions that pay attention to South Africa's historical context, and to the implications of the 4IR for distributive justice. Policies would also be required to address aspects of 4IR technologies such as the availability of data resources, and how these could influence decision-making. The imperative for public–private partnerships to propel South Africa into the 4IR also needs to be addressed.

In adopting 4IR technologies, therefore, policy measures should not only consider their scientific, technical and economic merits, but also their context suitability, particularly the historical context of South Africa. The key question that should guide policy formulation is whether the adoption of 4IR technology aids in addressing historical inequalities and ecological degradation. This chapter has argued for the prioritisation of addressing inequity, because care must be taken not to increase the already escalating social divides in the country through, for example, data and technology inequalities (UN Global Pulse, 2012; UN Data Revolution Group, 2014).

The following policy recommendations aim to be in line with this strong focus on ensuring that policymakers propel South Africa into a just low-carbon economy.

Policy recommendations regarding water resource management, water quality and WWTWs:

- There is an urgent need for a comprehensive green wastewater treatment strategy that draws on AI, nanotechnology and on-site renewable energy solutions
- A smart water quality system, and an integrated water management system that relies on 4IR technologies should be developed. These will assist in the distribution of water, guide water monitoring tools and assist in water governance.
- It is crucial to develop an organic pollution abatement strategy

as organic pollution is the leading cause of water resource deterioration.

- Water quality guidelines, which depict the water withdrawal quantities and water discharge quantities, should be updated. This would facilitate the better use of 4IR technologies, which could be used to manage water resources to avoid further environmental transgression.

Policy recommendations regarding the mining sector:
- Incentives should be offered for investment in minerals and metals that will assist the transition to a low-carbon future.
- A comprehensive green industrialisation strategy for structural transformation that does away with coal dependence and instead fosters renewable energy should be built; this should connect mining to upstream, side stream and downstream opportunities associated with products of the 4IR (such as the electric vehicle).
- It is important to decouple economic growth from environmental degradation by offering incentives for minimally invasive automated mining, and for reducing and then abandoning fossil fuel extraction.

As South Africa is well versed in policy aimed at reducing the GHG emissions and reducing climate change impacts, the recommendations above are aligned with those policies, specifically the Carbon Tax Act, National Development Plan and Green Transportation Strategies. The recommendations are aimed at supporting efforts to achieve a just transition to a low-carbon economy.

CONCLUSION

This chapter demonstrates that for South Africa to be on an environmentally sustainable path, investment needs to be made in developing 4IR technologies, together with the development of agile, polycentric, responsive governance and institutional framework within the context of a social-ecological system. It is argued that 4IR technologies hold enormous potential to address ecological, social and economic challenges in South Africa. A transition towards a low-carbon economy can be facilitated by 4IR technologies if these are partnered with the correct policy and governance conditions. The rolling out of 4IR technologies in South Africa needs to be coupled with insightful consideration of social, historical and economic variables. The transition should be fair to all; therefore employment, health, safety, knowledge sharing and inclusion should be at the centre.

There is significant potential for 4IR technologies to slow down the transgressions of some of the key sectors identified within South Africa's ecologies. In the water quality and management sector, 4IR technologies such as machine learning, nanotechnology and IoT can help to restore ecosystem integrity. The development of smart water systems can aid in water supply and distribution, and the reduction of WWTW dysfunctionality. The adoption of 4IR technologies in the past (for example, Cape Town and miniSASS) have proven to be successful and can be made applicable more broadly in the water sector. These same technologies can be applied to land system changes to boost the effectiveness of decision-making.

It was noted that the mining sector is a crucial one in South Africa, and that the future of renewable energies depends on the mining industry. In order to transform, mines should adopt a more sustainable *modus operandi*. The technologies for this include drones, AI, IoT, and hydrogen and electric vehicles. These will facilitate greater employment, safety and economic gains. When mining is operating sustainably, our GHG emissions will be significantly reduced, helping to address energy and climate change concerns. 4IR technologies can be used to support all activities – by government and other stakeholders – aimed at meeting South Africa's climate obligations. Technologies like CPS, blockchain, IoT, AI and others can assist in the maintenance of data and the fair distribution of ecosystem services and so help conserve and restore natural biodiversity, thus offering benefits for humanity now and in future generations.

Fourth Industrial Revolution technologies hold great potential to be a catalyst for justice, if introduced in an appropriate manner. However, uncritical adoption of these technologies may deepen and widen the already escalating inequalities in the country. This chapter therefore highlights some of the environmental and social risks associated with 4IR technologies as well their potential benefits for addressing environmental concerns. This means that policies are required to create strategies and interventions that ensure both an equitable distribution of ecosystem services and that further ecological transgressions are limited.

REFERENCES

African Development Bank (AfDB). 2019. 'Study report unlocking the potential of the fourth industrial revolution in Africa'. https://4irpotential. africa/wp-content/uploads/2019/10/AFDB_4IRreport_Main.pdf, accessed 5 May 2020

Amhez, N.E.H. 2019. 'Agrarian transition and food security in the village of Nabha, Central Bekaa'. American University of Beirut. https:// scholarworks.aub.edu.lb/bitstream/handle/10938/21612/st-6925. pdf?sequence=1, accessed 30 April 2020

Akcil, A. and Koldas, S. 2006. 'Acid mine drainage (AMD): Causes, treatment and case studies'. *Journal of Cleaner Production*, 14(12/13), 1139–1145.

Alabi, M., Telukdarie, A. and Jansen, N.V.R. 2019. 'Industry 4.0 and water industry: A South African perspective and readiness'. *American Society for Engineering Management (ASEM)*, 1–11.

Allen, R. 2012. 'Backward into the future: The shift to coal and implications for the next energy transition'. *Energy Policy*, 50, 17–23.

Archer, E.R. 2019. 'Learning from South Africa's recent summer rainfall droughts: How might we think differently about response?'. *Area*, 51(3), 603–608.

Atebe, B., Prinsloo, J. and Gawlik, R. 2019. 'The significance of electricity supply sustainability to industrial growth in South Africa'. *Energy Reports,* 5, 1324–1338.

Atkeson, A. and Kehoe, P. 2001. 'The transition to a new economy after the second industrial revolution (No. w8676)'. National Bureau of Economic Research. http://www.nber.org/papers/w8676, accessed 16 April 2020.

Bega S. 5 October 2019. 'Govt report paints a bleak picture of South Africa's rivers'. *Saturday Star.* https://www.iol.co.za/saturday-star/news/govt-report-paints-a-bleak-picture-of-south-africas-rivers-34111285, accessed 17 April 2020.

Bester, R., Blignaut, J.N. and Crookes, D.J. 2019. 'The impact of human behaviour and restoration on the economic lifespan of the proposed Ntabelanga and Laleni dams, South Africa: A system dynamics approach'. *Water Resources and Economics*, 26, 100126.

Bezdek, R.H. 2019. 'The hydrogen economy and jobs of the future'. *Renewable Energy and Environmental Sustainability*, 4, 1.

Blasco, J and Corsi, I. (eds). 2019. *Ecotoxicology of Nanoparticles in Aquatic*

Systems. Boca Raton: CRC Press. https://doi.org/10.1201/9781315158761.

Brent, A.C., Hietkamp, S., Wise, R.M. and O'Kennedy, K. 2009. 'Estimating the carbon emissions balance for South Africa'. *South African Journal of Economic and Management Sciences*, 12(3), 263–279.

Bwapwa, J.K. 2019. 'Analysis on industrial and domestic wastewater in South Africa as a water-scarce country'. *International Journal of Applied Engineering Research*, 14(7), 1474–1483.

Carels, C., Maroun, W. and Padia, N. 2013. 'Integrated reporting in the South African mining sector'. *Corporate Ownership and Control*, 11(1), 991–1005.

Casey, J.P. 20 May 2019. 'History of mining in South Africa'. *Mining Technology*. https://www.mining-technology.com/features/history-of-mining-in-south-africa/, accessed 9 July 2020.

Chauhan, A., Sillu, D. and Agnihotri, S. 2019. 'Removal of pharmaceutical contaminants in wastewater using nanomaterials: A comprehensive review'. *Current Drug Metabolism*, 20(6), 483–505.

Cinnamon, J. 2020 'Data inequalities and why they matter for development'. *Information Technology for Development,* 26(2), 214–233.

Constitution of the Republic of South Africa. 1996. https://www.justice.gov.za/legislation/constitution/SAConstitution-web-eng.pdf.

Cortès, U., Sànchez-Marrè, M., Ceccaroni, L., R-Roda, I. and Poch, M. 2000. 'Artificial intelligence and environmental decision support systems'. *Applied Intelligence,* 13(1), 77–91.

Dane, A. (no date). 'The potential of electric vehicles to contribute to South Africa's greenhouse gas emissions targets and other developmental objectives – how appropriate is the investment in electric vehicles as a NAMA?'. University of Cape Town. http://www.sll.uct.ac.za/sites/default/files/image_tool/images/119/Papers-2013/13-Dane-Electric_vehicles.pdf, accessed 10 May 2020.

Department of Environmental Affairs (DEA). https://www.environment.gov.za/, accessed 2 March 2020.

Department of Energy and Mineral Resources (DEMR). 2019. 'Integrated resource plan (IRP) 2019'. http://www.energy.gov.za/files/media/Pub/IRP-2019.pdf, accessed 10 May 2020.

Dong, J., Wang, G., Yan, H., Xu, J. and Zhang, X. 2015. 'A survey of smart water quality monitoring system'. *Environmental Science and Pollution Research,* 22(7), 4893–4906. DOI 10.1007/s11356-014-4026-x.

Duan, P., Wang, Y. and Yin, P. 2020. 'Remote sensing applications in monitoring of protected areas: A bibliometric analysis'. *Remote Sensing*, 12(5), 772.

Exposito-Alonso, M., Drost, H.G., Burbano, H.A. and Weigel, D. 2020. 'The Earth BioGenome project: Opportunities and challenges for plant genomics and conservation'. *The Plant Journal*, 102(2), 222–229.

Folke C. 2006. 'Resilience: The emergence of a perspective for social-ecological systems analyses'. *Global Environmental Change,* 16, 253–267.

Freire, G.R. and Cota, R.F. 2017. 'Capture of images in inaccessible areas in an underground mine using an unmanned aerial vehicle'. Proceedings of the First International Conference on Underground Mining Technology.

Gillson, L. Midgley, G. and Wakeling, J. 2012. 'Exploring the significance of land-cover change in South Africa'. *South African Journal of Science*, 108 (5/6), 3–5.

Graham, P.M., Dickens, C.W. and Taylor, R.J. 2004. 'miniSASS: A novel technique for community participation in river health monitoring and management'. *African Journal of Aquatic Science*, 29(1), 25–35.

Harvey, R. 2018. 'Greening South African mining through the Fourth Industrial Revolution'. In: Valiani, S. (ed.), *The Future of Mining in South Africa: Sunset or Sunrise?* Johannesburg: MISTRA, 144–178.

Harvey, R. 2019. 'Mining for a circular economy in the age of the Fourth Industrial Revolution: The case of South Africa'. South African Institute of International Affairs. https://media.africaportal.org/documents/Policy-Briefing181harvey.pdf, accessed 9 February 2020

Heffron, R. and McCauley, D. 2018. 'What is the "just transition"?'. *Geoforum*, 88, 74–77.

Herbig, F.J. and Meissner, R. 2019. 'Talking dirty-effluent and sewage irreverence in South Africa: A conservation crime perspective'. *Cogent Social Sciences*, 5(1), 1701359.

Herweijer, C., Combes, B., Johnson, L., McCargow, R., Bhardwaj, S., Jackson, B. and Ramchandani, P. 2018. 'Enabling a sustainable fourth industrial revolution: How G20 countries can create the conditions for emerging technologies to benefit people and the planet'. Economic Discussion Paper, No. 2018-32. Kiel Institute for the World Economy. http://www.economics-ejournal.org/economics/discussionpapers/2018-32, accessed 6 March 2020.

Hydrogen South Africa (HySA). 2020. 'Clean mining platform. Hydrogen in mining and tunneling'. HySA Infrastructure. https://hysainfrastructure.com/clean-mining-platform/, accessed 18 February 2020.

Jordaan, A., Bahta, Y.T. and Phatudi-Mphahlele, B. 2019. 'Ecological vulnerability indicators to drought: Case of communal farmers in Eastern Cape, South Africa'. *Jàmbá: Journal of Disaster Risk Studies*, 11(1), 1–11.

Kiersz, A. and Akhtar, A. 14 August 2019. '21 high-paying careers for people who want to save the planet – and also have job security'. *Business Insider*. https://www.businessinsider.com/high-demand-renewable-energy-jobs-of-the-future-2019-8?IR=T, accessed 10 July 2020.

Laurance, W. 2019. 'The Anthropocene'. *Current Biology*, 29(19), R953–R954.

Lewin, H., Robinson, G., Kress, W., Baker, W.J., Coddington, J. et al. 2018. 'Earth BioGenome Project: Sequencing life for the future of life'. *Proceedings of the National Academy of Sciences*, 115(17), 4325–4333.

Lewis, S. and Maslin, M. 2015. 'Defining the Anthropocene'. *Nature*,

519(7542), 171–180.

Madikizela, B.R. and Dye, A.H. 2003. 'Community composition and distribution of macroinvertebrates in the Umzimvubu River, South Africa: a pre-impoundment study'. *African Journal of Aquatic Science*, 28(2), 137–149.

Mamun, K., Islam, F., Haque, R., Khan, M., Prasad, A. et al. 2019. 'Smart Water Quality Monitoring System Design and KPIs Analysis: Case Sites of Fiji Surface Water'. *Sustainability*, 11(24), 7110.

Manickum, T. and John, W. 2014. 'Occurrence, fate and environmental risk assessment of endocrine disrupting compounds at the wastewater treatment works in Pietermaritzburg (South Africa)'. *Science of the Total Environment*, 468, 584–597.

Mckenzie, R., Siqalaba, Z. and Wegelin, W. 2012. 'The state of non-revenue water in South Africa (2012)'. Water Research Commission, http://www.wrc.org.za/wp-content/uploads/mdocs/TT%20522-12.pdf.

Mcsweeney, R. and Timperley, J. 2018. 'The carbon brief profile: South Africa'. https://www.carbonbrief.org/the-carbon-brief-profile-south-africa, accessed 7 March 2020.

Miller, A.R., Berg, G., Barnes, D.L., Eisele, R.I., Tanner, D.M. et al. 2012. 'Fuel cell technology in underground mining'. 5th Southern African Institute of Mining and Metallurgy (SAIMM) International Platinum Conference. Sun City, South Africa, 17–21 September, 533–545.

Minerals Council South Africa. 2019. 'Facts and Figures pocketbook. 2019'. https://www.mineralscouncil.org.za/downloads/send/18-current/871-facts-and-figures-2019-pocketbook.

Mokyr, J. 1998. 'The second industrial revolution, 1870–1914'. In: Castronovo, V. (ed.). *Storia dell'economia Mondiale*. Rome: Laterza Publishing, 219–245

National Planning Commission (NPC). 2013. 'National development plan vision 2030'. https://www.gov.za/issues/national-development-plan-2030, accessed 9 March 2020.

Niedertscheider, M., Gingrich, S. and Erb, K. 2012. 'Changes in land use in South Africa between 1961 and 2006: An integrated socio-ecological analysis based on the human appropriation of net primary production framework'. *Regional Environmental Change,* 12, 715–727.

Nikolakopoulos, G., Gustafsson, T., Martinsson, P. and Andersson, U. 2015. 'A vision of zero entry production areas in mines'. IFAC-Papers OnLine, 48(17), 66–68.

Nthunya, L. Gutierrez, L., Derese, S., Nxumalo, E.N., Verliefde, A. et al. 2019. 'A review of nanoparticle-enhanced membrane distillation membranes: Membrane synthesis and applications in water treatment'. *Journal of Chemical Technology and Biotechnology*, 94, 2757–2771. doi:10.1002/jctb.5977

Pearson, P. and Foxon, T. 2012. 'A low carbon industrial revolution? Insights and challenges from past technological and economic transformations'. *Energy Policy*, 50, 117–127.

Pennis, E. 24 February 2017. 'Biologists propose to sequence the DNA of all life on Earth'. *Science*. https://www.sciencemag.org/news/2017/02/biologists-propose-sequence-dna-all-life-earth, accessed 6 August 2020.

Petrillo, A., Felice, F.D., Cioffi, R. and Zomparelli, F. 2018. 'Fourth Industrial Revolution: Current practices, challenges, and opportunities'. In: Petrillo, A., Cioffi, R. and De Felice, F. (eds.). *Digital Transformation in Smart Manufacturing*. London: IntechOpen, 1–20.

Pham, T.H.Y., Shahrour, I., Aljer, A., Lepretre, A., Pernin, C. and Ounaies, S. 2020. 'Smart monitoring for urban biodiversity preservation'. *CIGOS 2019, Innovation for Sustainable Infrastructure: Proceedings of the 5th International Conference on Geotechnics, Civil Engineering Works and Structures*, 1123–1128. Singapore: Springer.

Poljak, D. 2018. 'Industry 4.0-New challenges for public water supply organizations'. Lean Spring Summit 2018.

PwC. 2017. 'Fourth industrial revolution for the earth – harnessing the 4th industrial revolution sustainable emerging cities'. https://www.pwc.com/gx/en/sustainability/assets/4ir-for-the-earth.pdf, accessed 10 May 2020.

Reiter, D., Meyer, W., Parrott, L., Baker, D. and Grace, P. 2018. 'Increasing the effectiveness of environmental decision support systems: Lessons from climate change adaptation projects in Canada and Australia'. *Regional Environmental Change*, 18(4), 1173–1184.

Reynolds, E. 23 September 2019. 'How smart meters saved water and money in drought-ravaged Cape Town'. *Innovate Africa*. https://edition.cnn.com/2019/09/23/business/cape-town-drought-water-meter-intl/index.html, accessed 5 March 2020.

Rockstrom, J., Steffen, W., Noone, K., Persson, A., Chapin F.S. et al. 2009. 'Planetary boundaries: Exploring the safe operating space for humanity'. *Ecology and Society*, 14(2), 32.

Rylnikova, M., Radchenko, D. and Klebanov, D. 2017. 'Intelligent mining engineering systems in the structure of Industry 4.0'. *E3S Web of Conferences*, 21, 1032.

Scherman, P.A. and Palmer, C.G. 2013. 'Critical analysis of environmental water quality in South Africa: Historical and current trends'. Water Research Commission deliverable, project No. K5/2184.

Schwab, K. 2017. *The fourth Industrial Revolution*. London, United Kingdom: Penguin Books Ltd.

Sengupta, P., Choudhury, B., Mitra, S. and Agrawal, K. 2019. 'Low carbon economy for sustainable development'. *Encyclopedia of Renewable and Sustainable Materials*, (3), 551–560.

Sharifi S, Behzadi S, Laurent S, Forest M, Stroeve P. and Mahmoudi M. 2012. 'Toxicity of nanomaterials'. *Chemical Society Review*, 41(6), 2323–2343.

Singh, N. 2017. 'Weathering the "perfect storm" facing the mining sector'. *Journal of the Southern African Institute of Mining and Metallurgy*, 117(3), 223–229.

Slaughter, A. and Hughes, D. 2013. 'A simple model to separately simulate point and diffuse nutrient signatures in stream flows'. *Hydrology Research*, 44(3), 538–553.

Snudden, J. 2019. 'Progression to the next industrial revolution: Industry 4.0 for composites'. *Reinforced Plastics*, 63(3), 136–142.

Statistics South Africa (Stats SA). 2017. 'General household survey, selected development indicators, 2016'. Statistical release P0318.2.

Statistics South Africa (Stats SA). 2019. 'Inequality trends in South Africa: A multidimensional diagnostic of inequality'. Stats SA Library Cataloguing-in-Publication (CIP) Data Report No. 03-10-19.

Tanger, S.M., Zeng, P., Morse, W. and Laband, D.N. 2011. 'Macroeconomic conditions in the US and congressional voting on environmental policy: 1970–2008'. *Ecological Economics*, 70(6), 1109–1120.

Tempelhoff, J. 2009. 'Civil society and sanitation hydropolitics: A case study of South Africa's Vaal River Barrage'. *Physics and Chemistry of the Earth*, 34, 164–175.

Thessen, A. 2016. 'Adoption of machine learning techniques in ecology and earth science'. *One Ecosystem*, 1, 8621.

Thirion C. and Jafta (eds). 2019. 'River Ecostatus Monitoring Programme State of Rivers Report 2017-2018. N/0000/00/REMP/2019'. Department of Water and Sanitation.

Trading Economics. No date. 'South Africa – Annual Freshwater Withdrawals, Agriculture (% Of Total Freshwater Withdrawal)'. https://tradingeconomics.com/south-africa/annual-freshwater-withdrawals-agriculture-percent-of-total-freshwater-withdrawal-wb-data.html, accessed 18 July 2020.

Union of Concerned Scientists (UCSUSA). 10 October 2019. 'Each country's share of CO2 emissions.' https://www.ucsusa.org/resources/each-countrys-share-co2-emissions, accessed 12 March 2020.

United Nations (UN). December 2015. 'Paris agreement'. http://unfccc. int/files/essential_background/convention/application/pdf/english_paris_agreement. pdf.

UN Data Revolution Group. 2014. 'A world that counts: Mobilising the data revolution for sustainable development'. Geneva: United Nations Secretary General's Independent Expert Advisory Group.

UN Global Pulse. 2012. 'Big data for development: Opportunities and challenges'. United Nations.

Valicek, P. and Fourie, F. 2014. 'Fuel cell technology in underground mining'. *The 6th International Platinum Conference, 'Platinum–Metal for the Future'*, The Southern African Institute of Mining and Metallurgy, 325–332.

Vetter, S., Goodall, V.L. and Alcock, R. 2020. 'Effect of drought on communal livestock farmers in KwaZulu-Natal, South Africa. *African Journal of Range & Forage Science*, 37(1), 93–106.

Vogel, C. and Olivier, D., 2019. 'Re-imagining the potential of effective drought responses in South Africa'. *Regional Environmental Change*, 19(6), 1561–1570.

Wang, H., Yang, Y., Keller, A.A., Li, X., Feng, S. et al. 2016. 'Comparative analysis of energy intensity and carbon emissions in wastewater treatment in USA, Germany, China and South Africa'. *Applied Energy*, 184, 873–881.

World Economic Forum (WEF). 2017a. 'PwC and Stanford Woods Institute for the Environment. Harnessing the Fourth industrial revolution for the earth'. *Fourth industrial revolution for the earth series*. World Economic Forum, Switzerland.

World Economic Forum (WEF). 2017b. 'The future of jobs and skills in Africa – preparing the region for the fourth industrial revolution'. http://www3.weforum.org/docs/WEF_EGW_FOJ_Africa.pdf, accessed 10 May 2020.

World Economic Forum (WEF). 2018. 'Harnessing the fourth industrial revolution for water'. World Economic Forum, Switzerland. https://www.weforum.org/reports/harnessing-the-fourth-industrial-revolution-for-water.

Ziervogel, G., New, M., Can Garderen, E., Midgley, G., Taylor A. et al. 2014. 'Climate change impacts and adaptation in South Africa'. *WIREs Climate Change*, 5, 605–620.

ELEVEN

Inclusive energy transitions: An analysis of the potential for a digital revolution in South Africa's electricity system

BLANCHE TING

INTRODUCTION

South Africa's electricity system has a state-owned, vertically integrated[1] incumbent, namely Eskom, which is responsible for the generation, transmission and distribution of electricity. Although South Africa has known since its 1998 White Paper on Energy that an electricity deficit would be reached by 2007, no new plants were commissioned in time to meet this deadline (DME, 1998). It has been said that the average age of Eskom's generation fleet is around 37 years, and hence reasonably mature (DOE, 2019). The integrated resource plan (IRP) has indicated that 75 per cent of Eskom's current coal fleet is scheduled for decommissioning by 2040, and some older plants are due to be closed

---

1 Vertical integration is an arrangement in which the supply chain of a company is managed by that company.

as early as 2021 (DOE, 2019). With an ageing electricity infrastructure and lack of maintenance due to low generation reserves, the country has undergone a series of 'load shedding' episodes since 2008. Under increasing pressure to diversify away from coal, decarbonise its green house gas (GHG) emissions and improve the reliability of electricity generation, the country is experiencing a period of rapid change involving the introduction of new sources of energy technologies and new actors. In response, to diversify its energy security, an integrated resource plan (IRP) was initiated and, in 2012, the plan was implemented in an ambitious Renewable Energy Independent Power Producer Procurement Programme (REIPPP). It has met with some indicators of success, albeit on a limited scale compared to the overall supply (DOE, 2015; DOE, 2019). By 2018, the country's renewable energy programme had seen swift procurement of approximately 6.4 gigawatts (GW), of which half were operationalised and synchronised to the grid (Ting and Byrne, 2020). Additionally, from 2015, a Gas Independent Power Producers Procurement Programme (Gas I4P) was also initiated, totalling 3.6 GWs for power generation (DOE, 2016; Ting, 2019). Despite this rapid introduction of renewables into the grid, South Africa continues to experience extensive electricity challenges, often bringing economic activities to a halt due to incessant load shedding. Nevertheless, the current electricity crisis may offer the potential for a digital revolution in the country's energy system.

A digitally enabled energy transition offers a unique opportunity for the current system to improve, or at least to minimise the damaging effects of what currently exists: a fossil fuel-based, centralised large-scale and unidirectional power system with relatively few actors and a load profile that is passive and predictable, as well as a monetary flow that favours upstream players (such as generators) (Figure 11.1a). An improvement would lead towards an emerging paradigm that is lower carbon-based, decentralised and utilises a bidirectional power system with a multitude of actors on the supply and demand sides (IRENA, 2019a; Wolsink, 2012). With this system, the load profile would be active and stochastic due to increased use in distributed energy resources (DERs) and there would be potential for revenues for both generators and consumers (US-DOE, 2015). (Figure 11.1b).

Figure 11.1: A comparison between South Africa's existing electricity system and a potential, emerging system engaging 4IR technologies

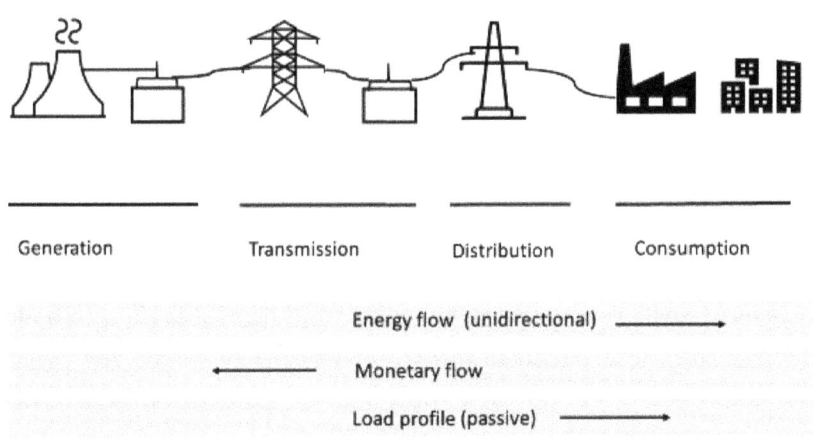

(a): Schematic illustration of South Africa's current electricity system: the energy flow is unidirectional; the load profile is passive, and monetary flow favours generators and distributors.

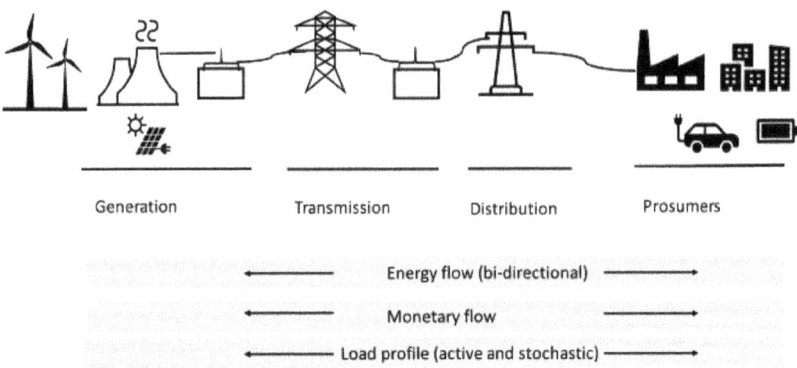

(b): Illustration of an emerging paradigm: the energy flow is bi-directional, the load profile is active and stochastic, and the monetary flow is across the whole value chain, including proactive consumers or 'prosumers'.

Source: Author's own

Digitalisation is a significant enabler in power sector transformation because it allows for the organisation of a considerable volume of data and for better management of increasingly complex systems. For the power sector, digitalisation is essentially converting data into value (IRENA, 2019b). The rising significance of digitalisation in the electricity sector is associated with two other innovation trends: decentralisation and electrification. Decentralisation is led by the increased implementation of distributed energy resources (DERs) (such as rooftop solar photovoltaic (PV) and wind), connected to the distribution grid. Electrification has enabled the integration of these resources into both vehicles and buildings, such as electric vehicles (EVs), heat pumps and electric boilers, which involve large and irregular quantities of energy loads. All these contemporary assets on the supply and demand sides are adding complexity to the power sector, making management and control of it critical in future. Digital technologies are understood to support such transformation, including better control, operation and maintenance of DERs assets. Digitalisation encompasses a range of digital-enabled technologies, including: 1) the Internet of Things (IoT); 2) Artificial Intelligence (AI) and big data; and 3) blockchain (IEA, 2017a; IRENA, 2019c).

Essentially, transformation in the electricity system is driven by the following (US-DOE, 2015; IEA, 2017a; IRENA, 2019b):

- shifting electricity generation from the traditional, centralised model with limited large, centralised station plants towards a model using integrated hybrid (centralised/decentralised) distributed energy resources (DERs);
- changing demand loads due to growing consumer participation and broader electrification;
- integrating smart grid technologies for dealing with complex power systems, due to the availability of innovative technologies;
- increasing the potential for a robust and reactive electricity grid (which will be of increasing importance, given the projections for more recurrent and powerful weather events) and linkages with critical infrastructure such as ICTs, water and transport; and
- developing a mature electricity infrastructure that requires new

technologies to provide dynamic real-time monitoring, control and detection of system conditions.

This chapter seeks to investigate the applicability of a digital transformation in South Africa's electricity system. To do so, the discussion uses an interdisciplinary lens, drawing on a combination of sustainability transitions and the energy justice framework in order to examine the potential for a digital revolution.

The sustainability transitions community has produced significant theories, concepts and debates concerning fundamental shifts away from fossil fuel energy systems towards low-carbon development (Markard et al., 2012; Schot and Kanger, 2018). The urgency for addressing sustainability challenges has grown over the last few decades because of the rapid depletion of natural resources, rising GHG emissions, air pollution and energy security and access, among other factors (Markard et al., 2012). Due to the inter-relatedness of the challenges faced in achieving sustainable development, the term 'transition' has been used to frame a need for systemic changes (Geels and Schot, 2010).

Sustainability transitions researchers have deliberated over the role that a digital transformation in electricity systems could play in cost and energy savings, as well in opening up new market opportunities (Sovacool et al., 2017; Mourik et al., 2020). However, there are concerns that these benefits may be uneven and, in some cases, reinforce inequality (Milchram et al., 2018). Through an energy justice lens, for a transition to be successful it should be inclusive, participative, and collaborative with consumers. All too frequently, consumers, particularly those who lack the capabilities to articulate their needs, are seen as largely passive actors, who could miss out on being future 'prosumers', or consumers who are involved in designing products aimed at them (Raimi and Carrico, 2016; Milchram et al., 2018). This chapter contributes to the ongoing debate about the country's electricity system by arguing that a technology fix found in digital solutions is simply too narrow and therefore attention must be given to a 'just' systemic change. Issues of inclusive transformation should therefore be part of the deliberation.

BACKGROUND LITERATURE: SOUTH AFRICA'S ELECTRICITY REFORM

South Africa has a historically energy-intensive economy with ambitious plans to lower the carbon emissions it produces (Ting, 2015). This vision includes a plan for a transition to a low-carbon economy[2] by implementing a just transitions pathway.[3] A 'just transition' is considered a comprehensive, sustainable development plan for the country and covers water, land, food security and energy issues, in order to shift into a low-carbon, climate-resilient economy and society by 2050. These plans have been recognised by the Department of Public Enterprises (DPE) roadmap for a reformed electricity sector. In this roadmap, the DPE (2019) has recognised the opportunity for small-scale, embedded generation to stimulate investment from the private sector and communities and so to stimulate job creation. Moreover, a transformed electricity sector is expected to help address spatial inequalities, as decentralisation of generation would have 'spill over' effects into socioeconomic activities. These would have the potential to relieve pressure on overburdened metropolitan areas and offer growth opportunities in peripheral urban areas (DPE, 2019). Thus, a reformed electricity sector offers numerous opportunities in line with the country's just transitions process. The following are outlined in the DPE plan (DPE, 2019):

- affordable and accessible energy, particularly for underprivileged and marginalised communities;
- job protection and creation, particularly in the downstream value chain;
- a decentralised energy grid infrastructure which would provide opportunities for a redress of spatial inequalities; and
- new kinds of ownership, which would facilitate participation in the economy.

---

2 National Planning Commission (NPC), Chapter 5, Transitions to a low-carbon economy. https://www.nationalplanningcommission.org.za/assets/Documents/NDP_Chapters/devplan_ch5_0.pdf.
3 Developing models and pathways for a low-carbon economy and resilient society. https://www.sustainable.org.za/project.php?id=55.

To achieve a just energy transition, it is important to note that, in trying to achieve sustainability, South Africa has the complex task of balancing salient trade-offs and synergies against multiple priorities and objectives in different sectors (Ting, 2020). Specifically, the urgency of environmental sustainability in the country is placed alongside other equally critical developmental goals, known locally as the triple challenge: reducing inequality, poverty reduction and job creation (NPC, 2012). In August 2019, the country had reached its highest unemployment rate – 29 per cent – in the last two decades.[4] Moreover, as a measure of South Africa's inequality, the Gini coefficient was 0.65 in 2015.[5] Strides have been made with regard to electricity access, however, more than 80 per cent of South African households in both urban and rural areas are connected to the grid (Stats SA, 2019; Stats SA, 2020b).

This means the impact of a digital revolution in the country's electricity system would be highly significant. However, in assessing this impact, one needs to consider the reality of South Africa's digital divide. In 2016, it was reported that almost two-thirds (59 per cent) of South African households had access to the internet, either from home, workplace, educational facilities or internet cafés (Stats SA, 2020b). Most users gained access through mobile phones, and fewest from home. One of the major barriers to internet access in the country is the unaffordability of data. South Africa's typical mobile prepaid data costs an average of over R100, or roughly $7 for 1GB, one of the highest rates on the African continent (Gillwald et al., 2018; Gillwald and Mothobi, 2019). Of further concern is the fact that for those who can afford access to the internet, most of the digital use is consumptive (Gillwald et al., 2018), for example social media such as Facebook, WhatsApp or Twitter. Most South Africans using the

---

4   StatsSA, Quarter Labour Force Survey (QLFS) for the second quarter of 2019, http://www.statssa.gov.za/?p=12376.

5   'How unequal is South Africa'. http://www.statssa.gov.za/?p=12930, accessed 1 June 2020. The Gini coefficient is a measure of income inequality. Briefly put, the range is from 0 to 1 where 0 reflects an equal distribution of income (where everyone has the same income) and 1 reflects absolute inequality (one person has all the income and no one else has anything). The closer to 1 the higher the inequality. Comparative numbers include: South Africa (0.6), Brazil (0.5), China (0.4), Japan (0.3). https://data.worldbank.org/indicator/SI.POV.GINI?locations=ZA-BR-IN-CN-TH-JP-GB, accessed 15 July 2020.

internet miss out on productive use of it for tasks such job searches, educational applications and e-commerce, among others (Gillwald and Mothobi, 2019). The stark challenge of affordable data was highlighted during the global COVID-19 pandemic in 2020 as students struggled to adapt to online learning.[6] These are some of the issues within which to contextualise the feasibility of a digital revolution in the country.

The next section discusses the research methodology used, justifying the need for an interdisciplinary lens, using both social and technical literature.

RESEARCH METHODOLOGY

Some scholars suggest that choices associated with technological innovation are not neutral, but are accompanied by certain ideas, preferences, values and interests (Chataway et al., 2014; Schot and Steinmueller, 2018). These subjective concerns are evident in relation to smart grids, for instance. Consumers are apprehensive about privacy (such as data collection of household energy use); safety (such as radiation emitted by wireless smart meters); affordability and security in relation to smart energy technologies (Milchram et al., 2018; Raimi and Carrico, 2016). Understanding these apprehensions is therefore important, as they have implications for the scale, scope and pace of the adoption of smart energy technologies.

Additionally, there are criticisms directed at unquestioningly positive expectations of the relationship between innovation and development. It is important to recognise that while some innovations lead to socioeconomic development, others contribute to exacerbating poverty and inequality. This is in part because contemporary innovation is reliant on high-quality networked infrastructure, capital-intensity, a good supply of skilled labour, and products and services which are typically accessible to affluent consumers (UNCTAD, 2017). This disadvantages the less well-off, and marginalised groups (for example, rural inhabitants), which in turn contributes to reinforcing poverty and inequality as these groups

---

6   'Covid-19 presents a curricula crunch for SA's universities'. https://www. iol.co.za/saturday-star/news/covid-19-presents-curricula-crunch-for-sas-universities-47191206, accessed 1 July 2020.

are excluded from the benefits of socio-technical change (UNCTAD, 2017; Chataway et al., 2014). It therefore becomes important to pause and question which kinds of innovation will promote social justice, while simultaneously discouraging harmful innovation. This chapter draws particularly on social and technical perspectives to develop the analytical framework. On the social dimensions, it can be argued that a distinction should be drawn between the old and emerging paradigm of a digitalised electricity system (Table 11.1). To elaborate, the emerging paradigm places more emphasis on actively engaging the consumers (as 'prosumers'), that decision-making is 'democratised', participation is inclusive, and access to energy is socially just. In terms of the technical dimensions, the emerging paradigm is towards distributed networks, stochastic generation (DERs), adaptive transmission and distribution, and bi-directional flow of electricity.

Table 11.1: A reflection on the differences between the old and emerging paradigms in decentralised and distributed electricity systems

Social dimensions		
	Old paradigm	**Emerging paradigm**
Access to energy	Income dependent	Fair and just
Consumer	Passive	Active (prosumers)
Decision	Technocratic	Democratic
Technological dimensions		
Grid	Centralised	Distributed
Generation	Dispatchable	Stochastic
Load profile	Passive and predictable	Engaged customers
Infrastructure	Static transmission and distribution structure	Adaptive T&D
Network	One directional flow	Bi-directional flow

Source: Author's own

Given these perspectives, this chapter uses an inclusive innovation lens through which to assess the potential digital revolution in South Africa's electricity system. The chapter uses two sources of literature, one considering sustainability transitions and another relating to energy justice. Sustainability transitions (ST) has gained significance in the last twenty years, mainly because of rising interest from those who are concerned with enabling the shift towards sustainable and inclusive economic growth (Markard et al., 2012; Schot and Kanger, 2018). Sustainability transitions are considered radical, not necessarily because of speed but rather because of the scope in changes of user practices, cultural meanings, industry structures, markets, policies, technologies, and supporting infrastructures (Geels and Schot, 2010).

In line with ST literature, it is important to appreciate that technologies are selected by the society in which they are embedded, and the selection mechanisms have material, political, cultural and sociological characteristics (Ting, 2020). In other words, socio-technical change is 'selected' not only by markets but also by social characteristics such as values, beliefs, expectations and visions among actors (Geels and Schot, 2010). In this case study, various technologies are 'selected' through social and technical dimensions. Moreover, once technologies are adopted in society, there is a process of retention and reproduction – or 'lock-in' – making changes away from the retained technologies difficult (Ting and Byrne, 2020) (Figure 11.2).

Since the potential for a digital revolution in South Africa's electricity system is in a nascent stage, this chapter's analytical framework is focused on the selection environment. Moreover, this chapter develops an argument for including an energy justice lens in a selection environment. The aim is that by combining two sources of literature (sustainability transitions and energy justice), it will be possible to assess whether the promise of a digital revolution in South Africa's electricity system is indeed inclusive or not.

The following dimensions in the selection environment form the analytical framework (Figure 11.2):

- Markets and distributional benefits. The markets refer to the economics, price instruments, customer preferences and practices associated with the selection of new technology. This also includes

Figure 11.2: The analytical framework used in the case study, outlining the multi-dimensional selection environment

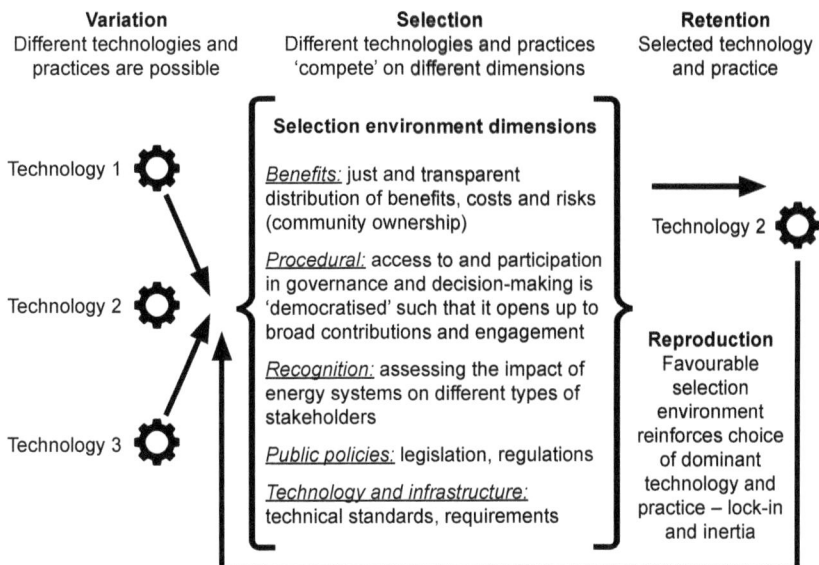

Source: Adapted from Ting and Byrne (2020: 4). In this example, the selection of Technology 2 is facilitated by the combination of the various dimensions of the selection environment which include material, political, cultural and sociological characteristics.

the distribution of benefits – which relate to costs – and obligations among participants in an energy system (Sovacool et al., 2019). Highlighting these issues would enable an analysis of the impact of new technologies on vulnerable groups of consumers such as low-income households, in comparison to the well-resourced middle class (Milchram et al., 2018).

- Procedural justice is concerned with impartial access to and participation in governance and decision-making such that the process is 'democratised' and respects due process and representation (Milchram et al., 2018). Democratisation is a key process in broadening out and opening up to broad participation of stakeholders in technological assessments. This has the effect of pluralising varying perspectives and thus encouraging inclusive and robust decision-making (Stirling, 2008; Ely et al., 2014).

- Equitable recognition refers to stakeholder groups involved in energy systems. Moreover, recognition refers to the inclusion of all actors in decision-making processes but goes beyond that since it also addresses issues of raising capabilities to articulate specific needs related to energy (Schot and Steinmueller, 2018; Sovacool et al., 2019).
- Public policies refer to legislation and regulations that govern and influence the adoption and diffusion of new technologies.
- Technology and infrastructure are technical standards and infrastructure requirements constitute a technical dimension of the selection environment.

THE CASE OF SOUTH AFRICA

This section discusses the total electricity value chain, from generation, transmission and distribution to consumption. Given that the emerging paradigm has an increasing presence of DERs, the resulting variability and uncertainty at different times scales will require several flexible solutions. This chapter evaluates the potential of digital technologies, which enable flexibility in the following sections: 1) supply-side, 2) grid, and 3) demand side (Figure 11.3). The next section starts with a discussion on generation.

GENERATION

South Africa's generation mostly comes from Eskom, a state-owned vertical monopoly electricity supplier that uses coal as its primary resource. Eskom has remained a monopoly responsible for electricity generation (>90 per cent), transmission (95 per cent) and distribution (>50 per cent) (DEFF, 2016; Eskom, 2016), deriving most of its revenue from three primary customers: redistributors such as municipalities; industry; and large mining companies. Together, these three consumer categories account for approximately 80 per cent of electricity consumption and revenues (Ting and Byrne, 2020).

South Africa has an installed generation capacity of approximately 40 GW. At present, electricity generation is mostly from coal (just over 83 per cent), followed by nuclear (almost six per cent) and independent power producers (IPPs) (four per cent) (Figure 11.4) (DOE, 2019).

Figure 11.3: Depicting Eskom's vertically integrated model and the different kinds of flexibility that are needed due to an increase in DERs.

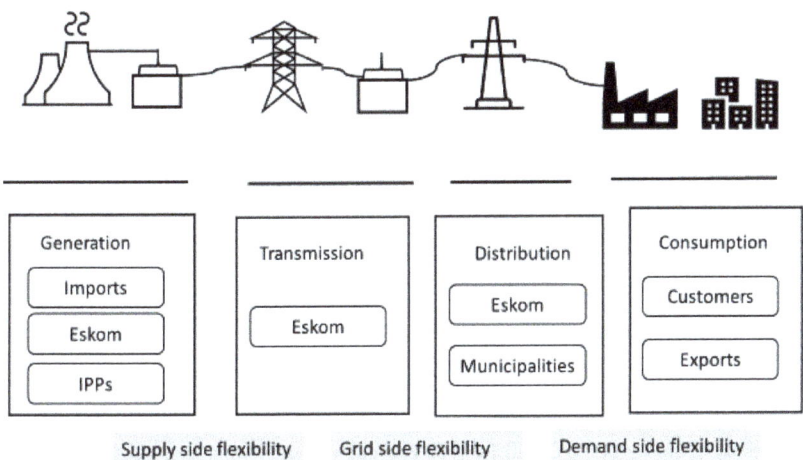

Source: Author's own

Figure 11.4: Installed capacity in South Africa's electricity system

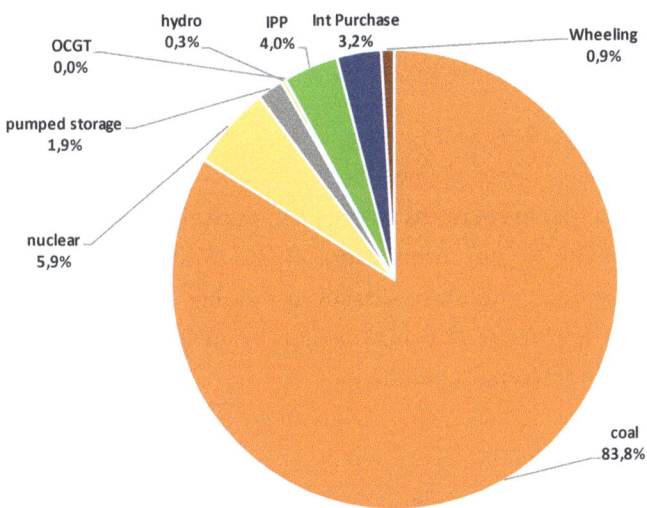

Source: DOE (2019)[7]

---

7   Wheeling here is defined as access between a non-Eskom generator (NEG) and Eskom to facilitate trading of energy. The NEG uses the national grid to transport electricity from where it is produced to a location of end use. http://www.eskom. co.za/Whatweredoing/Pages/Wheeling_Of_Energy.aspx. accessed 1 July 2020.

TRANSMISSION

The transmission network consists of high-voltage power lines[8] which connect large-scale generators to distribution (US-DOE, 2015). When the generated electricity exits the power station, the electricity is ramped by a step-up transformer to voltages around 132 kV or 400 kV (DPE, 2019). These large voltages are essential to drive the flow of electricity through the transmission lines and lessen expenses. System generation is expected as close as possible to the system load to ensure that system frequency is maintained at or very close to nominal levels (Ela et al., 2011). In South Africa, the nominal frequency of the National Integrated Power System (NIPS) is 50 Hz and is normally controlled within specific limits as defined in the Grid Code (NERSA, 2019).

Eskom has over 32,000 km of transmission lines, which service the domestic market, as well as regional markets through the Southern African Power Pool (SAPP) (DPE, 2019; Eskom, 2019b). However, Eskom transmission lines have an average age of more than 30 years, which is considered middle-aged for these types of assets (DPE, 2019). This ageing infrastructure is reflected in deteriorating network performance over the years.

SYSTEM SIDE FLEXIBILITY

The next set of discussions pertains to the main areas in which supply-side flexibility could be enhanced using digital tools. These areas pertain to transmission congestion, advanced weather forecasting and the potential of new revenue streams through balancing load services (see Figure 11.3 as a reference guide).

Transmission congestion

Renewable power systems with large amounts of variable generation (both wind and solar), which can increase or decrease output unexpectedly, have resulted in a diversity of power flows in the current

---

8   South Africa transmission system consist of power lines and substation equipment that operate at a nominal voltage of more than 132kV (DME, 2008).

transmission network. Transmission congestion has therefore become an issue among renewable energy developers. One of the solutions being developed is the advancement of capacity forecasting with reliable weather data prediction, using dynamic line rating (DLR). DLR provides real-time data on transmission lines on capacity ratings such as current and temperature. Usually, as transmission lines increase in heat, they tend to expand and sag, resulting in contact with trees and other objects, leading to outages. Ideally, high-voltage power lines are better able to conduct electricity through cooler conditions (Tomich, 2020). With the advent of smart grids, sensors are physically installed on transmission lines to monitor temperature, tension and sag, and to determine the thermal rating of the powerlines. This data is combined with weather forecasting and modelling, enabling network operators to gain optimised rating information, which can assist in variable renewable energy (VRE) integration into the grid.

Advanced weather forecasting

As the portion of VRE increases over time, the unpredictability in electricity generation also becomes regular and important, causing challenges to system operation. A challenge for system operators is increased sudden drops in electricity feed-in due to the continued expansion of wind and solar power. One of the digital enabling tools which can help reduce the uncertainty associated with VRE generation is advanced weather forecasting. These digital tools can generate real-time data (for example, they could track cloud movements, and forecast solar irradiation and wind speed variability) within a specified geographical location from meteorological devices, store them in a Cloud, and enable predictive models based on algorithms, all of which lead to a better understanding of VRE integration for systems operators (IRENA, 2019c). Moreover, accurate weather forecasting provides systems operators with improved ways of planning, as generation assets can be committed, and dispatched efficiently, thereby reducing the need for operating reserves (US-DOE, 2015).

Although AI is in the early stages of development for predicting advanced weather forecasting combined with power plant output data, nevertheless, as the power of data processing becomes cheaper, faster and more complex, so does the ability of machines to learn and thereby improve accuracy. In South Africa, meteorological data for renewable power plants (RPP) must be provided to the operator of the system as set out in the RPP grid code (NERSA, 2019). Depending on the type of RPP, the following meteorological data is required: wind speed, wind direction, air temperature, air pressure, air density and solar radiation relative to a specific location. Moreover, it is expected that the RPP generator provides the systems operator with a production forecast a day ahead and week ahead (NERSA, 2019). All these requirements would need accurate data generation, which can have an impact on systems operation and efficiency.

New revenues streams: Balancing load services

One of the ways in which new revenues can be generated is by providing balancing load services. These are new kinds of services that are becoming significant as power systems become more unpredictable, and flexibility turns into an asset. However, to balance load services, time-granular wholesale prices in energy markets are needed, which can be achieved through real-time price signals that are cost-reflective of short-term demand-and-supply conditions. Such ancillary services would need to be stepped up as variable renewable energy sources increase. However, system operators require incentives to participate in ancillary service markets (IRENA, 2019c).

Applying these concepts of new revenue streams to South Africa's vertically integrated monopoly model becomes challenging (Table 11.2). For instance, real-time wholesale tariffs are difficult if Eskom is the main electricity provider in the country. Moreover, if IPPs such as renewable energy developers were to compete, they would need to pay grid operator costs, which would raise tariffs to a premium. These complications are part of the current debate on the unbundling of Eskom into different divisions of generation, transmission and distribution (DPE, 2019). Therefore, it can be argued that realising a digital transformation is part of broader issues in the current electricity system, and these include Eskom's unbundling, just transitions, ageing infrastructure, and the decommissioning of older coal fleets, among others (Ting, 2020). Thus, there are numerous problems that offer context for a truly inclusive digital transformation (Table 11.2).

DISTRIBUTION

In South Africa, the responsibility of distribution (retail) of electricity is divided between Eskom, the municipalities and other licensed distributors. The distribution is currently split between municipalities (60 per cent) and Eskom (40 per cent), with municipalities accounting for most of the distribution into the country's 12 metros and the largest municipalities (DME, 2008; DPE, 2019). The country has over 45,000 km of distribution lines, 300,000 km of reticulation power lines, and more than 7,000 km of underground cables, which are owned by Eskom (DPE, 2019, Eskom, 2019b). The flow of electricity from transmission (400 kV or 275 kV) towards distribution networks is a stepped-down voltage (132 kV) where power is delivered to redistributors or end-users.[9] From there the voltage levels are a further step down from distribution substations towards reticulation lines (33 kV, 22 kV and lower) into a service connection. The main point to appreciate here is that the current distribution system is designed to be relatively passive, with a focus on unidirectional power flow, which includes distribution towards the consumer. While this design standard is adequate to provide customers with consistent electrical service, it can be contended that it is insufficient to meet future needs for greater resilience, more complex electricity load profiles and increasing consumer participation (US-DOE, 2015).

Currently, the task of municipalities is to buy bulk electricity from Eskom and act as reticulators of electricity and gas, upon which most of their revenues are based. In 2018, it was reported that municipalities had on average received 30 per cent of their income through a mark-up of electricity purchased from Eskom (Figure 11.5) (Stats SA, 2020a). Moreover, municipalities use the revenues derived from these sales to cross-subsidise other local activities (Montmasson-Clair et al., 2017). At present, there are numerous challenges within the legislative framework that governs the municipalities' task of reticulation of electricity within their specified boundaries. The next section discusses these challenges in further detail.

---

9   A distribution system operates at a nominal voltage of 132kV or less and subsequently a distributor is defined as a legal entity that owns, operates or distributes electricity through a distribution system (DME, 2008).

Table 11.2: A summary of an inclusive energy transitions lens related to the supply side

Inclusive energy transitions lens	Supply-side flexibility
Markets and distributional benefits	Due to an increase in VRE, rewarding flexibility will be key in the future. Furthermore, there is a need to properly remunerate grid balancing support services.
Procedural	An electricity reform would require fair and non-discriminatory access to the transmission grid. This would then, in turn, facilitate grid access irrespective of Eskom or IPPs.
Recognition	New kinds of role in a reformed electricity sector. These include the operator of the system, which would manage the supply and demand balances in actual time through a variety of minimum cost options. There is also the market operator which would contract with suppliers and distributors (DPE, 2019).
Public policies	Grid connection code for renewable power plants (RPP)
Technology and infrastructure	Digital tools would enable advanced weather forecasting (for example, wind speed, wind direction, air temperature, air pressure, air density and solar radiation) of variable renewable energy (VRE) generation. Dynamic line rating (DLR) – sensors and analytics to monitor and optimise transmission lines in near real-time, enabling them to carry increased loads under certain conditions. On a cooler, breezier day, a line can carry more than it can in the sweltering heat.

Source: Author's own

Figure 11.5: Total income received across all South African municipalities in December 2019

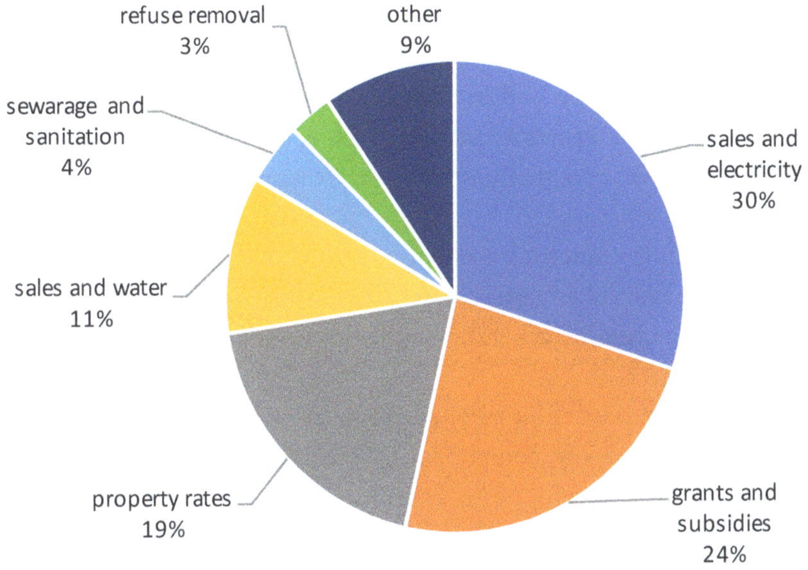

Source: Stats SA (2020a)[10]

GRID-SIDE FLEXIBILITY

The next set of discussions concerns the challenges and opportunities for grid-side flexibility in a digitalised electricity system. These pertain to the following: municipal legislative challenges, maintaining grid stability and reliability by using aggregators, grid infrastructure, and tariffs and municipal debt (see Figure 11.3 as a reference guide).

Municipal legislative challenges

According to DPE (2019), there are numerous legislative challenges for municipalities that have implications for the distribution model, and these include the following:

- different tariffs from Eskom and municipalities in the same municipal area;

10 'An update to municipal spending and revenue. December 2019'. http://www.statssa.gov.za/?p=13146 , accessed 1 June 2020.

- ownership of distribution infrastructure is regularly divided between Eskom and municipalities;
- municipalities have different capacities to distribute electricity, and a substantial number are in crisis;
- worsening quality and standards in municipal distribution with significant maintenance backlogs affecting the security of supply; and
- lack of investment in municipal infrastructure.

Given these numerous challenges, addressing the issues associated with municipalities will be necessary if South Africa intends to pursue and realise embedded generation at distribution and consumer levels.

Grid stability and reliability through aggregators

Distribution is a key component in a smart grid plan because the future electricity profile is likely to have a greater interface between distributors and distributed energy resources (DERs). This means that the function of distribution system operators needs to change, as they need to operate the assets connected to their grid for the assistance of both the grid and consumers (IRENA, 2019b). This potential of increase in DERs would result in a reduced need for high voltages, and consequently sales which require lower voltages will increase and sales to high-voltage customers could reduce significantly.[11] To mitigate this challenge, there is a need for better tools on low-voltage grid intelligence, which could enhance customer service. As articulated in South Africa's smart grid plan, the distribution grids/networks operating at 132 kV level and below will be crucial for future integration of the renewable energy and other intermittent loads (DOE, 2017). To do so, a 'smart grid' would require digital technologies, such as IoT, to manage, aggregate and control varying demand loads and provide balancing services to the grid (IRENA, 2019c).

Moreover, one of the emerging challenges in the introduction of renewable energy, particularly in a centralised, passive grid, typified in South Africa's current electricity system, is the volatility in grid load

---

11 Eskom's energy and revenue loss management, (2014) https://www.ee.co.za/article/eskoms-energy-revenue-loss-management.html.

and the integration of heterogenous DERs. These fluctuations need to be minimised in order to maintain a stable grid. One of the best examples of mitigating fluctuation is 'smoothing out' peak demand by matching available generation with distribution capacity (IRENA, 2019b). In countries where digitalisation of electricity has advanced (Finland, Denmark, Netherlands, and Germany to name a few), 'smoothing of peak demand' is an important value addition, through the provision of innovative ancillary services by key emerging actors called aggregators (US-DOE, 2015). Here, aggregators can decrease the need for expensive spinning reserves[12] and lessen the necessity for long-term investments in new generation plants and other capital expenditures (IRENA, 2019a:11; Ela et al., 2011).

An aggregator role is to group together physical devices (dispatchable and stochastic generators), creating sufficient capacity comparable to that of a conventional generator (IRENA, 2019b; Pasetti et al., 2018). This aggregation can also be called a virtual power plant (VPP) (see Box 11.1). Here, a VPP uses information technology, such as real-time data of current capacity, storage levels and standby status of assets, data visualisation such as information of consumer groups and geographical location, and full automation in control. Aggregators can then trade electricity or ancillary services (for example, frequency and voltage support) via an electricity exchange, in the wholesale market, or through procurement by the system operator (IRENA, 2019c). Aggregators would be key players in an emerging future in which distributed energy sources would be enabled to participate in electricity and ancillary service markets.

A study on the potential role of municipalities as an aggregator in which energy is traded has been conducted (Montmasson-Clair et al., 2017). Here the municipalities' role would be different from a distributor, as it operates as an aggregator of electricity and trade in a market. The crucial

---

12 In order to maintain grid stability, a utility must provide reserve capacity that is able to be deployed within a specified time, to mitigate load variation and unexpected electricity load. This reserve capacity is called the spinning reserve which is the amount of unused capacity in online energy assets that can compensate for power shortages or frequency drops within seconds or minutes (Pöller and Obert, 2017: 22).

role here is to provide a link between IPP generators of renewable energy resources and consumers. At present, there is only a single energy trading company in South Africa, called PowerX, which was granted a license by National Energy Regulator of South Africa (NERSA), and has been operational since 2006. However, this energy trader role for municipalities faces several hurdles, such as competing tariffs from Eskom and a wheeling agreement with Eskom transmission lines. Moreover, the municipalities funding model is currently mandated under the Municipal Systems Act (MSA), which prohibits them from entering commercial activities such as trading. Currently, the MSA directs the municipality to 'finance their affairs by 1) charging fees for services; and 2) imposing surcharges on fees, rates on their property' (Montmasson-Clair et al., 2017: 32). At the time of finalising this chapter, regulations had just been published to allow municipalities to develop or procure their own power – a decision that would have some positive implications for the issues raised in this chapter, at least for those with the financial status to qualify (Creamer, 2020).

All things considered, if DERs were to be implemented at scale, such that aggregators are needed, there would remain substantial obstacles, given the country's vertically integrated model, and municipalities' constraints within the existing legislative framework.

Box 11.1: An example of a VPP

> Next Kraftwerke's is a power trader and operator of a large virtual power plant (VPP) in Europe that aggregates 8,000 energy-producing and consuming units.[13] The VPP has a total networked capacity of over 7,000 megawatts, and trades aggregated power on varying spot markets. The VPP is able to integrate distributed energy resources (for example, solar, wind, biomass units), with flexible power consumer (such as demand-side response), as well as storage systems, in order to monitor, forecast, enhance and dispatch generation or consumption.[14]

---

13 Next-kraftwerke: https://www.next-kraftwerke.com/.
14 Next-kraftwerke: https://www.next-kraftwerke.com/vpp/virtual-power-plant.

Simply put, digitalisation has allowed Next Kraftwerke the ability to forecast a given electricity unit (for example, generated from a solar PV or combined heat and power plant) to trade where the demand is needed, thus creating flexibility and grid stability.

Source: Next-Kraftwerke, 2020

Grid infrastructure

It is expected that future smart grids will be resilient, reliable and safe, particularly enabled by the introduction of technologies related to IoT (US-DOE, 2015). IoT-enabled technologies are projected to provide fast and predictive analytics, such as grid stress, robustness, frequency and voltage instability (US-DOE, 2015). Further, IoT can obtain data, perform decision-support algorithms, avoid or minimise interruptions, dynamically control the flow of power and restore service swiftly (DOE, 2017). Moreover, there is an opportunity for predictive risk assessments, based on real-time analysis, that can recognise field assets, and power lines that are most likely to break down (US-DOE, 2015; DOE, 2017). In this way, the grid becomes 'self-healing' as it has corrective capabilities that can detect and analyse issues (For example, fluctuation in voltage and currents or poor power quality), anticipate potential threats from grid failure, and quickly restore grid components or network units (DOE, 2017). A self-healing grid is possible using IoT, as it ensures grid reliability, security, power quality, and efficiency.

In South Africa, the updated IRP expects that 2.6 GW of embedded electricity generation will be implemented by households and businesses for their own use, at a rate of 200 MW a year (DPE, 2019). This is expected to attract investment from consumers who will invest in DERs aimed at selling electricity back to the national grid. One of the initiatives by the Department of Minerals and Energy Resources (DMER), called 'the approach to distribution asset management (ADAM)', has highlighted a significant gap in electricity distribution infrastructure maintenance and a substantial refurbishment backlog for both Eskom and municipalities (DOE, 2017). The ADAM baseline report in 2008 estimated that R27 billion was needed for electricity infrastructure maintenance, rehabilitation, and strengthening backlog (DOE, 2012). However, by 2014 a follow-up review had estimated that R69 billion was needed, an indication that the situation has worsened over the years (DOE, 2017). Thus, major challenges in electricity distribution, particularly with municipalities, are insufficient investment in asset maintenance; lack of refurbishment investment, and low revenue collection due in part to ageing infrastructure and theft. In relation to smart grids, a lack of infrastructure for grid intelligence could lead to vulnerability to physical attacks such as frequent and intense weather events, as well as potential cyber threats.

Tariffs and municipal debt

In South Africa, the tariff structure reflects a vertically integrated business model, where larger customers (commercial and industrial) generally subsidise smaller ones (Eskom, 2009). There is, therefore, an inherent design in cross-subsidising the poor by richer consumers, which means some customers are not paying cost-reflective prices (Eskom, 2009; Lekoloane et al., 2018). One factor increasingly impacting on Eskom revenues is a drastic drop in the cost of embedded technologies, such as rooftop solar PV (Ting and Byrne, 2020; DPE, 2019). This means that an increasing number of consumers are defecting from the grid due to an improved business case for embedded generation. Consequently, Eskom's tariff setting methodology is becoming untenable, because as volumes of sales decreases, it must raise tariffs to recover costs. These was laid out by DPE (2019: 14–15):

> Eskom's operating business model is outdated and based on the era of excess electricity supply and captive customers ... With declining demand, Eskom is facing a utility death spiral. Modern systems have become increasingly decentralised, enabling the participation of multiple suppliers, including generation from renewables.

These weaknesses in Eskom contribute to making it unsustainable (Eskom, 2019a, DPE, 2019). In this state, the more Eskom pushes a spike in electricity tariffs, the greater the incentives for consumers to become more self-sufficient, and possibly to defect from the grid, thereby lowering Eskom sales volumes. Moreover, the business model of subsidising poorer customers could be disrupted, because it is likely that those who can afford embedded solutions will typically be the first adopters to do so (Lekoloane et al., 2018). In addition, there is growing concern about municipal debt, which has risen from almost R5 billion in 2015 to over R26 billion in 2019 (COGTA, 2019) (Figure 11.6). Some of the reasons for an increase in municipal debt are poor billing, poor debt management processes and a 'culture of non-payment' as customers ceased paying because there were no incentives to do so (COGTA, 2019). The implications of poor revenue management can hinder new types of market opportunities in a digitalised electricity system, such as incentive schemes in net billing and time of use tariffs which would facilitate prosumers playing a part in wholesale markets.[15]

The culture of non-payment is best demonstrated through the resistance and rejection of prepaid meters in 'hot spot' areas such as Tembisa, Soweto and other similar communities (Makonese et al., 2012). These are areas where electricity is highly politicised, and the root causes of non-compliance are complex and multifaceted (Fjeldstad, 2004). Some of the factors are due to an enduring resistance to paying for services which are expected at least to be affordable if not free (Mathe, 2019). Fjeldstad (2004) has noted that during apartheid non-compliance became the chief weapons against an illegitimate government. This strategy was used by people living mostly in townships and in rural areas in the so-called homelands, which had the intended effect of giving rise to an ungovernable state. Post-apartheid, non-compliance (such as non-payment) did not cease but instead became a norm as a

---

15 Time-of-use tariffs is a type of time-varying power prices available to consumers providing them options to shift their electricity consumption, allowing them to save on electricity bills.

Net billing is a type of market mechanism, which allows a prosumer to export surplus electricity into the public grid at a specified value per kWh relative to time.

DER – distributed energy resources are able to provide ancillary services or electricity to the grid, increasing flexibility in the system while being remunerated accordingly.

Figure 11.6: Escalating municipal debt owed to Eskom

Source: *COGTA (2019: 8)*

response to the government's failure to provide basic services such as electricity and water, among others (Fjeldstad, 2004).

In addition to municipal debts, there are challenges that can hamper digitalisation in distribution. Municipalities lack the new skills and capabilities needed for an electricity transformation. Furthermore, there is a need for reforms in legal (such as procurement from IPPs) and regulative frameworks (such as wheeling), and in financial processes (such as metering and billing). These are some of the issues that can serve as a major stumbling block in realising a digital revolution at distribution level (Table 11.3).

CONSUMPTION/PROSUMERS

The development of distributed energy resources (DERs) along with the extensive availability of 'smart' devices, has initiated new opportunities for consumers as active participants. Examples of DERs include electric vehicles, roof-top solar PV, battery storage, and power to heat, all of which contribute to decentralising the system (DPE,

2019; DOE, 2017; IRENA, 2019b) (see Box 11.2). Figure 11.7 illustrates the heterogenous types of consumers or prosumers in a digitalised electricity system, at different scales ranging from homes and large residences to smart cities, among a few examples. The significance here is that prosumers can self-generate, store or trade electricity as enabled by DERs.

Figure 11.7: A diagram of DERs at the consumer/prosumer level.[16]

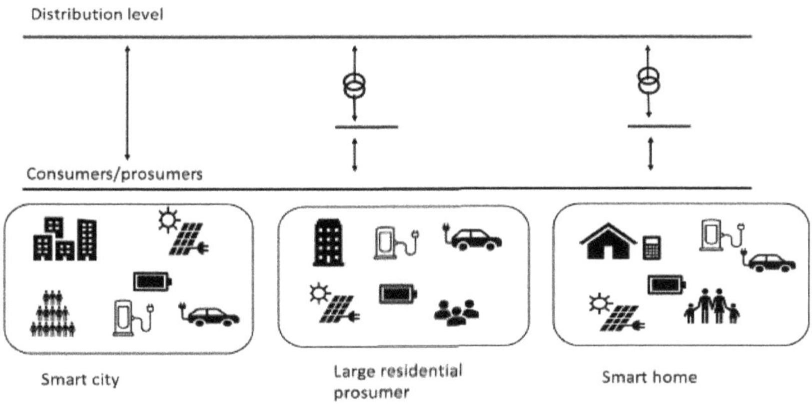

Source: Adapted from Pasetti et al. (2018: 6)

DEMAND-SIDE FLEXIBILITY

The next section discusses opportunities as well as challenges in demand-side flexibility in a digitalised electricity system by democratising participation through prosumers and community-based energy systems (see Figure 11.3 as a reference guide and Table 11.4 as a summary on the cautious steps that needs to be recognised concerning demand-side flexibility).

Democratising participation through prosumers
At the consumer level, one of the exciting opportunities in an electricity

---

16 The DERs represent smart EV, solar PV, and energy storage, which interface with various consumers at a scale of smart city, large residential prosumer or a smart home.

Table 11.3: A summary of an inclusive energy transitions lens related to grid-side flexibility

Inclusive energy transitions lens	Grid-side flexibility
Markets and distributional benefits	There are new business opportunities and revenue streams for municipalities. However, new roles and functions for municipalities are highly resisted (Montmasson-Clair et al., 2017). There are opportunities, such as net billing and time of use tariffs, which would enable the participation of prosumers, but these are not yet available.
Procedural	There is a need for new approaches to systems operation through strengthening collaboration between distribution and transmission system operators, as well as local government stakeholders, considering the evolving role of distribution system operators. This can include active partner engagements with stakeholders (SALGA, AMEU, municipalities, for example), discussion on revenue management, and technical interventions.
Recognition	• It is important to recognise the 'culture of non-payment', as reflected in the increase in municipal debt over the years. If smart grids are to be implemented, a question is who pays for the costs, and who faces the risks and benefits. • Digital revolution will require new skills in load management, new revenue streams, as well as technical know-how. It must be recognised that there is a need for new capabilities in data collection, data pre-processing, processing and testing. • Potential of new emerging players, for example, aggregators. Municipalities could play a role here, or other entities, but these are limited given the current legislative framework.

Inclusive energy transitions lens	Grid-side flexibility
Public policies	There are numerous limitations in the current legislative framework that governs electricity supply. There needs to be consideration of the role of municipalities in a reformed distribution market, since the current business model relies on municipality revenues having a mark-up from Eskom.
Technology and infrastructure	• A potential for grids to anticipate and automatically respond to system disturbances as a form of 'self-healing'. However, there is a lack of maintenance on ageing electricity distribution infrastructure. • There is a need for grid intelligence at low voltage (LV) networks (132 kV) in a system that has an increasing presence of DERs.

system transformed towards decentralisation and digitalisation is the changing of roles of consumers. Active energy consumers – who will be what are now termed prosumers – can self-generate, store and sell electricity, thereby saving costs and producing income. For instance, peer-to-peer trading (P2P) is a business model that allows for the use of mini grids, providing consumers and prosumers with a marketplace to trade electricity without the need for a retailer (IRENA, 2019a). P2P can use third-party digital platforms to trade with minimal involvement from suppliers. For instance, with blockchain, trade is possible using price signals and actual time DER production data all through the network (IEA, 2017a; IRENA, 2019a). Other uses of blockchain are in grid management and system operation, payment of charging EVs, and real-time green certificates (Figure 11.9). Most of the companies that use blockchain with an energy application were established between 2016 and 2018 (IRENA, 2019a). This, therefore, reflects the early stage of technology and its potential is still largely untested.

Box 11.2: Examples of distributed energy resources (DERs)

Distributed generation
Mini grids and microgrids
A mini grid refers to a contained combination of electricity generation, distribution, storage and consumption within a specified area (UNDESA, 2014). Mini grids are often locally managed and have a scale of less than 10 MW of installed capacity and serve a location of around 50 kilometres. Microgrids have a capacity of from 100 kW up to a few megawatts which serves an area with less than an eight-kilometre boundary (UNDESA, 2014). Microgrids can be found in large university campuses, military bases, hospitals and shopping malls. The IEA has suggested that off-grid solutions, such as solar PVs, are likely to be the most cost-effective solutions for universal electrification in sub-Saharan Africa (IEA, 2017b). This is because decentralised systems, which are smaller in scale, have been shown to be more appropriate for remote, inaccessible non-urban areas.

A good example of a microgrid in South Africa is Eskom's Smart Grid Centre of Excellence, which has a 32 kWp solar PV, located in Wilhelmina Farm in the Free State. Launched in 2018, the demonstration plant provides electricity to the 14 households that make up the Wilhelmina community. The plant also contains three sets of lithium-ion batteries, totalling 90 kWh of storage (Eskom, 2019b). The microgrid plant has an advanced metering infrastructure (AMI) and smart prepaid metering systems, which can detect theft and tampering as well as energy balancing. A hybrid system is used to manage communication between the home systems and the plant, with live information from the site being relayed to the nearest substation. In the control room, live feeds of weather are used to forecast the performance of the plant, predict threats ahead of time and manage load remotely (Eskom, 2019b).

Behind-the-meter batteries

Behind-the-meter (BTM) batteries are characterised by their bidirectional response capability to store and discharge electric power as needed (US-DOE, 2015). They are usually applied in homes and workplaces. BTM batteries have a range of fewer than 5 megawatts and are frequently accompanied by rooftop solar PV (IRENA, 2019c)

Contribution to VRE integration

Enables the storing of energy, and dispatching either for consumer needs or for participation in ancillary service market (smooths the peak load profile, for example)

Demand response

A process that enables consumers to alter their electricity consumption patterns and provide grid services, individually or through an aggregator (IRENA, 2019c).

Smart charging electric vehicles

The future of mobility will be autonomous, shared and electric. It has been reported that most personal vehicles remain in parking lots or garages more often than being on the road, but importantly near buildings with electrical power (Sovacool et al., 2017). There is a growing interest in vehicle-to-grid (V2G), which uses EVs as a bi-directional flow of electricity between a vehicle and power grid. This has the effect of using EVs as potential storage devices as it helps balance the grid by trading electricity during peak when not operational.

Power-to-heat

To support sector coupling, one of the emerging applications enables renewable power-to-heat, which is the use of renewable power to generate useful heat energy for buildings or industrial processes. Examples of these include thermal boilers, heat pumps and thermal storage. Power to heat is part of a trend in the electrification of end-use such as transport, which will increase the load of distribution networks (IRENA, 2019c).

The potential for active participation by the consumer offers a way of 'democratising' access to the grid. However, seen from an inclusive lens, it is important to caution that such opportunities could be uneven, depending on the ability of consumers to access the hardware (technology) and software (know-how). According to Milchram (2018), smart grids could highlight the inequitable distribution of benefits as they reinforce financial injustices faced by those who are less economically well off. In this section of the population, their energy choices may well be narrowed, and they may be excluded from the smart grid, due to affordability.

For instance, in South Africa, poorer households are dependent on government-subsidised, free basic electricity, which provides 50 kWh per month for a grid-based system (Eskom, 2009). Therefore, the concept of a truly smart home, where there is access to solar panels, smartphones, plugs and meters, as well EV charging points, may be more beneficial to an elite section of society. This points to one way in which digital solutions can exacerbate inequality if they are not applied inclusively. To elaborate: Those who currently have access to affordable, or free, electricity will still need to rely on subsidised electricity, while those who can afford to go off grid can do so, leaving behind fewer paying consumers to bear the cost of tariff subsidisation. This is a case of disproportionate allocation of costs, given the current tariff structure of cross-subsidies in a vertically integrated business model (Lekoloane et al., 2018). Perhaps a solution lies in a community-based energy system, where energy generation (such as mini grids) is collectively owned, and the benefits, costs and risks are shared.

Community-based energy system
In a community-based energy system, the technologies need to be bundled such that the scale reflects local demand. Moreover, a community-based logic facilitates opportunities that focus on the

Figure 11.9: Blockchain initiatives in the power sector

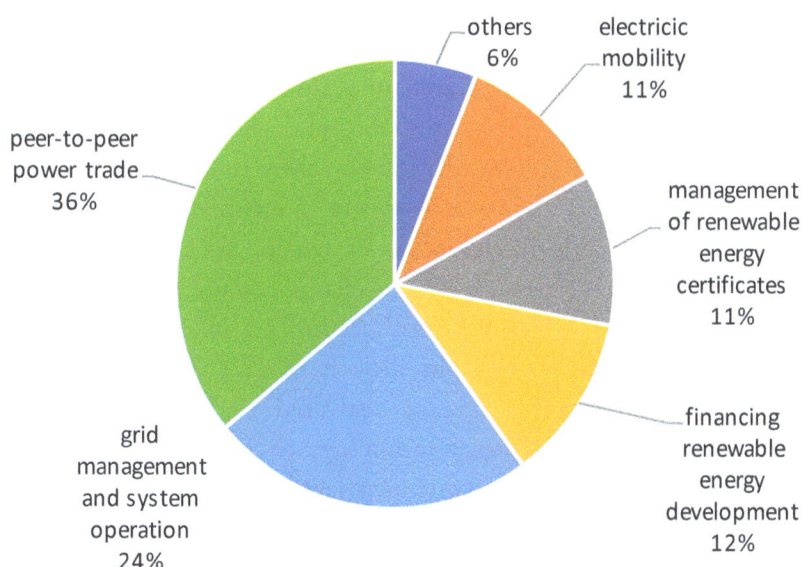

Source: IRENA (2019a:11)

local economy, create local jobs, embeddedness and acceptance of energy solutions (Mourik et al., 2020). Therefore, energy solutions become reflective of community needs, instead of a one-size-fits-all approach. Associated with the community-based energy system is the use of pay-as-you-go (PAYG) models, which allow payment of electricity as needed, as opposed to fixed payments (IRENA, 2019c). One such community logic initiative is called a participatory business model (PBM) (Lennon et al., 2019). The PBM places emphasis on renewable energy businesses that are fair and equitable to those living near such initiatives. Furthermore, the PBM promotes community approaches that entail democratic control, cooperatives, local ownership and community benefits beyond profit (social and environmental benefits, for example) (Lennon et al., 2019). Possibly, in South Africa, implementing a smart grid programme could use a policy mix approach, where there are different types of support depending on the scale of renewable energy to be implemented, and the type of households (urban vs semi urban and rural).

DISCUSSION AND CONCLUSION

Using an inclusive energy transition lens, this chapter revealed the numerous cautious steps required to realise a digital revolution in South Africa's electricity system (Table 11.5). Since decentralisation and digitalisation are still in their infancy, there is an opportunity for the country to 'select' the technologies in an inclusive manner, before systems become path-dependent or locked in. Several studies have shown that once systems are locked in, changes away from the retained technologies become difficult (Unruh, 2000; Ting and Byrne, 2020). A good example is the current tight 'lock-in' of a high-carbon pathway due to South Africa's dominant coal-fired electricity generation (Ting, 2015). Therefore, energy technology choices need to take into consideration the long-term, systemic effects of new technologies.

Reflecting on South Africa's overall electricity system, it looks as if the overdependence on coal-based electricity generation and its monopoly model is fast declining. This is not least because the pattern of electricity demand has changed over the years, particularly the waning support for energy-intensive mining, but there is also an emphasis on more sustainable energy sources. Furthermore, there are changes in the global electricity system towards transactive energy, which allows for consumers to be no longer passive but engaged (prosumers), as well as shifts from dispatchable to stochastic generation (for example, distributed), and growing linkages with digitisation (smart meters, automation and real-time data). All of these scenarios have the potential to shape South Africa's electricity system beyond the current monopoly model, in the direction of diverse energy sources and infrastructure that cater to flexible, modular and decentralised demand.

From the supply side, it is clear that increasing flexibility will be key if there is an increase in VRE in the system. At present, digital tools such as AI have been used in limited ways – for example, in meteorology, control and predictive maintenance. But with increasing digital connectivity, which creates a massive amount of granular data, AI could be used effectively in supporting decision-making, planning and condition monitoring (IRENA, 2019c). Likewise, many of these opportunities are dependent on the transparency and pace with

Table 11.4: A summary of an inclusive energy transitions lens related to demand-side flexibility

Inclusive energy transitions lens	Demand-side flexibility
Markets and distributional benefits	The smart grid would allow for new kinds of participation such as self-generation, store, trade (P2P) and pay-as-you-go (PAYG) for electricity. These are new opportunities for revenue streams available to consumers. However, the smart grid would also disrupt the current tariff structure of cross-subsidising poorer customers, while affluent customers can afford self-sufficiency measures, and perhaps grid defection. This leaves an imbalance between consumers who continue to rely on the government, while others can install their own DERs – thus, widening inequality to energy access, affordability and security.
Procedural	• Potential for integrated prosumers: consumers can contribute to and access the grid in real time. • If there is comprehensive electricity reform in the country, consumers will have more choices, which will allow them to optimise their energy uses.
Recognition	A potential risk of smart grids is that they require affordable access to the internet, which could exclude societal groups that are economically vulnerable (Milchram et al., 2018).
Public policies	• Due to the nascent stage of energy reform, there is a need to experiment with different types of business models; regulatory 'sandboxes' should be established to allow for this (IRENA, 2019a). • Regulatory change is dependent on whether Eskom is unbundled. Unless and until this occurs, developing a flexible retail market, which allows for proactive 'prosumers', is not feasible
Technology and infrastructure	• Smartphones • EV infrastructure • Behind-the-meter batteries • Mini grids

which Eskom is unbundled into separate parts, namely generation, transmission and distribution. This will enable inclusive access to transmission lines and enable real-time wholesale market prices. However, it must be noted that with its deep historical ties to energy-intensive mining activities, Eskom is an entrenched incumbent that prefers to maintain the status quo (Ting, 2015). It seems unbundling Eskom is less about the private sector or state ownership. Rather, the heart of the issue is more about removing the conflict of interest between the state and the private sector and improving governance so that Eskom is less able to act in its own narrow interests, and there is instead greater accountability, more transparency and public participation (Ting, 2020). Until there is certainty about Eskom's unbundling status, it is difficult to determine the nature of a transformed electricity system, as the market remains a monopoly, with a limited number of participants who have privileged influence over energy and economic policy, as well as access to, and distribution of, investments.

Municipalities have an enormous role to play in grid-side flexibility as they serve as an interface between consumers, or potential prosumers, and energy suppliers, particularly since VRE is argued to operate at lower voltages and will require integration into distribution networks. However, municipalities are at present overwhelmed with escalating debt due to a culture of non-payment, lack of critical skills, revenue losses due to ageing infrastructure and theft (COGTA, 2019). Additionally, the current business model of municipalities is reliant on marking up Eskom's wholesale prices in order to generate revenues. This means that realising new roles and functions for municipalities has major cost implications. In the medium term, a few of the municipalities will be able to take advantage of the right for self-generation or autonomous procurement.

At the level of consumers, an increased decentralisation of energy resources will enable local ownership of energy production, which in turn will change the ways that people engage with energy. There are opportunities, for example, to self-generate, store and trade (peer-to-peer trading) electricity depending on the type of VREs available. Moreover, using blockchain or pay-as-you-go services can enable new kinds of financial transactions, particularly with mini grids which focus on initiatives on a local scale. However, given the country's wide inequality, as well as increasing non-payment of electricity bills, questions must be raised on how the digital revolution enables the participation of economically vulnerable citizens. More importantly, if digitalisation of the country's electricity system is pursued in earnest, it should ensure inclusive participation by all sections of the population, and so provide an opportunity to uplift socioeconomic wellbeing. Simply put, the potential of digital solutions for the electricity system must be embedded in the communities these solutions are targeted at. Solutions cannot merely be narrowed down to technological choices. In a country that has high inequality, social values become important – particularly those related to energy access and affordability. Thus, social costs cannot be ignored nor dissociated from technological choices.

An inclusive energy transitions lens has shown that a more nuanced approach is needed to better understand the multidimensional selection environment in which new technologies are chosen (Table 11.5). In line with this perspective, energy justice scholars have noted that reducing the social dimensions of energy transitions to merely a question of public acceptance/rejection dichotomy – is insufficient (Lennon et al., 2019). Rather, there is a need to engage citizens to encourage a diversity of expression, agency and participation about the readiness of society for the technological solutions proposed.

This chapter has argued that the current, vertically integrated monopoly model by Eskom hinders efforts to realise a truly democratised grid. Moreover, the current tariff structure does subsidise poorer customers with tariffs from large, intensive energy users. Therefore, a potential disruption could result in unfair distributional justice, because more affluent customers would be able to afford to defect from the grid, but those who couldn't would be left to rely on the government, and to bear the burden of increasing electricity prices.

This chapter described the rejection of prepaid meters in some areas in the country. If the Fourth Industrial Revolution were to be applied in the electricity sector (smart meters and smart grids), there would be a need for regulatory frameworks and programmes to protect low income households, so that an electricity transformation becomes

Table 11.5: Summarising the overall potential of a digital revolution in South Africa's electricity system

Inclusive energy transitions lens	Overall discussion
Markets and distributional benefits	Given the number of emerging applications of digitised electricity solutions, there is a need to test these various applications to better understand potential risk, benefits and impacts before implementation on a wider scale.
Procedural	• The disruptive potential is still in the early stages and is far from being thoroughly understood. • At present, there is a limited understanding of how end-users will interact with the energy system.
Recognition	There is a need to gain greater insight into prosumers' needs and, more importantly, their agreement to the digital tools suggested to them.
Public policies	Changes to regulatory policy are needed to create regulatory certainty.
Technology and infrastructure	• At present, there is a lack of technological maturity and awareness about a digitalised electricity system. • It will be necessary to assess and develop the interoperability of communication systems and standards.

Source: Author

desirable, particularly to less well-off communities (Kambule et al., 2018).

Digital solutions in the energy system are only beginning to be explored, even though there are many promising applications. Significant hurdles that stand in the way include, among others, regulatory uncertainty, scalability, lack of interoperability and standards, and a lack of technological maturity and awareness (Table 11.5).

An inclusive energy transition will require 'democratised'

governance structures that will empower all stakeholders, and ensure that the risks and rewards are shared. Furthermore, selecting technologies will require an openness to experimentation, with different ways of dealing with problems, and there is a need to involve a diverse set of actors, with arrangements for collaboration across multiple sectors. If this is achieved, new technologies may be able to justly reflect societal needs in a way that addresses South Africa's triple challenge of minimising inequality, unemployment and poverty.

REFERENCES

Chataway, J., Hanlin, R. and Kaplinsky, R. 2014. 'Inclusive innovation: An architecture for policy development'. *Innovation and Development*, 4(1), 33–54.

Cooperative Governance and Traditional Affairs (COGTA). 2019. 'COGTA plans to resolve a municipal debt to Eskom; with deputy ministers'. https://pmg.org.za/committee-meeting/29506/, accessed 1 June 2020.

Department of Energy (DOE). 2015. 'State of renewable energy in South Africa'. http://www.energy.gov.za/files/media/Pub/State-of-Renewable-Energy-in-South-Africa.pdf, accessed 1 June 2020.

Department of Energy (DOE). 2016. 'Independent Power Producer Procurement Office – Information memorandum for new generation capacity for the LNG to power'. https://financialinstitutionslegalsnapshot.lexblogplatformthree.com/wp-content/uploads/sites/161/2016/10/Information-Memorandum_LNG-to-Power-IPPPP_Oct-2016.pdf.

Department of Energy (DOE). 2017. 'Strategic National Smart Grid Vision for the South African Electricity Supply Industry'. https://www.ee.co.za/wp-content/uploads/2017/12/Smart-Grid-Vision-Document-2017.pdf.

Department of Energy (DOE). 2019. 'Integrated Resources Plan (IRP)'. http://www.energy.gov.za/IRP/2019/IRP-2019.pdf, accessed 30 July 2020.

Department of Enterprise (DPE). 2019. 'Roadmap for Eskom in a reformed electricity supply'. https://dpe.gov.za/wp-content/uploads/2019/10/ROADMAP-FOR-ESKOM_0015_29102019_FINAL1.pdf.

Department of Environment, Forestry and Fisheries (DEFF). 2016. 'Renewable energy: Eskom and Minister of Environmental Affairs, Eskom presentation to the Portfolio Committee on Environmental Affairs'. https://pmg.org.za/committee-meeting/23506/, accessed 1 June 2020.

Department of Minerals and Energy (DME). 1998. 'White Paper on the Energy Policy'. http://www.energy.gov.za/files/policies/whitepaper_energypolicy_1998.pdf.

Department of Minerals and Energy (DME). 2008. 'Electricity pricing policy

(EPP) of the South African electricity supply industry'. http://www. eskom.co.za/CustomerCare/TariffsAndCharges/Documents/18671_ not13981.pdf, accessed 1 June 2020.

Ela, E., Milligan, M. and Kirby, B. 2011. 'Operating reserves and variable generation: A comprehensive review of current strategies, studies, and fundamental research on the impact that increased penetration of variable renewable generation has on power system operating reserves'. National Renewable Energy Laboratory, NREL. https://www.nrel.gov/, accessed 1 June 2020.

Ely, A., Van Zwanenberg, P. and Stirling, A. 2014. 'Broadening out and opening up technology assessment: Approaches to enhance international development, co-ordination and democratisation'. *Research Policy*, 43(3), 505–518.

Eskom. 2009. 'Tariffs and Charges, 2009/2010'. http://www.eskom.co.za/ CustomerCare/TariffsAndCharges/Pages/Tariff_History.aspx, accessed 1 June 2020.

Eskom. 2016. 'Integrated report'. http://www.eskom.co.za/OurCompany/ Investors/IntegratedReports/Pages/Annual_Statements.aspx, accessed 1 June 2020

Eskom. 2019a. 'Engagement on Eskom's Strategy Executive Forum'. Presentation by acting CEO, J. Mabuza to senior Eskom managers and executives, 22 August 2019. https://www.ee.co.za/wp-content/ uploads/2019/08/Executive-Forum-Strategy-Presentation-22-August-2019-v7.pdf, accessed 1 June 2020.

Eskom, 2019b. 'Integrated report, March 2019'. http://www.eskom.co.za/ OurCompany/Investors/IntegratedReports/Pages/Annual_Statements. aspx, accessed 1 June 2020.

Fjeldstad, O.H. 2004. 'What's trust got to do with it? Non-payment of service charges in local authorities in South Africa'. *The Journal of Modern African Studies,* 42, 539–562.

Geels, F.W. and Schot, J. 2010. 'The dynamics of transitions'. In: Grin, J., et al. (ed.). *Transitions to Sustainable Development: New Directions in the Study of Long Term Transformative Change.* New York: Taylor & Francis Group.

Gillwald, A. and Mothobi, O. 2019. 'After Access 2018. A demand-side view of mobile internet from 10 African countries'. Research ICT Africa. Policy Paper Series No. 5 After Access: Paper No. 7. https://cisp.cachefly.net/ assets/articles/attachments/78404_2019_after-access-africa-comparative-report.pdf.

Gillwald, A., Mothobi, O. and Rademan, B. 2018. 'The state of ICT in South Africa'. Research ICT Africa Policy Paper no. 5, Series 5. https:// researchictafrica.net/wp/wp-content/uploads/2018/10/after-access-south-africa-state-of-ict-2017-south-africa-report_04.pdf.

International Energy Agency (IEA). 2017a. 'Digitalization and energy'. *International Energy Agency.* https://www.iea.org/reports/digitalisation-and-energy.

International Energy Agency (IEA). 2017b. 'Energy Access Outlook 2017: From poverty to prosperity'. https://doi.org/10.1787/9789264285569-en.

International Renewable Energy Agency (IRENA), 2019a. *Enabling Technologies.* https://www.irena.org/media/Files/IRENA/Agency/Publication/2019/Sep/IRENA_Enabling-Technologies_Collection_2019.pdf.

International Renewable Energy Agency (IRENA), 2019b. *Innovation Landscape Brief: Internet of Things.* https://www.irena.org//media/Files/IRENA/Agency/Publication/2019/Sep/IRENA_Internet_of_Things_2019.pdf?la=en&hash=0576FFAC16E131D9AB017B118655933FD892A6C6.

International Renewable Energy Agency (IRENA). 2019c. *Innovation Landscape for a Renewable Powered Future: Solutions to Integrate Variable Renewables.* https://www.irena.org/-/media/Files/IRENA/Agency/Publication/2019/Feb/IRENA_Innovation_Landscape_2019_report.pdf.

Kambule, N., Yessoufou, K. and Nwulu, N. 2018. 'A review and identification of persistent and emerging prepaid electricity meter trends'. *Energy for Sustainable Development,* 43, 173–185.

Lekoloane, G., Wright, J. Carter-Brown, C. 2018. 'Municipal energy transition: Opportunities for new business models and revenue streams'. Presentation: AMEU Convention 2018, CSIR Convention Centre, Pretoria, 7–10 October 2018.

Lennon, B., Dunphy, N. P. and Sanvicente, E. 2019. 'Community acceptability and the energy transition: A citizen's perspective'. *Energy, Sustainability and Society,* 9, 35.

Makonese, T., Kimemia, D. and Annegarn, H. 2012. 'Assessment of free basic electricity and use of pre-paid meters in South Africa'. Proceedings of the 20th Domestic Use of Energy (DUE), 3–4 April 2012, Cape Town, South Africa.

Markard, J., Raven, R. and Truffer, B. 2012. 'Sustainability transitions: An emerging field of research and its prospects'. *Research Policy,* 41, 955–967.

Mathe, T. 18 October 2019. 'Why we don't pay for power'. *Mail & Guardian.* https://mg.co.za/article/2019-10-18-00-why-we-dont-pay-for-power/, accessed 16 July 2020.

Milchram, C., Hillerbrand, R., Van de Kaa, G., Doorn, N. and Künneke, R. 2018. 'Energy justice and smart grid systems: Evidence from the Netherlands and the United Kingdom'. *Applied Energy,* 229, 1244–1259.

Montmasson-clair, G., Kritzinger, K., Scholtz, l. and Gulati, M. 2017. 'New roles for South African municipalities in renewable energy'. A Review of Business Models Discussion Paper South African-German Energy Partnership (SAGEN).

Mourik, R.M., Breukers, S., Van Summeren, I.F.M. and Wieczorek, A.J. 2020. 'The impact of the institutional context on the potential contribution of new business models to democratising the energy system'. In: Lopes, M., Antunes, C.H. and Janda, K.B. (eds.). *Energy and Behaviour.* London: Academic Press.

National Energy Regulator of South Africa (NERSA). 2019. 'Grid connection code for renewable power plants (RPPs) connected to the electricity transmission system (TS) or the distribution system (DS) in South Africa. Version 3.0'. https://nersa.org.za/, accessed 1 June 2020.

National Planning Commission (NPC). 2013. 'National Development Plan (NDP), 2030. Our future make it work'. https://www.gov.za/sites/default/files/gcis_document/201409/ndp-2030-our-future-make-it-workr.pdf.

Pasetti, M., Rinaldi, S. and Manerba, D. 2018. 'A Virtual Power Plant Architecture for the Demand-Side Management of Smart Prosumers'. *Applied Sciences*, 8(3), 432.

Pöller, M., and Obert, M.2017. 'Assessing the impact of increasing shares of variable generation on system operations in South Africa (flexibility study)'. https://www.sagen.org.za/publications/24-assessing-the-impact-of-increasing-shares-of-variable-generation-on-system-operations-in-south-africa-flexibility-study/file, accessed 1 June 2020.

Raimi, K.T. and Carrico, A.R. 2016. 'Understanding and beliefs about smart energy technology'. *Energy Research & Social Science*, 12, 68–74.

Schot, J. and Kanger, L. 2018. 'Deep transitions: Emergence, acceleration, stabilisation and directionality'. *Research Policy*, 47, 1045–1059.

Schot, J. and Steinmueller, W.E. 2018. 'Three frames for innovation policy: R&D, systems of innovation and transformative change'. *Research Policy*, 47, 1554–1567.

Sovacool, B.K., Axsen, J. and Kempton, W. 2017. 'The future promise of vehicle-to-grid (V2G) Integration: A sociotechnical review and research agenda'. *Annual Review of Environment and Resources*, 42, 377–406.

Sovacool, B.K., Lipson, M.M. and Chard, R. 2019. 'Temporality, vulnerability, and energy justice in household low carbon innovations'. *Energy Policy*, 128, 495–504.

Statistics South Africa (Stats SA). 2019. 'Inequality trends in South Africa, a multidimensional diagnostic of inequality'. http://www.statssa.gov.za/publications/Report-03-10-19/Report-03-10-192017.pdf, accessed 1 June 2020.

Statistics South Africa (Stats SA). 26 March 2020a. 'An update to municipal spending and revenue (December 2019)'. http://www.statssa.gov.za/?p=13146, 1 July 2020.

Statistics South Africa (Stats SA). 2020b. 'General household survey. 2018. P0318'. http://www.statssa.gov.za/publications/P0318/P03182018.pdf, accessed 1 June 2020.

Stirling, A. 2008. '"Opening up" and "closing down": Power, participation, and pluralism in the social appraisal of technology'. *Science, Technology, & Human Values*, 33, 262–294.

Ting, M. B. 2015. 'Historical review of the relationship between energy, mining and the South African economy'. In: Mytelka, l., Msimang, V. and Perrot, R. (eds.). *Earth, Wind and Fire: Unpacking the Political, Economic and Security Implicaitons of Discourse on the Green Economy.* Johannesburg: MISTRA.

Ting, M.B. 2019. 'Multiple regime interactions, conversion, and South Africa's liquefied natural gas'. SPRU working paper, 08.

Ting, M.B. 2020. 'Socio-technical transitions in South Africa's electricity system'. Doctoral thesis (PhD), Science Policy Research Unit (SPRU), University of Sussex. http://sro.sussex.ac.uk/id/eprint/92279/, accessed 6 July 2020.

Ting, M.B. and Byrne, R. 2020. 'Eskom and the rise of renewables: Regime-resistance, crisis and the strategy of incumbency in South Africa's electricity system'. *Energy Research & Social Science*, 60, 101333.

Tomich, J. 27 January 2020. 'Grid congestion costs billion, stymies renewables'. *E&E News,* https://www.eenews.net/stories/1062188485, accessed 1 July 2020.

United Nations Conference on Trade and Development (UNCTAD). 2017. 'New innovation approaches to support the implementation of the sustainable development goals'. https://unctad.org/en/PublicationsLibrary/dtlstict2017d4_en.pdf.

United Nations Department of Economic and Social Affairs (UNDESA). 2014. 'Electricity and education: The benefits, barriers, and recommendations for achieving the electrification of primary and secondary schools'. https://sustainabledevelopment.un.org/index.php?page=view&type=400&nr=1608&menu=1515, accessed 1 June 2020.

Unruh, G.C. 2000. 'Understanding carbon lock-in'. *Energy Policy*, 28, 817–830.

United States Department of Energy (US-DOE). 2015. 'Chapter 3: Enabling modernization of the electric power system. Quadrennial technology review an assessment of energy technologies and research opportunities'. https://www.energy.gov/downloads/chapter-3-enabling-modernization-electric-power-system, accessed 1 July 2020.

Wolsink, M. 2012. 'The research agenda on social acceptance of distributed generation in smart grids: Renewable as common pool resources'. *Renewable and Sustainable Energy Reviews*, 16(1), 822–835.

TWELVE

Intellectual property and Fourth Industrial Revolution technologies[1]

Caroline Ncube and Isaac Rutenberg

INTRODUCTION

This chapter provides a discussion of the relevance of intellectual property (IP) law to this volume's critical examination of South Africa's Fourth Industrial future. It leverages the definition and framing of the Fourth Industrial Revolution (4IR) in preceding chapters and complements the governance, policy and legislature discussion by Gillwald. It is also complementarity to Oguamanam's chapter on (re) positioning indigenous knowledge for 4IR, which also discusses IP. Artificial Intelligence (AI) readiness and uptake in South Africa and a few other African countries has been discussed in the introductory

1 The research and editorial assistance of Desmond Oriakhogba, Paul Kithinji and Melissa Omino is gratefully acknowledged. All views expressed and errors, however, remain those of the authors. The research is supported by the Open African Innovation Research partnership (OpenAIR) which is currently funded by International Development Research Centre, Canada, the Social Sciences and Humanities Research Council of Canada, and Queen Elizabeth Scholars. The views expressed herein do not necessarily represent those of Open AIR's funders.

chapter and other parts of the book. Building upon, and drawing from, these aspects of this volume, the chapter focuses on understanding the significance of IP to 4IR with regard to three questions:

- What role has IP played in the early stages of development and deployment of 4IR technologies in South Africa and other parts of the continent?
- To what extent is IP likely to hinder or encourage the development and adoption of 4IR technologies in South Africa and the wider continent?
- Given the historical and expected rapid evolution of 4IR technologies in South Africa and elsewhere in Africa, are national IP systems fit for the purpose of encouraging further and equitable development of such technologies, specifically from a gendered perspective?

This examination of IP is important because IP rights (IPRs) protect technological inputs and outputs and are relied upon extensively to appropriate and monetise technologies. IPRs are a legal mechanism through which eligible work is protected, primarily through the creation of economic exclusivity through which the right-holder controls reproduction, adaptation and distribution, among other exclusive rights. These rights are typically classified into 1) copyright and related rights, and 2) industrial rights, including patents, designs, trademarks and trade secrets. The World Intellectual Property Organization (WIPO), the primary international norm-setting platform for IP, is still coming to grips with the intersection between the 4IR and IP through a public consultation process (WIPO, 2019a; 2019b; 2020). Hundreds of comments were received in response to the draft WIPO Issues paper (WIPO, 2019a; n.d.). An examination of the comments database reveals that many stakeholders, including scholars, practitioners and corporations across the globe, are engrossed in these discussions. Patent offices have also been conducting their own consultation processes (for example, USPTO, 2019). Normative IP law discussions have primarily focused on Artificial Intelligence (AI), hence this chapter's main focus on AI. Although not directly interrogated in this chapter, it should be noted that tech giants, including social media platforms of various kinds, collect and store personal data on individuals, including private information and behavioural trends. Who this data belongs to and how it is used and shared is a fundamental question of IP and the right to privacy.

A full reprisal of the history of AI is beyond the scope of this chapter, but it is important to indicate the key phases of development, so that the discussion of the role of IP in the stages of AI development is contextualised. These phases of development may be summarised into six stages (see Table 12.1):

Table 12.1: The phases of development for artificial intelligence

Period	Event
1956–1974	Emergence of the discipline
1974–1980	'AI winter': limited capacities lead to reduction in funding and research
1980–1987	Resurgence in interest and funding due to new successes
1987–1993	'Second AI winter': collapse of hardware industry and exposure of further AI limitations
1993 – 2011	Renewed optimism and further successes as AI becomes more reliant on data
2012 – present	'New wave': upsurge and exponential growth

Source: WIPO (2019: 19); Stone et al. (2016: 50)

This chapter considers all developments prior to the new wave, which began in 2012, as the early stage of AI development. It also considers IP in relation to other 4IR technologies such as 3D printing, robotics and blockchain technology, which have developed exponentially in the last decade. Further, their uptake in parts of Africa has spread within that period as shown by the establishment and growth of FabLabs, which uses 3D printing, in both South Africa and Kenya since 2010 (Schonwetter and Van Wiele 2018: 13–16). The chapter also considers 5G mobile technologies and mobile money because they are integral to 4IR technologies (Mauro, 2019). It is noteworthy that meaningful work on IP and 4IR at an international level has only taken hold from 2019.

Substantial harmonisation of global IP frameworks occurred in the 1990s and early 2000s. This took place with little regard for the regulation of AI-generated work because the development of those technologies was progressing in fits and starts. Many African states have ratified many of these IP treaties (De Beer et al., 2018) to the point that the current IP laws in most countries share many similarities though there are differences in implementation, institutions, enforcement and capabilities. To a very large degree, the globalised standard framework for IPR evolved long before the new wave of 4IR technologies. Consequently, there is very little clarity among stakeholders and scholars as to how IPR should be applied or should further evolve in the face of such technologies. Judicial and administrative decisions have, of necessity, attempted to clarify some of the unexplored and controversial issues, but there remain many questions as to the ultimate influence of IPR on the Fourth Industrial future.

An additional complication in many African countries is that, while the IPR framework exists in law, implementation of IPR systems is burdened with numerous challenges, including lack of technical capacity to examine patent applications, and inadequate record keeping (Mgbeoji, 2013). Given these challenges and uncertainties, the next section of this chapter explores the role that IP has played thus far in the early stages of development and adoption of 4IR technologies in South Africa and in some other African countries. Since technologies typically evolve more rapidly than law and policy, and particularly given the expected rapid evolution of 4IR technologies in South Africa and Africa, a section on the discussion of the IPR system's probable impact on the future development of 4IR technologies follows. The section reprises the brief historical context of the role of IP and uses it as a foundation to predict whether, and to what extent, IP is likely to hinder or encourage the development of 4IR technologies in South Africa and the rest of the continent. This is followed by a section that considers other aspects of the further and equitable development of such technologies, including the position of marginalised market-actors such as women and small businesses. The last section concludes by summarising the arguments developed in preceding sections.

IP AND THE EARLY DEVELOPMENT
OF 4IR TECHNOLOGIES

As indicated above, comprehensively defining and conceptualising 4IR is beyond the scope of this chapter. However, it is important to be clear about the 4IR technologies considered in this chapter. These will be drawn from the following definition (DTPST, 2018: 18):

> The 4IR is a collective term for technologies and concepts of value chain organisation which draw together cyber-physical systems, the Internet of Things (IoT) and the Internet of Services (IoS), together with other emerging technologies, such as cloud technology, big data, predictive analytics, artificial intelligence, augmented reality, agile and collaborative robots and additive manufacturing.

It is equally important to be precise about the definitions of these technologies so that policy and regulatory discussions proceed with a common understanding of the technologies being scrutinised. Consequently, one of the first agenda items at WIPO is to secure agreement on definitions and to build a common glossary (WIPO, 2020: 3–4). For purposes of alignment with the international norm-setting discussions, this chapter adopts the definitions in the WIPO Revised Issues Paper (WIPO, 2020). As already indicated IP discussions tend to focus on AI, as defined in WIPO's Revised Issue Paper (WIPO, 2020: 3–4):

> a discipline of computer science that is aimed at developing machines and systems that can carry out tasks considered to require human intelligence, with limited or no human intervention. … AI generally equates to 'narrow' AI, which is techniques and applications programmed to perform individual tasks. Machine learning and deep learning are two subsets of AI.

Significant attention has also been given to 3D printing, robotics and blockchain technologies. The question of what role IP has played in the early phases of the development of 4IR can be conceived of in two ways, namely: 1) how, and to what extent, have IP rights been used to protect 4IR technologies, and 2) which normative and regulatory issues have arisen in the IP framework as it pertains to 4IR? Each of these questions will be considered in turn.

Using IP to protect AI

The easiest marker to use for IP protection in relation to AI is to count patents because they protect the functionality of AI-related inventions and are registered. Trade secrets also protect functionality, but they are not registered, therefore their usage is difficult to quantify. A recent WIPO study on IP and AI considered patent data and scientific publications to distil trends on the extent to which IP has been used to protect AI. The study found limited patents in the period 1956–2011, followed by exponential growth from 2012 (WIPO, 2019). The study also found that machine learning was the most cited 4IR technology in patent applications (WIPO, 2019: 18). Most AI patent applications have been filed in the US Patents and Trademarks Office (USPTO) and the Chinese Patent Office, and similarly most of the scientific publications have been published by organisations in these two countries (WIPO, 2019: 85). The only African state mentioned in the report is South Africa, which ranks 25th in the list of 30 patent offices; these are ranked by overall number of AI-related patent applications (WIPO, 2019: 84–86). By 2018, South Africa had received less than 1,000 patent applications while the top-ranking country, the USA, had received in excess of 140,000 applications (WIPO, 2019: 84-86). China and Japan received above 120,000 and 80,000 applications respectively (WIPO, 2019: 84-86).

The primary reason for such low levels of patent protection of AI in South Africa is that it 'is not an obvious location for 4IR, given its economy is still rooted in farming, mining and the informal sector, and burdened with high levels of unemployment, while the vast majority of its citizens lack advanced skills' (Sutherland, 2020: 234). South Africa has 26 public universities plus the Council for Scientific and Industrial Research (CSIR) which have some capacity for 4IR. The levels of capacity among these institutions are different though, because of their historical development and resourcing by the state (Xing and Marwala, 2017).

Normative questions

As for the second question, various normative issues have been raised in relation to AI-generated work and IP protection (WIPO, 2020). The main issue is the appropriate protection of work completely generated by an AI application, without human intervention and based on the AI's own creative choices (Gervais, 2020: 117). IP law has long recognised and protected technology-assisted works. This is evidenced in the statutory provisions and case law on the protection of computer-assisted works in, for instance, the UK, Hong Kong, India and New Zealand (Iglesias et al., 2019: 13; Abbot, 2020). This position is also entrenched in African jurisdictions such as Kenya, South Africa and Zimbabwe (Ncube, 2002; Rutenberg, 2017; Rutenberg and Nzomo, 2017). For instance, section 1(h) of the South African Copyright Act provides for the authorship of computer-generated computer programmes as follows: 'the author of a computer program which is computer-generated, means the person by whom the arrangements necessary for the creation of the work were undertaken.' This definition is premised on the legislation's acceptance of the creation of computer-generated computer programmes. With significant technological developments – from computer-generated and computer-assisted works to 4IR technologies – IP law now has to resolve how, if at all, it regulates AI-generated work within the scope of existing IPRs.

The fundamental policy question is whether IPRs as we know them should apply to such works or whether an alternative protection system should be created. This is a current topic of debate (Linke and Petrlík, 2020) which South Africa, and other African nations seeking to regulate 4IR technologies, need to engage with. The main considerations here are whether incentivisation of these works is best achieved by a *sui generis* system or the standard IPR framework.

In the meantime, while this policy matter is receiving attention, works created by AI are already being protected by IPRs (Li and Koay, 2020: 2). This has led to a series of secondary issues pertaining to patents, copyright, design, trade secrets and trademarks that have been highlighted in relation to both AI and robotics (Bonadio et al., 2018: 657–662). Such issues include, in the application of patent law, whether inventorship and ownership of AI-generated inventions ought to be ascribed to an AI application; whether they should be excluded from patent-eligibility; whether any revision of patentability criteria is necessary; what level of disclosure would be required in patent applications; and whether patent examination guidelines need to be revised (Iglesias et al., 2019: 6–8; WIPO, 2020: 4-7).

Resolving these questions is by no means a simple task because the questions raised are multi-layered. In some jurisdictions, patent offices have made decisions on AI inventorship. The decisions made about Thaler's Device for the Autonomous Bootstrapping of Unified Sentience (DABUS) will be used to illustrate these developments. Thaler obtained a patent on DABUS in 2008,[2] towards the end of the early phase of AI development (Thaler, 2019). DABUS began inventing and patent applications were filed in several jurisdictions, naming DABUS as the inventor. The UK IP Office (UKIPO), the European Patent Office (EPO) and the USPTO have all rejected the applications because AI could not be designated as an inventor (UKIPO, 2019; Li and Koay, 2020: 400; Crouch, 2020; EPO, 2020; USPTO, 2020). This stance, of requiring a human inventor, is similar to one taken in copyright law, which requires a human creator for a work to be eligible for copyright protection.

In relation to copyright the following questions have been raised: whether authorship and ownership should be attributed and ascribed to AI-generated work; how the originality of such work should be evaluated/the standard of originality; whether the list of works that are eligible for copyright protection should be revised to allow for AI-generated work to exclude sound recordings, broadcasts and performances; whether the use of copyright-protected works by AI to generate more works should constitute copyright infringement; and whether deep fakes ought to be regulated by copyright law (Iglesias et al., 2019: 8–9; WIPO, 2020: 7–9). Similar questions have been raised pertaining to other non-human works, for instance the authorship of a

---

2   US Patent 7,454,388 issued 18 November 2008.

photograph by a monkey has received considerable scholarly attention, including how the matter would be determined under South African and Nigerian copyright law (Ncube and Oriakhogba, 2018). The current position across most jurisdictions is that human authorship is required.

Some of the common questions that have been raised regarding designs include authorship and ownership of AI-generated designs; whether unauthorised use of data in registered designs constitutes infringement, and the proper treatment of unregistered designs (WIPO, 2020: 11–12). Regarding trade secret protection, issues that have arisen include whether it is an appropriate form of protection for data and AI applications; whether it should be subject to some exceptions and limitations; and whether there should be some guidelines on securing confidentiality and proving the nature of this confidentiality to a court, should litigation arise (WIPO, 2020: 13). Trademarks do not raise the same authorship or inventorship issues because the concept of an eligible author does not arise under trademark legislation. However, there are several fundamental issues of concern such as whether AI selection of products and services will affect the established trademark doctrines of 'distinctiveness, recollection, likelihood of confusion' (WIPO, 2020: 12).

South Africa is a first mover on the African continent, with regards to readiness for, and the harnessing of, the 4IR for the country's developmental progress. Relevant national initiatives are set out in Ougamanam's chapter and in other parts of this volume, though none has expressly included IP or IPRs.

IP'S PROBABLE IMPACT ON THE DEVELOPMENT OF 4IR TECHNOLOGIES

During previous periods of industrialisation and technological development, now-developed countries used national IPR laws to promote their national development and competitive advantage. The United States in the 19th century, for example, used favourable national copyright and patent laws to encourage local production and imitation of creative works and technologies. Similar legal favouritism was used by Switzerland to appease local industry (Correa, 2015: 2).

Now-developing countries have also used IPR laws in their favour. For decades, India limited the scope of patent protection in order to encourage certain sectors, particularly the development of an industry capable of producing pharmaceutical generics (Horner, 2014). Until recently, China selectively or minimally enforced IPR laws for foreign companies in order to encourage local industry (Andrews, 2006; Cox and Sepetys, 2006; Li and Koay, 2020).

Prior to the development and introduction of commercially feasible 4IR technologies, developed countries had already largely succeeded in harmonising IPR laws in all but the least developed countries. The harmonised laws do not allow for the same degree of flexibility that many developed countries enjoyed in their earlier stages of development. Thus, the global development and adoption of 4IR technologies is occurring in a legal environment that is quite different from previous periods of industrialisation, at least with respect to IPR. This is also the case because of the nature of most 4IR technologies, which rely heavily on developments in software. The application of IPR (particularly patents) to software is challenging and controversial. This is partly due to frequent changes by both courts and legislatures, and because it has evolved considerably over the past decades. Considering the unique aspects of IP law applied to 4IR technologies, it remains an open question whether IPR will, on balance, hinder or encourage the adoption, deployment and development of such technologies in Africa.

The impact of Open Source

The ICT industry has always had a tumultuous relationship with IPR. In California in the 1970s and 1980s, the Silicon Valley area developed a thriving community of IP law practitioners in parallel with global technology companies (Hulsink et al., 2007). Copyright and patent lawsuits became relatively common in the industry (Chander, 2014). In 1991, however, Linus Torvalds released Linux, an open-source adaptation of the Unix operating system, which marked a major milestone in the development of open source as a movement. One definition of open source software is 'computer software that is made freely available for anyone to modify or redistribute, provided that the source code of the software is made freely available to others' (Speres, 2009: 175), although there are now many open source licences with a variety of terms and conditions of use. The open availability of the source code for an operating system in 1991 was a paradigm shift, as it meant that it could be improved upon by any software developer and that anybody could develop applications. Ultimately, this resulted in a large and loyal ecosystem of users, applications and service providers. The concept of openly licensing technology and/or not enforcing IPRs ('open sourcing') continued to gain acceptance in the ICT industry. Some two decades after the debut of Linux, the electric vehicle company Tesla announced that it would not initiate any patent lawsuits against good faith users of their technology, essentially open sourcing their electric vehicle developments. Tesla CEO Elon Musk acknowledged at the time that competing with an entrenched industry (traditional automobile manufacturers) required more than a single company, and a large ecosystem of users, manufacturers, researchers and service providers would stand a greater chance of success.

Open sourcing a technology is now a well-known and increasingly popular business strategy, even with the largest of technology companies. The open source movement has had a substantial impact on the development of some 4IR technologies, most notably in AI and 3D printing (Haroon et al, 2020).

Regarding AI, several major global technology companies, including Microsoft and IBM, have open-sourced suites of AI algorithms and other AI tools. Tutorials on how to use the tools (or develop new ones) are also widely available. AI developers, working on hundreds of projects, are spread across the African continent and rely on these tools (Mohamed, 2019). Although some African products and research rely on in-house proprietary algorithms, the impact of open source AI tools is significant. It thus appears that IPRs are not a major impediment to the development of AI in Africa. Those who wish to work with AI have feasible options and access to training for the first input that is necessary: AI algorithms and related tools.

A major challenge, though, relates to the second necessary input for effective AI implementation: datasets of sufficient size, variety, and quality to train AI algorithms. In any of the areas thought to be ideal for AI, Africa lags developed regions in terms of the quantity and quality of data available to train AI algorithms. For example, for a dataset to enable an AI application such as language processing, a significant amount and variety of that language must be available in machine-readable format. Many African languages fall short of this requirement (see chapter 4 by Marivate). Furthermore, critically, where there is data, uncertainties over the ownership and control of that data mean uncertainty and risk for would-be users. Some copyright laws in Africa, such as in Kenya and South Africa, provide rights over databases, and/or may fail to adequately clarify permissive uses of databases. Furthermore, large datasets acquired by governments (such as, census data or national level survey data) are frequently unavailable in machine-readable format or are subject to copyright law with unclear rights of use by citizens. To the extent, then, that data ownership remains unclear, the second component of AI development will remain a challenge.

As 3D printing became feasible and widely accessible in the 2000s, a concern emerged that copyright law would undermine the potential of the technology. The concern has not disappeared but has been significantly mitigated by the widespread availability of open source materials. Ultimately, the 3D printing industry evolved into a mirror image of the situation for operating systems. For the latter, a small number of proprietary systems dominate the market. For 3D printing, although proprietary systems exist, they are largely for niche applications. Instead, hundreds of websites, containing thousands of design files and control software files, are available under a variety of open source licences. As the cost of hardware to build or buy 3D printers declines, the technology will become increasingly accessible to users in Africa. A significant proportion of social entrepreneurs in Kenya and South Africa who use 3D printing already rely upon and use open-source 3D technology (Schonwetter and Van Wiele, 2018: 29–30). Ultimately, basic polymer-based additive manufacturing will be widely available in Africa, although niche applications and the ability to print with other materials will likely remain less available primarily due to cost factors.

Similarly with 3D printing, the fundamental components of

blockchain technology are available under open-source licences. Numerous toolkits and tutorials are freely available, and business activity is primarily in the area of developing highly specialised blockchain applications. It is therefore unlikely that IPR will act as a major hinderance to the development of this technology.

Patents and other IPR barriers in 4IR

Notwithstanding the prevalence of open-source technologies, a large number of patent applications in developed countries are aimed at various aspects of AI technology. Patent applications and granted patents in the US mentioning aspects of AI, such as machine learning and neural networks, rose from 2010, although the growth of patent activity is not uniform across the various fields of AI. Similarly, the importance of patents in AI is not geographically uniform across various countries and regions of Africa. Google and IBM are conducting AI research on the continent, with facilities in Ghana and Kenya, respectively. The research translates to patent applications, but such filings are not common in Africa apart from South Africa.[3] Furthermore, even in South Africa, the volume of patent applications relating to AI is far below the volume of applications in the US or Europe, and it is clear that the industry is not prioritising large patent portfolios. It remains unknown, however, whether the relatively few applications filed in South Africa are of exceptionally high value and importance, and therefore likely to be highly impactful on the direction of AI in the country (IEEE, 2017).

There is little doubt, however, that patents play a crucial role in 5G mobile telecommunications. In prior generations, development of mobile telecommunications standards resulted in groups of patents, referred to as standard-essential patents (SEPs), that were critical for companies building and accessing global mobile telecommunications networks. When a technology becomes part of an industry standard,

---

3   The WIPO Patentscope database was searched on 15 June 2020. A search for AI-relevant terms in the English description ('artificial intelligence', or 'machine learning', or 'neural networks') of ARIPO, South African, and Kenyan patents and applications, returned 180 non-PCT documents. Of those, 179 documents were in South Africa, and one document was in ARIPO.

and the associated patents become SEPs, many jurisdictions mandate that the patent holders must make the patents available to all interested parties under fair, reasonable and non-discriminatory (FRAND) terms. Because the technology is still under development, the scope of SEPs in 5G mobile telecommunications remains somewhat unsettled, although some companies likely to hold many of the SEPs for 5G are active in filing patent applications in Africa. Nokia, for example, has over 3,000 patents declared essential to 5G and has several hundred patent filings in South Africa and other African countries (Nokia, 2020). The existence of such IPRs is not likely to prevent the technology from eventual widespread adoption in Africa. Installation of 5G mobile telecommunications systems is carried out by large telecommunications companies, which have significant experience in negotiating for access to the technology and expertise needed to install such technology.

Beyond patents, interoperability is an issue for many 4IR technologies, and is most relevant to the Internet of Things (IoT) (Gonzalez-Usach et al., 2019). Through various forms of IPR, including copyright and trade secrets, device and network manufacturers intentionally or unintentionally hinder the widest development of an IoT ecosystem. The European Union has used various mechanisms to overcome this barrier, including regulations and standardisation, competition law, and a decompilation exception in the Software Directive (2009/24) (Pihlajarinne, 2017). In South Africa, the CSIR made an early move towards IoT interoperability by creating an Internet of Things Engineering Group (Coetzee and Eksteen, 2011). More globally, interoperability may be solved in part by the implementation of Open Radio Access Network (Open RAN, or O-RAN) technology (Singh e al., 2020), which utilises open standards in a critical component of the interface between user devices and the core network for telecommunications. The implementation of O-RAN would dramatically reduce a barrier to 5G and IoT systems. At least two major telecommunications companies, Orange and Vodafone, operated O-RAN pilot projects in Africa in 2020, signalling great interest in the open source technology.

Local 4IR technology

Although mobile money transfer systems are not typically included when 4IR technologies are defined and enumerated, there is no doubt that the technology has revolutionised business and many aspects of society across much of Africa. From the earliest systems, such as the system piloted in Kenya in 2006, to the deeply penetrating and highly diversified systems of 2020, mobile money transfer has brought banking services to vast unserved and underserved regions and populations. The challenges that prevent such systems from succeeding in any specific jurisdiction are primarily due to a combination of inflexible banking laws and an established banking sector unwilling to allow competition from mobile operators. There are no patents on the fundamental technology in the countries that most readily embrace it, although it is not uncommon to find patents that relate to various applications of the technology. Copyright protection of the enabling software is also not a barrier, as these systems are typically developed in-house by the telecommunications companies, although there have been some minimal activity in copyright litigation over applications on mobile money transfer systems (*Faulu Kenya Deposit Taking Microfinance Limited v Safaricom Limited*, 2013). It was widely reported in 2019–2020 that the Kenyan and South African mobile telecommunications companies Safaricom and Vodacom, respectively, purchased the IPR and associated support services for the MPesa mobile money transfer technology from the UK company Vodafone.

FURTHER AND EQUITABLE DEVELOPMENT OF 4IR TECHNOLOGIES

It is important to preface this section with an acknowledgment of the fact that Africa has not yet fully leveraged the previous industrial eras (Oguamanam, 2019; Sutherland, 2020). Therefore, optimism about the harnessing of the 4IR has to have a heavy dose of reality. However, the possibility of leapfrogging and realising some significant returns from the 4IR is made possible by the existence of 'more than 400 tech hubs across the continent with Lagos, Nairobi and Cape Town emerging as internationally recognised technology centres' (Oguamanam, 2019). AI's potential as an opportunity for growth, development and democratisation is a well discussed theme that resonates through this volume (Access Partnership, 2020). A key intervention is to upskill the workforce and to train students appropriately to prepare Africa for the Fourth Industrial future (Marwala, 2019).

Recent research across several African countries has shown that collaborative dynamics are prevalent and require an IP framework that supports collaboration and open approaches to knowledge appropriation (De Beer et al., 2013). A previous section has demonstrated that open approaches are prevalent in the development and deployment of some 4IR technologies. Inclusive and equitable development are major themes in IP generally, where the focus has been on how to build an IP framework that enables and supports innovation by typically marginalised constituencies such as the informal sector, women and indigenous peoples and local communities (IPLC) (De Beer et al., 2013). These marginalised constituencies are likely to be vulnerable to exclusion in the IPR protection of 4IR technologies. Oguamanam has already discussed indigenous knowledge (IK) and the position of IPLC. This section will consider AI, gender and IP. This is an important area of exploration because the lack of a gendered analysis has been flagged as an area of concern (Thirukesan, 2019).

A gendered analysis of IP frameworks has also highlighted that women are generally underrepresented in the IPR system, particularly with regard to IPR registrations (De Beer et al., 2017: 2). There are limited scholarly works on the gendered analysis of IP frameworks in Africa (De Beer et al., 2017: 2). However, seminal work has been authored by Foster from a 'feminist decolonial technoscience' perspective on the Hoodia plant and its CSIR patent as it relates to indigenous people in South Africa (Foster, 2017). This chapter conceives of the 'gender gap' in relation to IP in a broad sense, not limited to a male/female binary but one that appreciates the complexities of gender as well as national contexts (Rushton et al., 2019). However, much of the scholarship drawn up for this section focuses on the binary conception of gender and thus concerns itself with the plight of women in IP and innovation contexts. Consequently, the language refers to binary divides in the singular. The gender gap exists in both the Global South and Global North, as evidenced by studies in Finland on design rights, utility models and trademarks (Heikkilä, 2018) and sustained scholarship on IP and gender in United States and Canadian legal literature (Swanson, 2016). The main areas of focus of this scholarship are the analysis of 1) gender disparity; 2) application of IP doctrines to gendered and sexualised subject matter; and 3) IP doctrines as gendered (Swanson, 2016). Studies on female inventors and authors in South Africa highlight the same focal concerns with respect to both high-tech and IK-based innovations. With respect to the latter, studies have demonstrated that the IP system has a tendency to disempower women, 'especially rural African women crafters whose works [are] created in collaborative communities' (Oriakhogba, 2020: 146). An example of the application of IP to such crafts is the possibility of seeking design protection for some crafts.

These gender-specific dynamics and concerns manifest in AI-related inventions, first in relation to the parties to whom inventorship is ascribed, and secondly in the design of 4IR technologies that have inherent gender bias (Gomez, 2019). The first issue, of underrepresentation of women, may be partially attributable to a skills deficit. However, this is only one factor in myriad considerations, so it would be glib to assert that training women in 4IR technologies specifically, and in STEM fields more generally, would remedy this underrepresentation. It would be more accurate to assert that it would partially contribute to addressing the issue. Other causes of under-representation that need to be addressed include the impenetrability (to women) of social structures and networks that support innovation and the socialisation of IPR and innovation systems (Burk, 2011; 2018).

Gender inequity and other forms of exclusion will therefore impact development of 4IR industries in Africa for the foreseeable future. Such impacts, as described previously, range from a higher frequency of gender biases in 4IR products to an overall slower growth of the sector. Furthermore, questions have been raised about the fitness for purpose of IPR systems in Africa, particularly in view of the extent to which most African IPR systems are under-resourced (Mgbeoji, 2013; Rutenberg, 2017). Notwithstanding these obstacles, the potential for wide adoption and integration of certain 4IR technologies remains promising, and the challenges of low-functioning IPR systems may ultimately be of minimal impact. We have previously described how open-source software is pervasive in AI, blockchain, virtual reality and 3D printing, and this widespread availability is instrumental in the ability of African researchers to use and improve upon such technologies. The availability of open-source resources encourages the development of further resources, as was seen in the community of technologies that developed around Linux. As a result, it appears unlikely that IPR will present an insurmountable one to the future development of most 4IR technologies in Africa.

Such a conclusion must be further qualified for 5G mobile telecommunications and the technologies that 5G enables or enhances, such as the Internet of Things. As noted previously, 5G mobile telecommunications is currently an exception to the importance of open source in 4IR technologies. Substantial IPRs exist to protect 5G mobile telecommunications, and implementation of the technology is likely to remain limited to large infrastructure projects involving the IPR owners or licensees. The experience of earlier generations of mobile telecommunications provides reason for both optimism and concern. Despite substantial protection of IPR in 3G and 4G technologies in more developed regions in Africa, such as South Africa, Nigeria and Kenya, access to 3G and 4G mobile telecommunications is widespread or rapidly improving. Nevertheless, and due to a variety of factors, 4G coverage in the whole of Africa remains far behind that of developed regions, and coverage or access is substantially inequitable, with bias against women and rural populations (GSMA, 2018). Thus, 5G mobile telecommunications will inevitably reach the African market, albeit inequitably and lagging behind implementations of the platform elsewhere.

CONCLUSION

Although previous industrial revolutions are yet to be fully realised in Africa, internet penetration is expanding rapidly based on a relatively well-developed mobile telecommunications sector and an expanding fibre optic backbone; broadband penetration via mobile phones reached 30 per cent of the whole of Africa in 2019 (ITU and UNESCO, 2019).

In the many tech hubs of Africa, internet access is already sufficient such that those who wish to develop 4IR technologies are able to access online, open-source tools. Equitable access will remain a barrier, but not an insurmountable one for many people and regions. The impacts of inequitable access to 3IR technologies is well documented and studied, and the impact of inequitable access to 4IR technologies will likely be equally problematic (Mozumder and Marathe 2007).

Utilisation of IPR systems in Africa has consistently lagged behind that in highly developed countries, but in most African countries IPR systems exist and function reasonably well, particularly in South Africa and Kenya. Copyright and trademark systems are highly developed and well utilised but are unlikely to present a substantial barrier to the development and implementation of 4IR technologies. In fact, as with other aspects of the ICT industry, trademarks are likely to benefit companies that seek to provide 4IR products and services. Branding and customer engagement have traditionally been critical components of business success in the ICT sector, and this is likely to remain true in 4IR technologies.

Data and data ownership remain a large unknown in the future evolution of 4IR technologies. Data and, especially, datasets pertaining to human activities, have rapidly evolved into one of the most critical components of the development of emerging technologies. Some technologies, such as AI, rely heavily on the availability of diverse data in order to function properly and effectively. Large datasets are collected from a variety of sources, including internet activities and social media. The IPR position with respect to datasets is not always clear or easily determined; copyright law applies in many (but not necessarily all) instances, and many datasets are regulated by private contracts. As the importance of data is further exposed, data ownership will become an increasingly important issue. Evolution of open-access datasets, similar to the evolution of open-source software, may become a critical component of 4IR technologies in Africa.

Ultimately, then, open-source approaches are among the most influential aspects of IPR with respect to emerging 4IR technologies. Throughout the last decade, the wide and deep availability of open-source platforms and toolkits have had a substantially positive impact on the growth and development of many 4IR technologies in Africa. This impact is expected to remain and grow in importance throughout the next decade, particularly through the growth of a community of open-source 4IR tools and tutorials.

REFERENCES

Access Partnership. 2020. 'Artificial intelligence for Africa: Opportunity for growth, development and democratisation'. *Access Partnership*. Pretoria: University of Pretoria. https://www.up.ac.za/media/shared/7/ZP_Files/ai-for-africa.zp165664.pdf, accessed 13 June 2020.

Andrews, J. 2006. 'Pfizer's Viagra patent and the promise of patent protection in China'. *Loyola of Los Angeles International and Comparative Law Review*, 28(1), 1–35.

Bonadio, E., McDonagh, L. and Arvidsson, C. 2018. 'Intellectual property aspects of robotics'. *European Journal of Risk Regulation,* 9, 655–676.

Burk, D.L. 2011. 'Do patents have gender?'. *American University Journal of Gender Social Policy and Law,* 19(3), 881–919.

Burk, D.L. April 2018. 'Bridging the gender gap in intellectual property'. *WIPO Magazine.* https://www.wipo.int/wipo_magazine/en/2018/02/article_0001.html, accessed 13 June 2020.

Chander, A. 2014. 'How law made Silicon Valley'. *Emory Law Journal*, 63(3), 662.

Coetzee, L. and Eksteen, J. 2011. 'The Internet of Things: Promise for the future? An introduction.' In: Cunningham, P. and Cunningham, M. (eds). *IST-Africa 2011 Conference Proceedings.*

Correa, C.M. 2015. 'Intellectual property: How much room is left for industrial policy?' UNCTAD Discussion Paper No. 223. http://unctad.org, accessed 17 July 2020.

Cox, A. and Sepetys, K. 2006. 'Intellectual property rights protection in China: Litigation, economic damages, and case strategies'. https://www.nera.com/content/dam/nera/publications/archive1/PUB_IPR_Protection_China_IP1138.pdf, accessed 18 July 2020.

Crouch, D. 27 April 2020. 'USPTO rejects AI-invention for lack of a human inventor'. *PatentlyO.* https://patentlyo.com/patent/2020/04/rejects-invention-inventor.html, accessed 13 June 2020.

De Beer, J., Baarbé, J. and Ncube, C.B. 2018. 'Evolution of Africa's intellectual property treaty ratification landscape'. *The African Journal of Information and Communication (AJIC),* 22, 53–82.

De Beer, J., Degendorfer, K., Ellis, M. and Gaffen, A. 2017. 'Open AIR briefing note – Integrating gender perspectives into African innovation research'. *Open AIR.* https://openair.africa/wp-content/uploads/2018/11/Briefing-

Note-Gender-2017-09-21.pdf, accessed 20 June 2020.

Department of Telecommunications and Postal Services (DTPST). 2018. 'Concept Document: Establishment of the Presidential Commission on the Fourth Industrial Revolution'. Government Gazette No. 42078.

European Patent Office (EPO). 2020. 'Grounds for the EPO Decision of 27 January 2020 on EP 18275163'. https://register. epo.org/application?documentId=E4B63SD62191498&number =EP18275163&lng=en&npl=false, accessed 2 June 2020.

Flynn, S., Geiger, C., Quintais, J.P., Margoni, T., Sag, M. et al. 2020. 'Implementing user rights for research in the field of artificial intelligence: A call for international action'. PIJIP Research Paper Series, 48. https:// digitalcommons.wcl.american.edu/research/48, accessed 12 June 2020.

Foster, L.A. 2017. *Reinventing Hoodia: Peoples, Plants and Patents in South Africa*. Seattle: University of Washington Press – Feminist Technoscience Series.

Gervais, D. 2020. 'Is intellectual property law ready for artificial intelligence?' *GRUR International*, 69(2), 117–118.

Gomez, E. 11 March 2019. 'Women in artificial intelligence: Mitigating the gender bias'. Joint Research Centre. https://ec.europa.eu/jrc/communities/ en/community/humaint/news/women-artificial-intelligence-mitigating-gender-bias, accessed 13 June 2020.

Gonzalez-Usach, R., Yacchirema, D., Julian, M. and Palau, C.E. 2019. 'Interoperability in IoT'. In: Gurjit, K. and Pradeep, T. (eds). *Handbook of Research on Big Data and the IoT*, 149–173. http://doi:10.4018/978-1-5225-7432-3.ch009.

GSM Association (GSMA). 2019. 'Mobile internet connectivity 2019 sub-Saharan Africa factsheet'. GSMA. https://www.gsma.com/ mobilefordevelopment/wp-content/uploads/2019/07/Mobile-Internet-Connectivity-SSA-Factsheet.pdf, accessed 15 June 2020.

Haroon, S.S., Viswanathan, A., Alyamkin, S. and Shenoy, R. 2020. 'Acceleration of 4IR driven digital transformation through open source: Methods and parallel industries knowledge reapplication in the field'. Offshore Technology Conference. doi:10.4043/30760-MS.

Heikkilä, J. 2018. 'IPR gender gaps: A first look at utility model, design right and trademark filings'. *Scientometrics*, 118(3), 869–883. https://doi. org/10.1007/s11192-018-2979-0.

Horner. 2014. 'The impact of patents on innovation, technology transfer and health: A pre- and post-TRIPs analysis of India's pharmaceutical industry'. *New Political Economy*, 19(3), 384–406. doi: 10.1080/13563467.2013.796446.

Hulsink, W., Manuel, D. and Bouwman, H. 2007. 'Clustering in ICT: From Route 128 to Silicon Valley, from Dec to Google, from Hardware to Content'. ERIM Report Series Reference No. ERS-2007-064-ORG.

https://ssrn.com/abstract=1032751.

Iglesias, M., Shamuilia, S. and Anderberg, A. 2019. 'Intellectual property and artificial intelligence – A literature review'. *EUR 30017 EN,* Publications Office of the European Union. doi:10.2760/2517, JRC119102.

International Telecommunication Union (ITU) and United Nations Educational, Scientific and Cultural Organization (UNESCO). 2019. 'Connecting Africa through broadband: A strategy for doubling connectivity by 2021 and reaching universal access by 2030'. Report by the Working Group on Broadband for All. https://www.broadbandcommission.org/Documents/working-groups/DigitalMoonshotforAfrica_Report.pdf.

Li, N. and Koay, T. 2020. 'Artificial intelligence and inventorship: an Australian perspective'. *Journal of Intellectual Property Law and Practice,* 15(5), 399–404.

Linke, D. and Petrlík, D. 2020. 'Copyright work and its definition with regard to originality and AI – Conference report on the Fourth Binational Seminar of TU Dresden and Charles University in Prague'. *GRUR International*, 69(1), 39–45.

Marwala, T. November 2019. 'Preparing Africa for the Fourth Industrial Revolution'. *WIPO Magazine*, Special issue, 11/2019. https://www.wipo.int/wipo_magazine/en/2019/si/article_0006.html, accessed 2 August 2020.

Mauro, I. 2019. '5G for the Fourth Industrial Revolution'. World Economic Forum. https://www.gsma.com/spectrum/wp-content/uploads/2019/05/1-Isabelle-Mauro-Director-Head-of-Telecoms-Digital-Communications-Industry-WEF.pdf, accessed 20 June 2020.

Mgbeoji, I. 2013. 'African patent offices not fit for purpose'. In: De Beer, J., Armstrong, C., Oguamanam, C. and Schonwetter, T. (eds.). *Innovation and Intellectual Property: Collaborative Dynamics in Africa*. Cape Town: University of Cape Town Press, 234–247.

Mohamed, S. 4 August 2019. '#Sautiyetu: Raising our voice in artificial intelligence'. Deep Learning Indaba Blog. https://deeplearningindaba.com/blog/2019/08/sautiyetu-raising-our-voice-in-artificial-intelligence/, accessed 27 June 2020.

Mozumder, P. and Marathe, A. 2007. 'Role of information and communication networks in malaria survival.' *Malaria Journal*, 6(136). https://doi.org/10.1186/1475-2875-6-136.

Ncube, C.B. 2002. 'Copyright protection of computer programs, computer-generated works and databases in Zimbabwe'. *The Journal of Information, Law and Technology (JILT)*, 2. http://elj.warwick.ac.uk/jilt/02-2/ncube.htm, accessed 3 June 2020.

Ncube, C.B. and Oriakhogba, D.O. 2018. 'Monkey selfie and authorship in copyright law: The Nigerian and South African perspectives'. *Potchefstroom Electronic Law Journal/Potchefstroomse Elektroniese Regsblad* , 21(1). http://dx.doi.org/10.17159/1727-3781/2018/v21i0a4979.

Nokia. 24 March 2020. 'Nokia announces over 3,000 5G patent declarations'. https://www.nokia.com/about-us/news/releases/2020/03/24/nokia-announces-over-3000-5g-patent-declarations/, accessed 13 June 2020.

Oguanaman, C. 19 November 2019. 'African innovation in the Fourth Industrial Revolution'. *Open AIR Africa*. https://openair.africa/african-innovation-in-the-fourth-industrial-revolution/, accessed 13 June 2020.

Oriakhogba, D. O. 2020. 'Empowering rural women crafters in KwaZulu-Natal: The dynamics of intellectual property, traditional cultural expressions, innovation and social entrepreneurship'. *South African Law Journal*, 137, 145–172.

Pihlajarinne, T. 2017. 'Internet of Things and Intellectual Property Rights'. Abstract, 36th Annual ECTA Conference. http://budapest2017.ecta.org/IMG/pdf/pihlajarinne_abstract_final-2.pdf, accessed 17 July 2020.

Rushton, A., Gray, L., Canty, J., and Blanchard, K. 2019. 'Beyond binary: (Re)defining "gender" for 21st century disaster risk reduction research, policy, and practices'. *International Journal of Environmental Research and Public Health*, 16(20), 3984. https://doi.org/10.3390/ijerph16203984.

Rutenberg, I. and Nzomo, V. 2017. 'Patenting the un-patentable? A review of patentable subject matter in granted patent claims: Kenya as a case study.' *South African Intellectual Property Law Journal*, 5, 58–74.

Schonwetter, T. and Van Wiele, B. 2018. '3D printing: Enabler of social entrepreneurship in Africa? The roles of FabLabs and low-cost 3D Printers'. Open AIR Working Paper 18. https://openair.africa/wp-content/uploads/2020/05/WP-18-3D-Printing-Enabler-of-Social-Entrepreneurship-in-Africa.pdf.

Singh, S.K., Singh, R. and Kumbhani, B. 2020. 'The evolution of radio access network towards open-RAN: Challenges and opportunities'. 2020 IEEE Wireless Communications and Networking Conference Workshops (WCNCW), Seoul, Korea (South), 1–6. doi: 10.1109/WCNCW48565.2020.9124820.

Speres, J. 2009. 'The enforceability of open source software licences: Can copyright licences be granted non-contractually?'. *South African Mercantile Law Journal*, 21, 174–187.

Stone, P., Brooks, R., Brynjolfsson, E., Calo, R., Etzioni, O., et al. 2016. 'Artificial intelligence and life in 2030: One hundred year study on artificial intelligence: Report of the 2015–2016'. Study Panel, Stanford University, Stanford, CA. http://ai100.stanford.edu/2016-report, accessed 12 June 2020.

Sutherland, E. 2020. 'The Fourth Industrial Revolution: The case of South Africa'. *Politikon*, 47(2), 233–252. doi: 10.1080/02589346.2019.1696003.

Swanson, K.W. 2016. 'Intellectual property and gender: Reflections on accomplishments and methodology'. *American University Journal of Gender, Social Policy and the Law*, 24(1), 175–198.

Thaler, S. 2019. 'DABUS in a nutshell'. *The American Philosophical Association Newsletter*, 19(1), 40–42.

Thirukesan, A. 1 December 2019. 'The absence of gender analysis in AI and its implications for Africa: With perspectives from WomENG'. OpenAIR. Africa. https://openair.africa/the-absence-of-gender-analysis-in-ai-and-its-implications-for-africa-with-perspectives-from-womeng/, accessed 13 June 2020.

UK Intellectual Property Office (UKIPO). 2019. 'Re Stephen L Thaler (2019). BL O/741/19'. https://www.ipo.gov.uk/p-challenge-decision-results/o74119.pdf

United States Patent and Trademark Office, Department of Commerce (USPTO). 2019. 'Request for Comments on Patenting Artificial Intelligence Inventions'. *Federal Register,* 84(166), 44889. https://www.federalregister.gov/documents/2019/08/27/2019-18443/request-for-comments-on-patenting-artificial-intelligence-inventions, accessed 3 March 2020.

World Intellectual Property Organization (WIPO). n.d. 'Artificial intelligence and intellectual property policy: Database of submissions'. https://www.wipo.int/about-ip/en/artificial_intelligence/policy.html#submissions, accessed 13 June 2020.

World Intellectual Property Organization (WIPO). 2019a. 'Draft issues paper on intellectual property policy and artificial intelligence'. WIPO/IP/AI/2/GE/20/1. https://www.wipo.int/edocs/mdocs/mdocs/en/wipo_ip_ai_2_ge_20/wipo_ip_ai_2_ge_20_1.pdf

World Intellectual Property Organization (WIPO). 2019b. 'Technology trends 2019: Artificial intelligence'. https://www.wipo.int/edocs/pubdocs/en/wipo_pub_1055.pdf.

World Intellectual Property Organization (WIPO). 2020. 'Revised Issues Paper on Intellectual Property Policy and Artificial Intelligence'. WIPO/IP/AI/2/GE/20/1 REV. https://www.wipo.int/meetings/en/doc_details.jsp?doc_id=499504, accessed 23 June 2020.

Xing, B. and Marwala, T. 2017. 'Implications of the Fourth Industrial Age for higher education'. *The Thinker*, 73(3). https://ssrn.com/abstract=3225331, accessed 5 June 2020.

Final reflections: Towards an inclusive and sustainable Fourth Industrial Revolution in Africa

Zamanzima Mazibuko-Makena and
Erika Kraemer-Mbula

The notion of the Fourth Industrial Revolution (4IR) holds great promise for technological innovations that can grow the economy while improving people's lives. However, unravelling the practicalities to fulfil such potential in a country like South Africa requires a systematic approach to understanding the 4IR in a developmental context. Countries' capabilities to develop, diffuse and use innovations are shaped by a range of historical, cultural, social, political and economic factors. These capabilities will manifest in innovations that are embedded in what is known as an innovation system, which is unique to each territory. Therefore, policy responses to the 4IR need to be tailored to a country's unique resources, productive capacity and socioeconomic needs. In South Africa in particular, and indeed the continent at large, some of the needs that require attention include the ability of the 4IR to create digital jobs to reduce unemployment, and the reduction of inequality through equitable access to the internet.

Chapters in this book illuminate the socioeconomic contexts

of developing countries that 4IR technologies must contend with to produce the outcomes they are purported to enable. Income inequalities, for instance, which are disproportionately prevalent in developing countries, have been shown to determine which population groups enjoy the fruits of innovations and which do not, including innovations involving 4IR technologies. This volume acknowledges that the magnitude of change is likely to be unprecedented and fast. However, the direction of such change remains uncertain. There is an increasing realisation that if the direction of change is not carefully steered, the 4IR will exacerbate existing social challenges. Therefore, the authors in this volume show the crucial points at which strategic interventions are required to harness 4IR technologies to create an inclusive, human-centred future. The authors draw largely from the South African context, and in some cases from other African and international countries to derive lessons and provide policy analysis and insights. The chapters provide recommendations that can be incorporated into 4IR strategies and complementary policies in South Africa, and beyond.

The authors apply a critical lens to the notion of the 4IR. The volume highlights the limitations of transplanting the World Economic Forum's (WEF's) notion of the 4IR (Schwab, 2016) to a developing country context, without adequately addressing the preconditions that need to be in place for implementation of the 4IR. Many of such preconditions are still lacking in African countries. For instance, policy strategies to accelerate Artificial Intelligence (AI) adoption in South Africa cannot be the same as those in countries with universal, reliable and fast internet access. Instead, a strategic approach to AI in South Africa ought to work in tandem with policies that prioritise support for basic information and communications technology (ICT) infrastructure. Authors in the book, therefore, argue for deliberate action to create the conditions that will enable the efficient application of 4IR technologies in such a way that they foster inclusion rather than exacerbate inequalities.

The book acknowledges the potential of 4IR technologies to tackle major global challenges, such as those encapsulated in the Sustainable Development Goals (SDGs). A recent report by WEF (2020) suggests that 70 per cent of the 169 targets underpinning the SDGs could be enabled by existing 4IR technology applications. Despite various interventions, Africa is underperforming in core areas that can help redesign its economies towards a more sustainable future. There is still a heavy reliance on commodities and the continent is still grappling with the Second and Third Industrial Revolutions and their adverse effects, compounded by persistent socioeconomic disparities. Advanced technologies are therefore not yet adequately accommodated in suitable regulatory frameworks, with some technologies operating without adequate government registration, regulation and intellectual property rights. It is, therefore, important to consider the local environment to formulate a 4IR strategy that dovetails with local realities.

The book provides recommendations for South African stakeholders to engage more effectively with the notion of the 4IR and to implement policies that lead to more sustainable and equitable outcomes. Through exploring the gaps in awareness and governance, the chapters in this volume reveal specific focus areas that policies should address. This chapter organises the recommendations under such focus areas, namely equitable access to current and advanced technologies; inclusive, ethical and sustainable innovation; governance and regulation of 4IR technologies; and skills, education and public sector capacity required for the 4IR.

EQUITABLE ACCESS TO TECHNOLOGIES TO PREVENT EXACERBATION OF EXISTING INEQUALITIES

There is an unequivocal technology divide on the African continent. Africa is experiencing the fastest growth in internet usage in the world, with the percentage of people using the internet increasing from 2.1 per cent in 2005 to 28.2 per cent in 2019 (ITU, 2019). However, Africa is still the region with the lowest internet usage rates, compared with more advanced regions where about 87 per cent of individuals currently use the internet (ITU, 2019). This technology divide reflects the inequality across the globe, which is increasing within African countries, albeit to varying degrees. The technological divide in South

Africa, which manifests very differently in other African countries and regions, is articulated by Gillwald, Gastrow, Lorenz and Kraemer-Mbula and Mazibuko-Makena in chapters 1, 5, 6 and 9, respectively. South Africa contends with several fundamental challenges to take advantage of emerging technologies, including in improving infrastructure, connectivity and competencies. The authors point to these foundational improvements that are necessary for the diffusion of advanced technologies in the ICT, education, manufacturing and healthcare sectors throughout the country, and for the benefits of 4IR technologies to be widely experienced.

Some of the major hindrances on the supply-side are linked to inadequate ICT infrastructure and policies for the implementation of emerging technologies. Stable, high-speed, widespread and affordable connectivity is a requirement for increasingly advanced 4IR technologies to be adopted and used extensively. At the same time, on the demand side, human capacity (ICT and other technical skills) and innovation systems need to be enhanced to be able to harness the improved infrastructure. Currently, income inequalities between populations based on geographical location (rural vs urban), race, gender, literacy levels and others, regulate access to technologies and, in turn, entrench these inequalities.

In chapter 1 Gillwald critiques the almost unquestioning adoption of 4IR strategies that are transferred from developed countries in to developing contexts. These imported strategies do not adequately take into consideration the deep-rooted socioeconomic challenges in South Africa. The chapter argues that despite the policies and strategies that have been put in place to rectify the socioeconomic inbalances in the country, sharp inequalities persist. Gillwald implies that a narrow, technology-focused strategy in South Africa will only serve to benefit those who are already at an advantage and, therefore, deepen imbalances. Similarly, Gastrow (chapter 5) paints a picture of the unequal education landscape in South Africa, in which the private schooling system already enjoys advanced technology, digital connection and high digital literacy and the public, less-resourced schooling system is often technologically under-developed and without digital connection. The equivalent conditions can be seen in South Africa's healthcare system, where there are unequal private and public sectors with disparities in access to good quality healthcare. Mazibuko-Makena (chapter 9) illuminates the conditions in the healthcare system that, if not addressed, would result in further inequitable access to 4IR-enabled healthcare. With deteriorating and insufficient infrastructure, particularly in the rural areas, a shortage of healthcare workers, inadequate funding and unsatisfactory leadership, the South African healthcare system certainly requires more than a technology-focused strategy. Lorenz and Kraemer-Mbula (chapter 6) describe the adoption of 4IR technologies in manufacturing activities, highlighting that such technologies are concentrated in a few firms and a few productive activities. The authors also explore the various impacts that 4IR technologies may have on employment and skills development, with special reference to low-skilled communities that are more vulnerable to unemployment and poverty.

The chapters, however, acknowledge the strides that the South African government has made over the years to redress inequalities across the various sectors, although much still needs to be done. In the health sector, for example, policies aimed at redistributing healthcare workers towards primary healthcare have been implemented. This has led to an overall increase in the use of health facilities by the most disadvantaged groups in society, most of whom rely on primary healthcare (Di Paola and Vale, 2019). However, the sector still contends with shortages in medical and technological supplies. Essentially, South Africa is still struggling to carry out some of the crucial elements of previous revolutions, while, at the same time, contending with unrelenting socioeconomic inequalities. 4IR strategies must, therefore, be informed by the socioeconomic realities of each country that they are implemented in, as barriers to access are unique. Policies need to be put in place that will create an enabling environment supportive of investment and the diffusion of 4IR technologies, while concurrently focusing on essential basic advancements in provision of basic service and infrastructure (including technology infrastructure).

THE NEED FOR INCLUSIVE, ETHICAL
AND SUSTAINABLE INNOVATION

The design and development of advanced technologies intended to be used in developing countries should involve the people they are designed to benefit. Population groups that have previously been excluded from technology development should strive, and be enabled, to participate fully in technological innovation. At different points in the book, this sentiment is expressed through terms such as 'co-creation', 'co-production' and 'co-evolution', and authors illustrate different areas that require inclusivity as well as sustainability in social, environmental and economic spheres. Technological innovation on the African continent cannot be confined to conventional research and development (R&D) processes and R&D-based scientific outputs. As exemplified in some of the existing empirical literature (Kraemer-Mbula and Wunsch-Vincent, 2016; Adesida et al., 2016), innovation in Africa encompasses informal, low-tech processes that also incorporate indigenous knowledge. Oguamanam (chapter 2) interrogates the exclusion of African indigenous knowledge systems (IKS) from science and technology policies, including 4IR strategies. Opportunities should be identified in local approaches to R&D that are complementary to advanced technologies. Further to this discussion is the failure to develop 4IR technologies that can efficiently recognise black African faces or use indigenous African languages. The design of 4IR technologies takes place in Western countries, therefore African languages are left out from the conception stage. In addition, the low numbers of Africans using digital technologies and the internet renders African language groups less able to benefit from the opportunities and information that they offer.

Marivate (chapter 4) emphasises that the 4IR in Africa should not take place without adequate recognition and representation of African languages. Engaging with intelligent machines in local languages should become standard, to enhance inclusion and advance the broad 'decolonisation' project in South Africa and Africa at large. The chapter outlines how the lack of data in Africa makes it harder to include African languages in natural language processing (NLP). The development of

AI outside the continent means that its use in Africa is open to biases and discrimination (also discussed in chapter 4 by Marivate), and places constraints on the ability of AI to improve the socioeconomic conditions of the majority of the population. Using South African languages as a case study, the chapter presents compelling evidence for the need to strengthen NLP in African languages and, more so, the need to develop these languages through increasing the amount of data available in these languages. Effectively, this entails improving local language data creation, collection, curation and annotation to create African NLP task datasets.

Under the 4IR, the conventional linear view of innovation driven by scientific discovery is challenged by complementary perspectives emphasising the importance of more interactive modalities of 'open innovation' (Chesbrough 2003), 'networked innovation' (Valkokari et al., 2012), and 'co-creation' (Prahalad and Ramaswamy, 2004; Ramaswamy and Ozcan, 2018). Through co-creation, various types of ecosystem stakeholders – such as corporations, government agencies, universities and colleges, not-for-profit entities in local communities and individual users – create value together by interacting, sharing ideas and ultimately shaping the design and implementation of new technologies. This collaborative process reduces the chances of developing a technology that may not be useful or accessible to the targeted population. Mokoele, Moyo and Mahlangu touch on this aspect in chapter 3 and briefly analyse the Collingridge dilemma: the effects of emerging technologies are not fully known until they are already embedded in society and it is usually not possible to reverse their adverse effects (Collingridge, 1980). The authors provide a framework for promoting an iterative intervention throughout the lifecycle of an AI technology, from design to deployment. The concept of co-creation is also explored by Mazibuko-Makena for the development of technologies in healthcare. The chapter discusses the advantages of co-creating with patients and allowing them to actively participate in the development of technologies meant to improve their access to healthcare. For such collaborative and co-creation processes to take place, Ncube and Rutenberg (chapter 12) discuss the importance of open approaches to the development and deployment of 4IR technologies (such as open access to datasets and open software). These authors discuss openness in the context of intellectual property rights (IPR) systems, arguing that it is essential to build IPR systems that enable and support typically marginalised communities, including women.

Putting innovation at the service of 'transformative change' (Schot and Steinmueller, 2018) entails integrating social issues (especially those affecting the most vulnerable) and environmental sustainability with considerations related to innovation, including those related to advanced technologies. Such dimensions are explored in chapters 10 and 11 by paying attention to socio-ecological and energy justice, respectively. Gibixego and Odume (chapter 10) deliberate on the prospects of a 'just' transition to a low-carbon economy using the 4IR. They present a case for 4IR technologies that have the potential to reverse the environmental damage caused by previous industrial revolutions and the potential to limit future devastation. The consequences of climate change are disproportionately felt by the most vulnerable population groups, who typically contribute the least towards environmental degradation. People living in impoverished areas where there is food insecurity, limited access to clean and reliable water and sanitation and possibilities of displacement are often vulnerable to climate change as they lack socioeconomic protection from extreme weather events (Matthews and Nel, 2019). Unprecedentedly heavy rains and drought, now more intense as a consequence of global warming, often exacerbate these conditions.

The responsible development and application of advanced technologies should thus not only ensure climate resilience and environmental sustainability, but also a just society. The authors, therefore, argue that moving into a 4IR world should be a just process that results in a change in the structure of the economy, into a low-carbon economy. This would entail a shift in the current system of production, distribution and consumption. The way energy is produced, the way work is conducted, and the way goods and services are provided would have to change while prioritising the needs of working and poor people (Barrett et al, 2017).

Continuing with the notion of a just transition, Ting (chapter

11) applies sustainable transitions and the energy justice framework to investigate the applicability of a digital transformation in South Africa's electricity system. Sustainability transitions suggest that technologies are often 'selected' from the society in which they are embedded. There is apprehension, though, about the benefits of energy reform being uneven and, in some cases, reinforcing inequality. Smart grids, for example, have associated costs which would worsen financial inequalities and/or exclude certain groups due to affordability (Milchram et al., 2018). The chapter, therefore, argues for a transition that is inclusive, participative and collaborative with consumers. It further states that, at the level of consumers, there should be an increased decentralisation of energy resources that will enable local ownership of energy production, which in turn would change the ways that people engage with the energy. There are opportunities, for example, to self-generate, store and trade (peer-to-peer trading) electricity. Moreover, using blockchain or pay-as-you-go services can enable new kinds of financial transactions, particularly with mini-grids that focus on localised scale initiatives. However, given the country's wide income inequalities, as well as increasing non-payment of electricity bills, questions are raised about how the digital revolution enables the participation of economically vulnerable citizens. More importantly, if digitalisation of the country's electricity system is pursued in earnest, it should ensure inclusive participation by all sections of the population, such that it provides an opportunity to improve citizens' socioeconomic wellbeing. Fundamentally, if the potential of digital solutions for electricity systems is to be realised, then these digital technologies must be embedded in the communities at which they are targeted. Solutions cannot merely be narrowed down to technological choices.

GOVERNANCE AND REGULATION OF 4IR TECHNOLOGIES AND THEIR IMPACT IN SOCIETY

As Africa becomes an 'adopter' of 4IR technologies, the impact of the 4IR on the continent will be contingent on the nature and quality of policies and regulation, and the extent to which they are implemented.

Ayentimi and Burgess (2019) are of the view that there is a need for both proactive and reactive policies in relation to 4IR. Proactive policies should outline strategies for what needs to be achieved, whereas reactive policies respond to particular issues and problems. Policies set the institutional and regulatory framework for the actions of key stakeholders (Amankwah-Amoah 2018; Badiane et al., 2018) and signal the region's readiness for the 4IR (Ayentimi and Burgess, 2019).

South Africa and the rest of the continent are confronted with significant policy developments and implementation gaps. Foundational policies, required to manage basic operations of governance, need to be put in place to regulate science, technology and innovation (STI), including the provision of support for emerging technologies. Policies on ICT, broadband, data protection, intellectual property and other relevant issues ought to be updated regularly to suit current needs. It is essential for these policies to be established to capitalise on the 4IR: a statement that is reiterated throughout the book. Moreover, the absence of regulation may exacerbate the negative impacts of the 4IR on vulnerable populations. Policies should be strategic and adapted to match South Africa's requirements and resources. In this regard, local expertise (from public sector, private sector and broader civil society) must be included in the development of policies and regulatory frameworks, rather than merely importing these frameworks from developed countries and implementing them wholesale. Imported policies often do not have the baseline data of a particular country and do not consider the provision of basic infrastructure as key to a successful strategy (Markowitz, 2019). Local experts can provide this necessary context and data to facilitate formulation of informed and evidence-based strategies. Furthermore, this would allow for endorsement from local officials and society.

The Presidential Commission on the Fourth Industrial Revolution (4IR Presidential Commission) in South Africa is an example of a strategic, multi-stakeholder approach aimed at coordinating work related to 4IR across all government. The 30-member commission was established in 2019, drawing in experts from different sectors of society, to assist the government in taking advantage of the opportunities presented by the digital industrial revolution. Coordination between, and systems thinking among, the public sector, the private sector, research institutions and broader civil society are crucial for STI policies, regulations, skills development and equitable access to produce the required outcomes for participation in a 4IR world. Oguamanam (chapter 2) and Lorenz and Kraemer-Mbula (chapter 6) call for a more visible representation from the indigenous knowledge and manufacturing sectors in such platforms. In order to develop an inclusive 4IR strategy, it will be crucial to take advantage of the informal sector, indigenous knowledge systems and any other local expertise in production, as well as science, technology and innovation.

Adapting STI policies to the local context becomes even more important for advanced 4IR technologies. For instance, AI technologies are currently being used for digital business models in the provision of services such as renting homes, food delivery and transport. In chapter 8, Mabasa and Qobo highlight the urgency of developing a suitable regulatory framework for the application of algorithm-based technologies. Mabasa and Qobo argue for the need to assess how the way in which a digitised business model is structured impacts on different sectors in the labour market. They consider how to adapt or amend labour legislation so it protects employees in the transition towards a digital economy. Using Uber as a case study, they reveal how little labour regulations are able to adapt to a digital economy organised around AI. Uber offers an algorithm-driven platform for drivers to generate income and does not employ them in a standard employment relationship (SER); this allows for loopholes in regulations about how drivers are treated. The chapter illustrates the adverse socioeconomic outcomes that are caused by the inadequate regulation of Uber operations including income inequalities, imbalanced sector competition, driver exploitation and violence. Mabasa and Qobo provide evidence that shows that Uber operates as an employer in the transport sector and has, in some countries, subsequently been legally defined as such. The authors thus propose an amendment of general transport taxi regulations to address the socioeconomic benefits and costs of digital e-hailing services. Furthermore, they advise the implementation of measures by local and provincial authorities to improve efficiencies in permit and licensing systems, which heavily impact competitiveness in the sector and on the costs of doing business. In addition to this, it is crucial for competition in the sector that regulations are reviewed, particularly those relating to pricing, licensing and area restrictions. The chapter also speaks to issues of ownership in the e-hailing sector as part of a just transition to a digital political economy. It is argued that there should be a national strategy that seeks to support local e-hailing operators, which should include incentives and the removal of market entry barriers.

SKILLS, EDUCATION AND PUBLIC SECTOR CAPACITY REQUIRED FOR THE 4IR

The 4IR requires a new and transferrable set of skills that are able to adapt to the knowledge economy. Good quality science, technology, engineering and mathematics (STEM) education, ICT skills development and a well-functioning innovation ecosystem are crucial for this era. In South Africa, the infrastructure required to harness the 4IR urgently requires improvement, development and regulation. However, the demand side should equally be prioritised. This is needed to ensure that the country has the capability to drive innovation and technological development; to equip its population with the skills needed for advanced technologies; and to create legislation and innovation ecosystems.

Curricula need to be updated and remain agile to respond to changing skills requirements. This requires intensified coordination and a multi/transdisciplinary approach to education and skills development. Gastrow (chapter 5) discusses the need for adaptability in education, skills development and capacity building. The chapter points out that adaptability requires constant research with an agenda focused on monitoring technological change and the impact it has on society. This would allow stakeholders in government, basic education, higher education, the private sector and broader civil society to respond more rapidly to technological change. Encompassed in this is the reality of workers having to contend with the advent of automation which might displace them and what is required to prepare for this. Gastrow discusses how multidisciplinary curricula are needed to equip students for the 4IR, including social sciences, engineering and natural sciences.

Lorenz and Kraemer-Mbula (chapter 6) further explore the impacts that the adoption of 4IR technologies in advanced manufacturing is having on skills. Focusing on firms manufacturing auto and mining equipment, their chapter identifies new skills and competencies, increasingly demanded by firms, which are scarce in South Africa. Although the needs for skills vary across sectors and individual firms, strong connections to local training institutions appear essential. Problem-solving skills are in high demand, pointing to the need for improving the links between vocational training institutions and industry, as well as increasing opportunities for experience-based learning through work-integrated study.

Furthermore, the development of skills that can easily be replaced by machines, such as those used in memorisation and repetition, should be adapted in favour of skills that machines are unlikely to replicate such as empathy, creativity, innovation, social skills, and managing ethical and cultural intricacies. The importance of developing these skills is reinforced by Molopyane (chapter 7) who illustrates the impact of automation and current technologies in mining and banking. The mining and banking sectors account for a substantial number of jobs in South Africa, thus it is important to strike a balance between the introduction of new technologies and the retention of jobs. In an economy that is becoming increasingly knowledge-based, however, the realities of the future of work cannot be avoided. Molopyane recommends upskilling current workers while ensuring there is a match between skills developed and those required in the workplace. Essentially, what is required to harness the potential of new and emerging technologies are agile policies that allow for shorter time periods for drafting, implementing and assessing relevant strategies; the development of appropriate curricula with more relevant pedagogy (including lifelong learning and online self-learning); and dynamic partnerships between public and private sectors in order to harness the potential of new and emerging technologies.

Additionally, ICT skills development should not be limited to formal educational institutions. There are large sections of the population who are not in the formal education system, with low levels of digital literacy, who should be equipped to benefit from the introduction of advanced technologies, making it necessary for digital skills development to reach informal settings (Kraemer-Mbula and Wunsch-Vincent, 2016). Studies have shown that people with low levels of digital literacy are significantly hampered in their use of digital platforms, as are internet users who cannot use digital tools beyond basic functions (Markowitz, 2019). A far wider range of ages, literacy levels and geographical locations need to be engaged in relation to skills development, and various organisations should form innovative partnerships to address this need across the board.

Similarly, rigorous capacity building within government institutions is required. Mazibuko-Makena's chapter shows that one of the reasons the implementation of the electronic medical record (EMR) system has been unsuccessful is the lack of training in ICT skills for officials in the public health sector. This has resulted in various iterations of the EMR system, and a majority of medical records remaining in paper format. Therefore, there is no efficient central digital record-keeping system for patients' medical information that can be accessed anywhere, at any time. Intentional and extensive capacity building is required; one study (Markowitz, 2019) found that it can take up to four years of training and capacity building for the effective development of regulations and skills development. Additionally, research and development and educational institutions should be part of capacity building to ensure institutionally sustainable capacity (Markowitz, 2019).

With these measures, South Africa can position itself to make the 4IR work for the country's national imperatives of economic growth, social justice and historical redress. It is important to participate in the 4IR in a way that truly takes cognisance of the African context. This means acknowledging all the factors that make it possible that the 4IR will bypass Africa, or, entrench its existing extreme inequalities: poor infrastructure; low levels of skills and education; peripheral participation in global resources, technologies and trading systems. But it also means drawing on the richness of what is available in the African context: the diversity of languages; the indigenous knowledge; the rich mineral endowments; the youthfulness of the population; and Africa's resilience and ability to innovate. Everything that comes with being

African should be included in Africa's approach to and experience of the Fourth Industrial Revolution. In this way, it will be possible to ensure that this, the Fourth Industrial Revolution, is inclusive of the people that have been left behind in the preceding three. With the right policies and infrastructure, South Africa and other African countries can turn the futuristic socio-technical imaginaries inspired by the 4IR into different imaginings – one of opportunity, social inclusion and economic wellbeing for all of their citizens.

REFERENCES

Adesida, O., Karuri-Sebina, G. and Resende-Santos, J. (eds). 2016. *Innovation Africa*. Bingley, UK: Emerald Group Publishing Limited.

Ayentimi, D.T. and Burgess, J. 2019. 'Is the fourth industrial revolution relevant to sub-Sahara Africa?' *Technology Analysis and Strategic Management*, 31(6), 641–652.

Barrett, J., Nzimande, F., Cock, J. and Naudé, L. 2017. 'A just transition to a low-carbon and climate resilient economy.' COSATU policy on climate change: A call to action, https://www.sagreenfund.org.za/wordpress/wp-content/uploads/2017/05/Naledi_A-just-transition-to-a-climate-resilient-economy.pdf, accessed 4 August 2020.

Chesbrough H.W. 2003. *Open Innovation: The New Imperative for Creating and Profiting from Technology*. Boston: Harvard Business School Press.

Collingridge, D. 1980. *The Social Control of Technology*. London: Frances Pinter Ltd.

Di Paola, M. and Vale, B. 2019. 'Knowledge, power and the role of frontline health workers for South Africa's epidemic preparedness. In: Mapungubwe Institute for Strategic Reflection (MISTRA), *Epidemics and the Health of African Nations*. Mazibuko, Z (ed). Johannesburg: MISTRA, 229–269.

Heeks, R., Foster, C. and Nugroho, Y. 2014. 'New models of inclusive innovation for development'. *Journal of Innovation and Development Strategy*, 4(2), 175–185.

Kraemer-Mbula, E. and Wunsch-Vincent, S. (eds). 2016. *The Informal Economy in Developing Nations*. Cambridge: Cambridge University Press.

Markowitz, C. 2019. 'Harnessing the 4IR in SADC: Roles for policymakers.' South African Institute of International Affairs. https://media.africaportal.org/documents/Occasional-Paper-303-markowitz.pdf, accessed 7 July 2020.

Matthews, N. and Nel, D. 2019. 'Climate change hits vulnerable communities first and hardest.' IISD. https://www.iisd.org/blog/climate-change-hits-vulnerable-communities-first-and-hardest, accessed 4 August 2020.

Milchram, C., Hillerbrand, R., Van de Kaa, G., Doorn, N. and Künneke, R. 2018. 'Energy justice and smart grid systems: Evidence from the Netherlands and the United Kingdom'. *Applied Energy*, 229, 1244–1259.

Prahalad C.K. and Ramaswamy, V. 2004. 'Co-creation experiences: The next practice in value creation'. *Journal of Interactive Marketing*, 18(3), 5–14.

Ramaswamy, V. and Ozcan, K. 2018. 'What is co-creation? An interactional creation framework and its implications for value creation'. *Journal of Business Research*, 84, 196–205.

Schot, J. and Steinmueller, W.E. 2018. 'Three frames for innovation policy: R&D, systems of innovation and transformative change'. *Research Policy*, 47(9), 1554–1567.

Schwab, K. 2016. *The Fourth Industrial Revolution*. New York: Crown Business.

Valkokari, K., Paasi, J. and Rantala, T. 2012. 'Managing knowledge within networked innovation.' *Knowledge Management Research & Practice*, 10, 27–40

INDEX

Lightning Source UK Ltd.
Milton Keynes UK
UKHW021839240321
380936UK00007B/41

9 781928 509165